W9-DBI-241

Plumber's
Standard
Handbook

Plumber's Standard Handbook

R. Dodge Woodson

McGraw-Hill

New York   San Francisco   Washington, D.C.   Auckland   Bogotá
Caracas   Lisbon   London   Madrid   Mexico City   Milan
Montreal   New Delhi   San Juan   Singapore
Sydney   Tokyo   Toronto

Library of Congress Cataloging-in-Publication Data

Woodson, R. Dodge (Roger Dodge), date.
  Plumber's standard handbook / R. Dodge Woodson.
    p.   cm.
  ISBN 0-07-134244-3 (book).—ISBN 0-07-134386-5 (set)
  1. Plumbing—Handbooks, manuals, etc. I. Title.
  TH6125.W56324   1999
  696'.1—dc21   98-47578
                                                          CIP

McGraw-Hill

A Division of The McGraw·Hill Companies

1 2 3 4 5 6 7 8 9 0   DOC/DOC   9 0 4 3 2 1 0 9

P/N 134244-3
PART OF
ISBN 0-07-134386-5

*The sponsoring editor for this book was Larry Hager, the editing super-
visor was David E. Fogarty, and the production supervisor was Pamela
A. Pelton. It was set in Century Schoolbook by Terry Leaden of
McGraw-Hill's Professional Book Group composition unit.*

Printed and bound by R. R. Donnelley & Sons Company.

*This book was printed on recycled, acid-free paper containing
a minimum of 50% recycled, de-inked fiber.*

McGraw-Hill books are available at special quantity discounts to use
as premiums and sales promotions, or for use in corporate training pro-
grams. For more information, please write to the Director of Special
Sales, McGraw-Hill, 11 West 19th Street, New York, NY 10011. Or contact your local bookstore.

To Afton and Adam, the two best children a father could ever ask for

Contents

viii   Contents

Part 2   Installations

Chapter 12. Design Criteria   12.1

Chapter 13. Sensible Pipe Sizing   13.1

Part 3   Troubleshooting and Repairs

Preface

Many books have been written about plumbing, but none have come close to covering the variety of topics that you will find in this book. Many books are geared to one aspect of the plumbing trade. For example, you could buy a book on code issues. Then you could buy a book written about troubleshooting and repairing plumbing systems and devices. And, you could buy a book to learn how to size and install plumbing systems. Here's the question you have to ask yourself. Why buy three or more books when this single book gives you all the information you need for codes, installations, troubleshooting, and repairs? The information contained here is more than enough to solve your day-to-day plumbing problems.

After more than 20 years in the plumbing trade, I've learned a lot. The price of my education has been steep. Mistakes are costly. Fortunately for you, the advice, suggestions, and details between these covers can help you to avoid mistakes. Used properly, this guide will save you time, and that means more time to make more money or to enjoy your favorite hobby.

I have arranged the chapters of this book in parts. The first part covers all major code issues. Part 2 is all about installations. Troubleshooting and repairs are covered in Part 3. If you are trying to size a plumbing system, look at the table of contents and find the appropriate chapter. It's Chapter 13. Solving your plumbing problems is that simple. Nearly every type of potential problem is covered in the troubleshooting section. There you will find step-by-step instructions for determining the cause of a problem and tips on how to correct the deficiency.

It's rare to find a single resource that is comprehensive enough to depend upon for all of your questions, but this book is an exception. It truly does hold the answers you need in your plumbing career. This may, in fact, be the only book of its kind. Written for professionals, by a professional, the text and illustrations are easy to understand, simple to use, and very effective in making the plumbing process more productive and more profitable.

Take a few moments and look over the table of contents. There are 30 fact-filled chapters to explore. Even if you have been in the trade for decades, you are likely to find helpful tidbits that have escaped you up until this time. Anyone new to the trade will be amazed at the wealth of information and experience offered here. But don't just take my word for it. See for yourself. Check out some of the chapters right now. It will only take a few minutes for you to see the superior value to this indispensable tool. If you want to waste less time, learn more about your trade, make more money, and enjoy your workday more, you need this one-of-a-kind book.

Are you tired of dealing with tons of paperwork, having to constantly look for a form to protect you and your business, and having to put up with generic, cheap-looking standard forms that do nothing for your business image? If so, check out the CD-ROM in the back of this book. It contains dozens and dozens of user-friendly forms that you can load on your computer and customize. (The files are in Rich Text Format.) The forms are there. All you have to do is add your company information, and the logos if you like. The multitude of forms on the CD-ROM were created by a master plumber who has been in the trade for more than 22 years and who has run his own plumbing company for more than 19 years. All of the forms have been reviewed by an attorney, but you should have your own legal counsel check the forms for compatibility with the laws of your state.

What kind of forms will you find on the CD-ROM? If you need a business form, it's probably on the disk. Would you like easy access to subcontractor agreements to use with your piece workers and other independent contractors? Are you interested in having forms you need to keep employee files updated? Does inventory in your trucks and stock give you a fit? Well, all of these needs, and a lot more, can be met with the forms provided on the enclosed CD-ROM. Just check the complete listing of all forms to see how the nearly 100 professional, ready-to-go forms will make your life easier and your business more profitable.

R. Dodge Woodson

Acknowledgments

I would first like to acknowledge and thank my parents, Maralou and Woody, for always being there when they're needed.

The following companies are thanked for their art contributions to this book:

Amtrol, Inc.

A. O. Smith Water Products Co.

A. Y. McDonald Manufacturing Company

CR/PL, Inc. (Crane Plumbing Fixtures)

Delta Faucet Company

EBCO Manufacturing Company

Goulds Pumps, Inc.

Hellenbrand Water Conditioners, Inc.

Moen, Inc.

Ridge Tool Company

Southern Building Code Congress International

Symmons Industries, Inc.

Universal-Rundel Corporation

UNR Home Products

Vanguard Plastics, Inc.

Woodford Manufacturing Company

Plumbing Code

Administrative Policies and Procedures

Administrative policies and procedures are what make the plumbing code effective. Without the proper procedures and administration, the plumbing code would be little more than an organized outline for good plumbing procedures. To be effective, the code must be enforced. To be fair, the rules for the administration of the code must be clear to all who work with it. Administrative policies dictate the procedure for code enforcement, interpretation, and implementation.

There is a fine line between administration and enforcement. This chapter deals with administration. Chapter 2 addresses the issue of code enforcement. While the two matters are broken into two separate chapters, there will be some overlap of material. Administration and enforcement are so closely related it is necessary to commingle the two from time to time.

The rules, regulations, and laws used to make the plumbing code are structured around facts. These facts are the result of research into ways to protect the health of our nation. As a plumber, you are responsible for the health and sanitation of the public. A mistake or a code violation could result in widespread illness or even death.

In some jurisdictions the plumbing code comprises rules. In other areas it is a compilation of laws. There is a big difference between a rule and a law. When you are working with a rule-based code, you are subject to various means of punishment for violating the rules. The punishment may mean the suspension or revocation of your license. There may be cash fines required for violations, but there is no jail time. For jurisdictions using a law-based code you could find yourself behind bars for violating the plumbing code.

The procedure required to obtain a plumber's license is not easy. Some people feel there are too many restraints in the licensing requirements for plumbers, but these people are not aware of the heavy responsibility plumbers must bear. The public often perceives a plumber as someone who works in sewers, has a poor education, and is slightly more than a common laborer. This is a false perception.

Professional plumbers are much more than sewer rats. Today's plumbers are generally well educated and have the ability to perform highly technical work. The mathematical demands on a plumber could perplex many well-educated people. The mechanical and physical abilities of plumbers are often outstanding.

While drain cleaning is a part of the trade, so is the installation of $2500 gold faucets. The plumbing trade is not all sewage and water. Plumbers are known for their fabled high incomes, and it is true that good plumbers make more money than many white collar professionals. The trade can offer a prosperous living, but it must be worked for.

Whether you are a plumber or an apprentice, you have much to be proud of. The plumbing trade is more than a job. As a professional plumber you will have the satisfaction of knowing you are helping to maintain the health of the nation and the integrity of our natural resources.

It was not long ago when there was no plumbing code or code enforcement. People could pollute our lakes and streams with their ineffective cesspools and outhouses. The plumbing code is designed to stop pollution and health hazards. The code is part of any plumber's life and career. Learning the code can be a laborious task, but the self-satisfaction obtained when you master the code is well worth the effort.

When you earn your master's license, you will have the opportunity to run your own shop. Having your own plumbing business can be quite profitable. To realize these profits you must first learn the code and the trade. Then, you must pass the tests for your journeyman and master's licenses. During your learning stages you are earning a good wage and providing a vital service to the community.

There are few professions that allow you to earn a good living while you are gaining the skills necessary to master your craft. When you attend college, you must pay for your education. With plumbing, you get paid to learn. After your training you are unlimited in the wealth you can build from your own business.

The first step toward financial independence as a plumber is a clear understanding of the plumbing code. Unlicensed plumbers are not allowed to work in many jurisdictions. Your license is your ticket to respectable paychecks and a solid future. Now, let's see how the administrative policies and procedures for the plumbing code will affect you.

What Does the Plumbing Code Include?

The plumbing code includes all major aspects of plumbing installations and alterations. Design methods and installation procedures are a cornerstone of the plumbing code. Sanitary piping for the disposal of waste, water, and sewage is controlled by the code. Potable water supplies fall under the jurisdiction of the plumbing code. Storm water, gas piping, chilled-water piping, hot-water piping, and fire sprinklers are all dealt with in the plumbing, mechanical, and building codes.

The plumbing code is meant to ensure the proper design and installation of plumbing systems and to ensure public health and safety. It is intended to be interpreted by the local code enforcement officer. The interpretation of the code officer may not be the same as yours, but it is the code officer's option to determine the meaning of the code under questionable circumstances.

How the Code Pertains to Existing Plumbing

The plumbing code requires any alterations or repairs to an existing plumbing system to conform to the general regulations of the code, as they would apply to new installations. No alteration or repair shall cause an existing plumbing system to become unsafe. Further, the alterations or repairs shall not be allowed to have a detrimental effect on the operation of the existing system.

For example, if a plumber is altering an existing system to add new plumbing, the plumber must make all alterations in compliance with code requirements. It would be a violation of the code to add new plumbing to a system that was not sized to handle the additional load of the increased plumbing.

There are provisions in the codes to allow existing conditions that are in violation of the current code to be used legally. If an existing condition was of an approved type prior to the current code requirements, that existing condition may be allowed to continue in operation so long as it is not creating a safety or health hazard.

If the use or occupancy of a structure is being changed, the change must be approved by the proper authorities. It is a violation of the plumbing code to change the use or occupancy of a structure without the proper approvals. For example, it would be a breach of the code to convert a residential dwelling to a professional building without the approval of the code enforcement office.

Small Repairs

Small repairs and minor replacements of existing plumbing may be made without bringing the entire system into compliance with the current plumbing code standards. These changes must be made in a safe and sanitary method and must be approved. For example, it would be permissible to repair a leak in a ½-inch (in) pipe without changing the ½-in pipe to a ¾-in pipe, even if the current code required a ¾-in pipe under the present use. It would also be allowed to replace a defective S-trap with a new S-trap, even though S-traps are not in compliance with the current code requirements. In general, if you are only doing minor repair or maintenance work, you are not required to update the present plumbing conditions to current code requirements.

It is incumbent upon the owner of a property to keep the plumbing system in good and safe repair. The owner may designate an agent to assume responsibility for the condition of the plumbing, but it is mandatory that the plumbing be kept safe and sanitary at all times.

Relocation and Demolition
of Existing Structures

If a building is moved to a new location, the building's plumbing must conform to the current code requirements of the jurisdiction where the structure is located. In the event a structure is to be demolished, it is the owner's, or the owner's designated agent's, responsibility to notify all companies, persons, and entities having utilities connected to the structure. These utilities may include, but are not limited to, water, sewer, gas, and electrical connections.

Before the building can be demolished or moved, the utilities having connections to the property must disconnect and seal their connections in an approved manner. This applies to water meters and sewer connections, as well as other utilities.

Materials

All materials used in a plumbing system must be approved for use by the code enforcement office. These materials shall be installed in accordance with the requirements of the local code authority. The local code officer has the authority to alter the provisions of the plumbing code so long as the health, safety, and welfare of the public is not endangered.

A property owner, or that owner's agent, may request a variance from the standard code requirements when conditions warrant a hardship. It is up to the code officer to decide whether or not the variance should be granted. The application for a variance and the final decision of the code officer shall be in writing and filed with the code enforcement office.

The use of previously used materials shall be open to the discretion of the local code officer. If the used materials have been reconditioned and tested and are in working condition, the code officer may allow their use in a new plumbing system.

Alternative materials and methods not specifically identified in the plumbing code may be allowed under certain circumstances. If the alternatives are equal to the standards set forth in the code for quality, effectiveness, strength, durability, safety, and fire resistance, the code officer may approve the use of the alternative materials or methods.

Before alternative materials or methods are allowed for use, the code officer can require adequate proof of the properties of the materials or methods. Any costs involved in testing or providing technical data to substantiate the use of alternative materials or methods shall be the responsibility of the permit applicant.

Code Officers

Code officers are responsible for the administration and enforcement of the plumbing code. They are appointed by the executive authority for the community. Code officers may not be held liable on a personal basis when working for a jurisdiction. Legal suits brought forth against code officers, arising from on-the-job disputes, will be defended by the legal representative for the jurisdiction.

The primary function of code officers is to enforce the code. Code officers are also responsible for answering questions pertinent to the materials and installation procedures used in plumbing. When application is made for a plumbing permit, the code officer is the individual who receives the application. After reviewing a permit application, the code officer will issue or deny a permit.

Once a permit is issued by the code officer, it is the code officer's duty to inspect all work to ensure it is in compliance with the plumbing code. When code officers inspect a job, they are looking for more than just plumbing. These inspectors will be checking for illegal or unsafe conditions on the job site. If the safety conditions on the site or the plumbing arc found to be in violation of the code, the code officer will issue a notice to the responsible party.

Code officers normally perform routine inspections personally. However, inspections may be performed by authoritative and recognized services or individuals instead of the code officers. The results of all inspections shall be documented in writing and certified by an approved individual.

If there is ever any doubt as to the identity of a code officer, you may request to see the inspector's identification. Code officers are required to carry official credentials while discharging their duties.

Another aspect of the code officer's job is the maintenance of proper records. Code officers must maintain a file of all applications, permits, certificates, inspection reports, notices, orders, and fees. These records are required to be maintained for as long as the structure they apply to is still standing, unless otherwise stated in other rules and regulations.

Plumbing Permits

Most plumbing work, other than minor repairs and maintenance, requires a permit. These permits must be obtained prior to the commencement of any plumbing work. The code enforcement office provides forms to individuals wishing to apply for plumbing permits. The application forms must be properly completed and submitted to the code enforcement officer.

The permit application shall give a full description of the plumbing to be done. This description must include the number and type of plumbing fixtures to be installed. The location where the work will be done and the use of the structure housing the plumbing must also be disclosed.

The code officer may require a detailed set of plans and specifications for the work to be completed. Duplicate sets of the plans and specifications may be required so that copies can be placed on file in the code enforcement office. If the description of the work deviates from the plans and specifications submitted with the permit application, it may be necessary to apply for a supplementary permit.

The supplementary permit will be issued after a revised set of plans and specifications have been given to the code officer and approved. The revised plans and specifications must show all changes in the plumbing that are not in keeping with the original plans and specifications.

Plans and specifications may not be required for the issuance of a plumbing permit. However, if plans and specifications are required, they may require a riser diagram and a general blueprint of the structure. The riser diagram must be very detailed. The diagram must indicate pipe size, direction of flow, elevations, fixture-unit ratings for drainage piping, horizontal pipe grading, and fixture-unit ratings for the water distribution system.

If the plumbing to be installed is an engineered system, the code officer may require details on computations, plumbing procedures, and other technical data. Any application for a plumbing permit to install new plumbing might require a site plan. The site plan must identify the locations of the water service and sewer connections. The location of all vent stacks and their proximity to windows or other ventilation openings must be shown.

In the event new plumbing is being installed in a structure served by a private sewage disposal system, there are yet more details to be included in the site plan. When a private sewage system is used, the plan must show the location of the system and all technical information pertaining to the proper operation of the system.

When a plumbing permit is applied for, the code officer will process the application in a timely manner. If the application is not approved, the code officer will notify the applicant, in writing, of the reasons for denial. If the applicant fails to follow through with the issuance of a permit within 6 months from the date of application, the permit request can be considered void. Once a permit is issued, it may not be assigned to another person or entity. Permits will not be issued until the appropriate fees are paid. Fees for plumbing permits are provided by the local jurisdiction.

Plumbing permits bear the signature of the code officer or an authorized representative. The plans submitted with a permit application will be labeled as approved plans by the code officer. One set of the plans will be retained by the code enforcement office. A set of approved plans must be kept on the job site. The approved plans kept on the job must be available to the code officer or an authorized representative for inspection at all reasonable times.

It is possible to obtain permission to begin work on part of a plumbing system before the entire system has been approved. For example, you might be given permission to install the underground plumbing for a building before the entire plumbing system is approved. These partial permits are issued by the code officer with no guarantee the remainder of the work will be approved. If you proceed to install the partial plumbing, you do so at your own risk in regard to the remainder of the plumbing not yet approved.

There are time limits involved after a plumbing permit is issued. If work is not started within 6 months of the date a permit is issued, the permit may become void. If work is started, but then stalled or abandoned for a period of 6 months, the permit may be rendered useless. A permit may be revoked if the code officer finds that the permit was issued under false information. Misrepresentation in the application for a permit or on the plans submitted for review is reason for the revocation of a plumbing permit.

All work performed must be done according to the plans and specifications submitted to the code officer in the permit application process. All work must be in compliance with the plumbing code. Code officers are required to conduct inspections of the plumbing being installed during the installation and upon completion of the installation. The details of these inspections are described in Chapter 2.

Multiple Plumbing Codes

There are many plumbing codes in use. Each local jurisdiction generally takes an existing code and amends it to local needs. There are three primary plumbing codes, the codes for zones one, two, and three. These three codes are similar in many ways and very different in others. In addition to the three primary codes, there are other codes in existence. To be sure of your local code requirements, you must check with the local code enforcement office.

This book is written to explain good plumbing procedures. However, various jurisdictions have different opinions of what good plumbing procedures are. Some states, counties, or towns adapt an existing code without much revision. Other areas make significant changes in the established code that is used as a model. It would not be unheard of to find a jurisdiction working with regulations from multiple plumbing codes. In light of these facts, always check with your local authorities before performing plumbing work.

2

Regulations, Permits, and Code Enforcement

When working with the plumbing code, you must be aware of the requirements and procedures involved with regulations, permits, and enforcement. These three elements work together to ensure the proper use of the plumbing code. This chapter is going to teach you about all three key elements, but it will also do much more. You are going to learn the facts about pipe protection, health, safety, pipe connections, temporary toilet facilities, and more. There will even be tips on how to work with plumbing inspectors, instead of against them.

Regulations

What are regulations? They are the rules or laws used to regulate activity, in this case, the performance of plumbing. The plumbing code is governed by rules in some states and laws in others. If you violate the regulations in a rule-base state, you may face disciplinary action, but not jail. In states where the plumbing regulations are laws, you risk going to jail, if you violate the regulations. It may be difficult to imagine going to jail for violating a plumbing regulation, but it could happen. Certain violations could result in personal injury or death. To protect yourself, and others, it is important to understand and abide by the regulations governing the plumbing trade.

Existing conditions

The first regulations we are going to examine have to do with existing conditions. This is an area where many people have difficulty in dctcrmining their responsibility to the plumbing code. Generally, any existing condition that is not a hazard to health and safety is allowed to remain in existence. However, when existing plumbing is altered, it may have to be brought up to current code requirements.

While the code is normally based on new installations, it does apply to existing plumbing that is being altered. These alterations may include repairs, renovations, maintenance, replacement, and additions. The question of when work on existing plumbing must meet code requirements is one that plagues many plumbers. Let's address this question and clear it up.

The code is only concerned with changes being made to existing plumbing. As long as the existing plumbing is not creating a safety or health hazard, and is not being altered, it does not fall under the scrutiny of most plumbing codes. If you are altering an existing system, the alterations must comply with the code requirements, but there may be exceptions to this rule. For example, if you were replacing a kitchen sink and there was no vent on the sink's drainage, code would require you to vent the fixture. Where undue hardship exists in bringing an existing system into compliance, the code officer may grant a variance.

In the case of the kitchen sink replacement, such a variance may be in the form of permission to use a mechanical vent (Fig. 2.1). When-

Figure 2.1 Mechanical vent.

ever you encounter a severe hardship in making old plumbing come up to code, talk with your local code officer. The code officer should be able to offer some form of assistance, either in the form of a variance or as advice on how to accomplish your goal.

Since the code does come into play with repairs, maintenance, replacement, alterations, and additions, let's see how it affects each of these areas. If you are repairing a plumbing system, you must be aware of code requirements. If no health or safety hazards exist, non-conforming plumbing may be repaired to keep it in service.

Do you need to apply for a plumbing permit to replace a faucet? No, faucet replacement does not require a permit, but it does require the replacement to be made with approved materials and in an approved manner. Remember this, you do need a permit to replace a water heater. Even if the replacement heater is going in the same location and connecting to the same existing connections, you must apply for a permit and have your work inspected. An improperly installed water heater can become a serious hazard, capable of causing death and destruction. Your failure to comply with the code in these circumstances could ruin your life and the lives of others.

Routine maintenance of a plumbing system must be done according to the code, but it does not require a permit. Alterations to an existing system may require the issuance of a permit, depending upon the nature of the alteration. In any case, alterations must be done with approved materials and in an approved manner. When adding to a plumbing system, you will normally need to apply for a permit. Adding new plumbing will come under the authority of the plumbing code and will generally require an inspection.

You can get yourself into deep water when adding to an existing system. When you add new plumbing to an old system, you must be concerned with the size and ability of the old plumbing. Increasing the number of fixture units entering an old pipe may force you to increase the size of the old pipe. This can be very expensive, especially when the old pipe happens to be the building drain or sewer. Before you install any new plumbing on an old system, verify the size and ratings of the old system. If your new work will rely on the old plumbing to work, the old plumbing must meet current code requirements.

Beware of the change-in-use regulations. If you will be performing commercial plumbing, this regulation can have a particularly serious effect on your plumbing costs and methods. If the use of a building is changed, the plumbing may also have to be changed. The change-in-use regulations come into play most often on commercial properties, but they could affect a residential building.

Assume for a moment that you receive a request to install a three-bay sink in a convenience store. You discover that the store's owner is

having the sink installed in order to perform food preparation for a
new deli in the store. This store has never been equipped for food
preparation and service. What complications could arise from this situation? First, zoning may not allow the store to have a deli. Second, if
the use is allowed, the plumbing requirements for the store may soar.
There could be a need for grease traps, indirect wastes, and a number
of other possibilities. When you are asked to perform plumbing that
involves a change of use, investigate your requirements before committing to the job.

The remaining general regulations of the plumbing code are easily
understood. Many of them were covered in the last chapter. By reading your local code book you should have no trouble in understanding
the regulations. Now, let's move on to permits.

Permits

Permits are generally required for various plumbing jobs. When a
permit is required, it must be obtained before any work is started.
Minor repairs and drain cleaning do not require the issuance of permits. Permits were discussed in the last chapter, but we will go over
them again and add some to what you have already learned.

In most cases plumbing permits can only be obtained by master
plumbers or their agents. In some cases, however, homeowners may
be allowed to receive plumbing permits for work to be done by themselves on their own homes. Permits are obtained from the local code
enforcement office and that office provides the necessary forms for
permit applications.

The information required to obtain a permit will vary from jurisdiction to jurisdiction. You may be required to submit plans, riser diagrams (Figs. 2.2 and 2.3), and specifications for the work to be performed. At a minimum, you will likely be required to adequately
describe the scope of work to be performed, the location of the work,
the use of the property, and the number and type of fixtures being
installed.

The degree of information required to obtain a permit is determined by the local code officer. It is not unusual for the code officer to
require two sets of plans and specifications for the work to be performed. The detail of the plans and specifications is also left up to the
judgment of the code officer. Requirements may include details of pipe
sizing, grade, fixture units, and any other information the code officer
may deem pertinent.

If your work will involve working with a sewer or water service,
expect to be asked for a site plan. The site plan should show the locations of the water service and sewer (Fig. 2.4). If you will be working

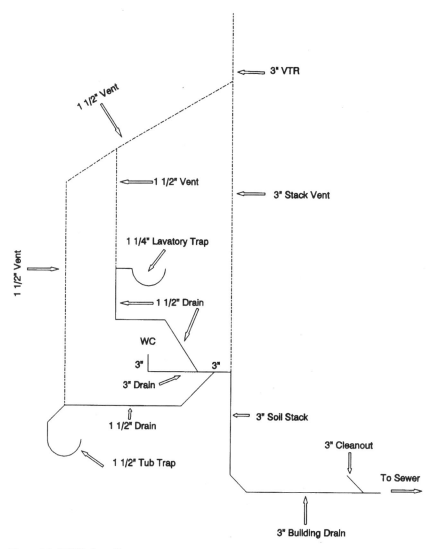

Figure 2.2 DWV riser diagram.

with a private sewage disposal system, its location should be indicated on the site plan (Fig. 2.5). Once your plans are approved, any future changes to the plans must be submitted to and approved by the code officer.

Plumbing permits must be signed by the code officer or an authorized representative. If you submitted plans with your permit application, the plans will be labeled with appropriate wording to prove they

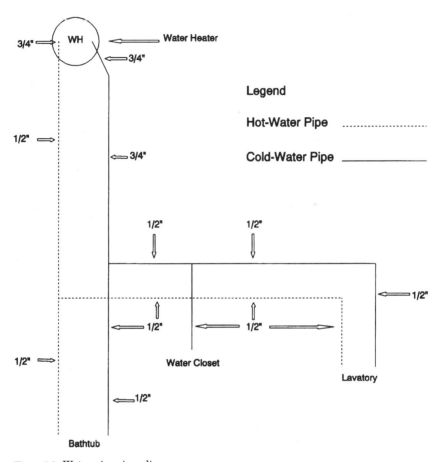

Figure 2.3 Water pipe riser diagram.

have been reviewed and approved. If it is later found that the approved plans contain a code violation, the plumbing must be installed according to code requirements even if the approved plans contain a nonconforming use. Most jurisdictions require a set of approved plans to be kept on the job site and available to the code officer at all reasonable times.

If plans are required, they must be approved before a permit is issued. All fees associated with the permit must be paid prior to the issuance of the permit. These fees are established by local jurisdictions. After a permit is issued, all work must be done in the manner presented during the permit application process.

It is possible to obtain a partial permit. This is a permit that approves a portion of work proposed for completion. When time is

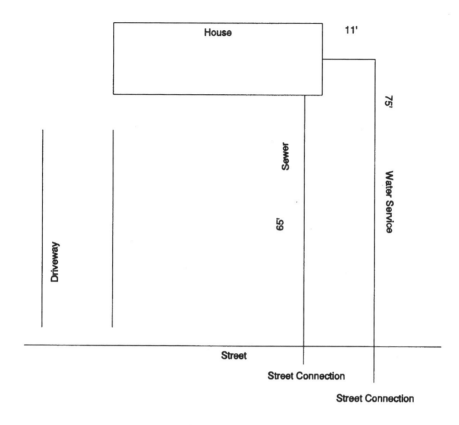

Figure 2.4 Site plan with city water and sewer.

important, it may be possible to obtain these partial permits, but there is risk involved. Assume you obtained permission and a permit to install your underground plumbing but had not yet been issued a permit for the remainder of the plumbing. With freezing temperatures coming on, you decide to install your groundworks so that the concrete floor can be poured over the plumbing before freezing conditions arrive. This is a good example of how and why partial approvals are good, but look at what could happen.

You have installed your underground plumbing and the concrete is poured. After awhile you are notified by the code officer that the proposed above-grade plumbing is not in acceptable form and will require major changes. These changes will affect the location and size of your underground plumbing. What do you do now? Well, you are probably going to spend some time with a jackhammer or concrete

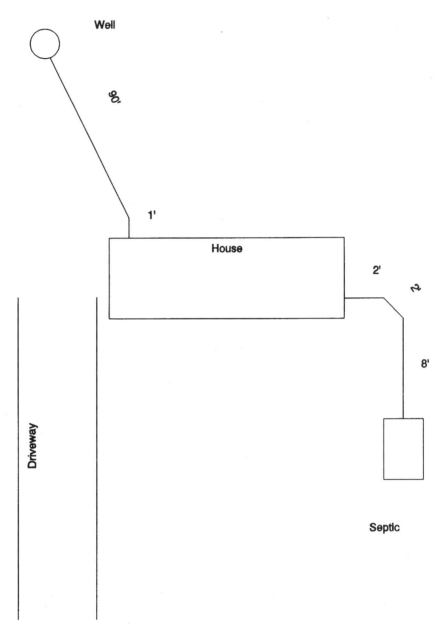

Figure 2.5 Site plan with private water and sewer.

saw. The underground plumbing must be changed, or the above-ground plumbing must be redesigned to work with the groundworks. In either case, you have trouble and expense that would have been avoided if you had not acted on a partial approval. Partial approvals have their place, but use them cautiously.

How can a plumbing permit become void? If you do not begin work within a specified time, normally 6 months, your permit will be considered abandoned. When this happens, you must start the entire process over again to obtain a new permit. Plumbing permits can be revoked by the code official. If it is found that facts given during the permit application were false, the permit may be revoked. If work stops for an extended period of time, normally 6 months, a permit may be suspended.

Code Enforcement

Code enforcement is generally performed on the local level, by code enforcement officers. These individuals are frequently referred to as inspectors. It is their job to interpret and enforce the regulations of the plumbing code. Since code enforcement officers have the duty of interpreting the code, there may be times when a decision is reached that appears to contradict the code book. The code book is a guide, not the last word. The last word comes from the code enforcement officer. This is an important fact to remember. Regardless of how you interpret the code, it is the code officer's decision that is final.

Inspections

Every job that requires a permit also requires inspection. Many jobs require more than one inspection. In the plumbing of a new home there may be as many as four inspections. One inspection would be for the sewer and water-service installation. Another inspection might be for underground plumbing. Then you would have a rough-in inspection for the pipes that are to be concealed in walls and ceilings. Then, when the job is done, there will be a final inspection.

These inspections must be done while the plumbing work is visible. A test of the system is generally required, with pressure from either air or water. Normally the inspection is done by the local code officer, but not always. The code officer may accept the findings of an independent inspection service. Before independent inspection results will be accepted, the inspection service must be approved by the code officer. Independent inspection services are commonly used to inspect prefabricated construction.

Plumbing inspectors are generally allowed the freedom to inspect plumbing at any time during normal business hours. These inspectors cannot enter a property without permission unless they obtain a search warrant or other proper legal authority. Permission for entry is frequently granted by the permit applicant when the permit is signed.

What inspectors look for

When plumbing inspectors look at a job, they are looking at many aspects of the plumbing. They will inspect to see that the work is installed in compliance with the code and in a way that the plumbing will be likely to last for its normal lifetime. Inspectors will check to see that all piping is tested properly and that all plumbing is in good working order.

What powers do plumbing inspectors have?

Plumbing inspectors can be considered the plumbing police. These inspectors have extreme authority over any plumbing-related issue. If plumbing is found to be in violation of the code, plumbing inspectors may take several forms of action to rectify the situation.

Normally, inspectors will advise the permit holder of the code violations and allow a reasonable time for correction of the violations. This advice will come in the form of written documents and will be recorded in an official file. If the violations are not corrected, the code officer will take further steps. Legal counsel may be consulted. After a legal determination is made, action may be taken against the permit holder in violation of the plumbing code. This could involve cash fines, license suspension, license revocation, and in extreme cases, jail.

Code officers have the power to issue a stop work order. This order requires all work to stop until code violations are corrected. These orders are not used casually; they are used when an immediate danger is present or possible. Code officers do have a protocol to follow in the issuance of stop work orders. If you ever encounter a stop work order, stop working. These orders are serious and violation of a stop work order can deliver more trouble than you ever imagined.

When code officers inspect a plumbing system and find it to be satisfactory, they will issue an approval on the system. This allows the pipes to be concealed and the system to be placed into operation. In certain circumstances code officers may issue temporary approvals. These temporary approvals are issued for portions of a plumbing system when conditions warrant them.

When a severe hazard exists, plumbing inspectors have the power to condemn property and force occupants to vacate it. This power would only be used under extreme conditions, where a health or safety hazard was present.

What can you do to change a code officer's decision?

If you feel you have received an unfair ruling from a code officer, you may make a formal request to have the decision changed. In doing this, you must make your request to an appeal board. Your reasons for an appeal must be valid and must pertain to specific code requirements. Your appeal could be based on what you feel is an incorrect interpretation of the code. If you feel the code does not apply to your case, you have reason for an appeal. There are other reasons for appeal, but you must specify why the appeal is necessary and how the decision you are appealing is incorrect.

Tips on Health and Safety

Health and safety are two key issues in the plumbing code. These two issues are, by and large, the reasons for the plumbing code's existence. The plumbing code is designed to assure health and safety to the public. Public health can be endangered by faulty or improperly installed plumbing. Code officers have the power and duty to condemn a property where severe health or safety risks exist. It is up to the owner of each property to maintain the plumbing in a safe and sanitary manner.

When it comes to safety, there are many more considerations than just plumbing pipes. Most safety concerns arise in conjunction with plumbing but not from the plumbing itself. It is far more likely that a safety hazard will result from the activities of a plumber on other aspects of a building. An example could be cutting so much of a bearing timber that the structure becomes unsafe. Perhaps a plumber removes a wire from an electric water heater and leaves it exposed and unattended; this could result in a fatal shock to someone. The list of potential safety risks could go on for pages, but you get the idea. It is your responsibility to maintain safe and sanitary conditions at all times.

A part of maintaining sanitary conditions includes the use of temporary toilet facilities on job sites. It is not unusual for the plumbing code to require toilet facilities to be available to workers during the construction of buildings. These facilities can be temporary, but they must be sanitary and available.

Pipe Protection

It is the plumber's responsibility to protect plumbing pipes. This protection can take many forms. Here we are going to look at the basics of pipe protection. You will gain insight into pipe protection needs that you may have never considered before.

Backfilling

When backfilling over a pipe, you must take measures to prevent damage to the pipe. The damage can come in two forms, immediate and long-term. If you are backfilling with material that contains large rocks or other foreign objects, the weight or shape of the rocks and objects may puncture or break the pipe. The long-term effect of having large rocks next to a pipe could result in stress breaks. It is important to use only clean backfill material when backfilling pipe trenches.

Even the weight of a large load of backfill material could damage the pipe or its joints. Backfill material should be added gradually. Layers of backfill between 6 and 12 inches deep are typically recommended as the average interval for filling a trench. Each layer of this backfill should be compacted before the next load is dumped.

Flood protection

If a plumbing installation is made in an area subject to flooding, special precautions must be taken. High water levels can float pipes and erode the earth around them. If your installation is in a flood area, consult your local code officer for the proper procedures in protecting your pipes.

Penetrating an exterior wall

When a pipe penetrates an exterior wall, it must pass through a sleeve. The sleeve should be at least two pipe sizes larger than the pipe passing through it. Once the pipe is installed, the open space between the pipe and the sleeve should be sealed with a flexible sealant. By caulking around the pipe you eliminate the invasion of water and rodents.

Freezing

Pipes must be protected against freezing conditions. Outside, this means placing the pipe deep enough in the ground to avoid freezing. The depth will vary with geographic locations, but your local code officer can provide you with minimum depths. Above-ground pipes, in unheated areas, must be protected with insulation or other means of protection from freezing.

Corrosion

Pipes that tend to be affected by corrosion must be protected. This protection can take the form of a sleeve or a special coating applied to the pipe. For example, copper pipe can have a bad reaction when placed in contact with concrete. If you have a copper pipe extending through concrete, protect it with a sleeve. The sleeve can be a foam insulation or some other type of noncorrosive material. Some soils are capable of corroding pipes. If corrosive soil is suspected, you may have to protect entire sections of underground piping.

Pipe Connections

Pipe connections can require a variety of adapters when combining pipes of different types. It is important to use the proper methods when making any connections, especially when you are mating different types of pipe together. There are many universal connectors available to plumbers today. These special couplings are allowed to connect a wide range of various materials.

Male and female adapters have long been an acceptable method of joining opposing materials, but today the options are much greater. You can use compression fittings and rubber couplings to match many types of materials to each other. Special insert adapters allow the use of plastic pipe with bell-and-spigot cast iron.

Working with the System Instead of against It

Code officers are expected to enforce the regulations set forth by the code. Plumbers are expected to work within the parameters of the code. Naturally, plumbers and code officers will come into contact with each other on a regular basis. This contact can lead to some disruptive actions.

The plumbing code is in place to help people, not to hurt them. It is not meant to ruin your business or to place you under undue hardship in earning a living. It really is no different than our traffic laws. The traffic laws are there to protect all of us, but some people resent them. Some plumbers resent the plumbing code. They view it as a vehicle for the local jurisdiction to make more money while they, the plumbers, are forced into positions to possibly make less money.

When you learn to understand the plumbing code and its purpose, you will learn to respect it. You should respect it; it shows the importance of your position as a plumber to the health of our entire nation. Whether you agree with the code or not, you must work within the constraints of it. This means working with the inspectors.

When inspectors choose to play hardball, they hold most of the cards. If you develop an attitude problem, you may be paying for it for years to come. Even if you know you are right on an issue, give the inspector a place to escape; nobody enjoys being ridiculed.

The plumbing code is largely a matter of interpretation. If you have questions, ask your code officer for help. Code officers are generally more than willing to give advice. It is only when you walk into their offices with a chip on your shoulder that you are likely to hit the bureaucratic wall. Like it or not, you must learn to comply with the plumbing code and to work with code officers. The sooner you learn to work with them on amicable terms, the better off you will be.

Little things can mean a lot. Apply for your permit early. This eliminates the need to hound the inspector to approve your plans and issue your permit. Many jurisdictions require at least a 24-hour (h) advance notice for an inspection request, but even if your jurisdiction doesn't have this rule, be considerate, and plan your inspections in advance. By making life easier for the inspector, you will be helping yourself.

3

Approved Materials and Their Connection

In order to do plumbing in the best and most cost-effective manner, you must choose the proper materials. Today's plumbers have so many materials to choose from that the decision of which material is best suited to the job can become perplexing. Take the water-distribution system as an example. When you are trying to decide what type of pipe to run your potable water through, you will have many options available. You could choose polybutylene (PB), chlorinated polyvinyl chloride (CPVC), or copper, just to name a few. If you decide to use copper, you must determine which copper to use. Will you use type M, type L, or type K? Can you use polyethylene (PE) or polyvinyl chloride (PVC)? As you can see, the choices can be confusing. Another solid choice for water distribution is a relatively new pipe. It's called cross-linked polyethylene. This pipe is known in the trade as PEX pipe. Since problems have occurred and major lawsuits have arisen around polybutylene pipe, some jurisdictions frown on its use. PEX piping is basically replacing polybutylene in many areas.

This chapter is going to give you a tour of the materials approved for use and the uses they are approved for. A material that is approved for above-ground use may not be allowed below ground. A pipe suitable for cold water may not work with hot water. This chapter will explain the different materials and give suggestions about which materials are best suited for specific uses. In addition, you will learn about approved connections.

Choosing the proper materials will help you in several ways. By working with only approved materials, you will not have to do the same job twice. Failure to use approved materials can result in code officers requiring you to rip out work already installed so that it can be replaced with proper materials. Effective material selection can save you money or make your bid for a job more competitive. By assessing all circumstances surrounding a job, your material selection can help you avoid problems later with callbacks and warranty work. Now, let's get down to business.

What Is an Approved Material?

An approved material is a material approved for specific uses, as determined by the local code enforcement office. Many approved materials and their approved uses are described in plumbing code books, but not all approved materials are listed. The materials detailed in code books represent the most commonly encountered approved materials. However, in certain cases other materials may be approved for use. This is frequently the case when a material exceeds the requirements listed in a code book.

The standards set in code books are normally the minimum acceptable standards. A product that exceeds these minimums may be allowed for use but not mentioned in the code. For general use, the materials listed in code books will be sufficient for your needs. Approved materials should be marked in a manner to identify themselves. Normally this identification will take the form of an embossed, molded, or indelible marking.

The type of identification marking used is frequently determined by the type of material being identified. Brass and copper fittings are often stamped. Plastic fittings usually have a molded marking. Pipe will typically carry a colored stripe and indelible letters to identify itself. For example, type M copper will have an indelible red stripe. Type L copper will have a blue stripe, and type K copper will have a green stripe. A yellow stripe will indicate a drain waste and vent (DWV) copper. By glancing at these color-codes, a code enforcement officer can quickly identify the type of copper being worked with.

A material is subject to local approval. For example, the water quality in a certain location may have an adverse effect on a particular type of pipe. In such a case, the local code enforcement office may deny the use of a material identified as an approved material in the code book. A similar deviation from the code may be made for pipe used below grade. The soils in some areas are not compatible with certain materials. Adjustments for local conditions may be made by the local authorities and must be checked to ensure the legality of material usage.

Materials approved for carrying potable water must not have a lead content of more than 8 percent. This applies to pipe and fittings. Solder used to join pipe and fittings for potable use may not contain more than 0.2 percent lead. There are several brands of lead-free solder available. The 50/50 solder, once common to the industry, is no longer approved for potable water systems.

When materials are approved, they are often approved for specific uses. It is not enough to say that PE pipe is approved for potable water use. In a sense, this is a true statement, but it has exceptions. PE pipe may not be used to convey hot water; it is not rated to convey hot water. Hot water is water with a temperature of at least 110°F. However, PE is approved to carry potable water as a water-service pipe. So, you see, PE is approved for potable water use, but its range of use is limited. This type of regulation confuses many people, but this chapter will clear the confusion for you. Now, let's look at approved materials and their allowed uses.

Water-Service Pipe

A water-service pipe is a pipe extending from a potable water source, a well or municipal water main, to the interior of a building. Once inside the building, the water-service pipe becomes a water-distribution pipe. This distinction is important, especially in zone three. When working in zone three, water-service pipe may not extend more than 5 feet (ft) into a building, unless it is of a material approved for water distribution. For example, a PE water service would have to be converted to an approved water-distribution material (Fig. 3.1) once it

Figure 3.1 Plastic insert adapter in PE pipe with a copper female adapter attached.

was in the building. But, a PB water service would not have to be converted because PB is an approved water-distribution pipe.

A water-service pipe must be rated for a pressure compatible with the pressure produced from the water source. Typically, a water service should be rated for 160 pounds per square inch (psi) at a temperature of 73.4°F. This rating will not always be applicable. If the pressure present at the water source is higher than 160 psi, the pipe's rating must be higher. On the other hand, if the water source is a well, and the water pressure from the well is below 160 psi, the rating of the pipe may be allowed to be lower. It is a good work principal to use pipe with a minimum working pressure of 160 psi even if a lower rating is approved. This allows for future changes that may increase the water pressure. The pipe must be rated with a pressure that is capable of withstanding the highest pressure developed from the water source.

Water-Service Materials

Now that you know what water service is, we are going to look at the materials approved for use for water service. The materials will be addressed in alphabetical order. Not all of the materials listed will be commonly used, but they are all approved materials. Remember, plumbing codes change and local jurisdictions have the authority to alter the code to meet local criteria. Before using these materials, or any other information in this book, consult your local code enforcement office to confirm the present status of your local code requirements.

Acrylonitrile butadiene styrene (ABS)

ABS pipe is a plastic pipe. It is normally used as a pipe for drains and vents, but if properly rated, it can be used for water service. It must meet certain specifications for pressure-rated potable water use and be approved by the local authorities. Since ABS is almost never used as a water-service pipe, it will be difficult to locate this material in a rating approved for water-service applications.

Asbestos cement pipe

Asbestos cement pipe has been used for municipal water mains in the past, but it is not used much today. It is, however, still an approved material.

Brass pipe

Brass pipe is an approved material for water-service piping, but it is rarely used. The complications of placing this metallic, threaded pipe below ground discourage its use. Brass pipe can also be used as a water-distribution pipe.

Cast-iron pipe

Cast-iron water pipe is approved for use for water service, but it would not be used for individual water supplies. Sometimes called ductile pipe, this pipe would be used in large water mains.

Copper pipe

Copper tubing is often referred to as copper pipe, but there is a difference. Copper pipe can be found with or without threads. Copper pipe is marked with a gray color code. Copper pipe is approved for water-service use, but copper tubing is used more often.

Copper tubing

Copper tubing is the copper most plumbers use. It can be purchased as soft copper, in rolls, or as rigid copper, in lengths resembling pipe. Copper tubing, frequently called copper pipe in the trade, has long been used as the plumber's workhorse. It is approved for water-service use and comes in many different grades.

The three types of copper normally used are type M, type L, and type K. The type rating refers to the wall thickness of the copper. Type K is the thickest, and type M is the thinnest, with type L in the middle. Type L is generally considered the most logical choice for water service. It offers more thickness than type M and is less expensive than type K. All three types are approved for water-service applications.

For a water-service pipe, soft, rolled copper is normally used. By using the coiled copper, plumbers can normally roll the tubing into the trench in a solid length, without joints. This reduces the risk of leaks at a later date. Copper water services are not as common as they once were. PE and PB are quickly reducing the use of copper. The reason is twofold. The plastic pipes are less expensive and are generally less affected by corrosion or other soil-related problems. Copper can also be used as a water-distribution pipe.

Chlorinated polyvinyl chloride

CPVC is a white or cream-colored plastic pipe allowed for use in water services when it is rated for potable water service. This pipe is not commonly used for water service, but it could be. CPVC is fairly fragile, especially when cold, and it requires joints if the run of pipe exceeds 20 ft. It is advisable to avoid joints in underground water services. CPVC can also be used as a water-distribution pipe.

Galvanized steel pipe

Galvanized steel pipe is an approved material for water services, but it is not a good choice. The pipe is joined together with threaded fittings. Over time, this pipe will rust. The rust can occur at the threaded areas, where the pipe walls are weakened, or inside the pipe. When the threaded areas rust, they can leak. When the interior of the pipe rusts, it can restrict the flow of water and reduce water pressure. While this gray, metal pipe is still available, it is rarely used in modern applications. If desired, galvanized pipe can be used as a water-distribution pipe.

Polybutylene

PB is available in rolled coils and in straight lengths. PB received mixed reviews when it was introduced to the plumbing trade.

PB is approved as a water-service pipe. It is available as a water-service-only pipe and as a water-service–water-distribution pipe. If the pipe is blue in color, it is intended only for water-service use. If the pipe is gray, it can be used for water service or water distribution. Before PB pipe can be used for potable water, it must be tested by a recognized testing agency and approved by local authorities. PB is relatively inexpensive, very flexible, and a good choice for most water-service applications. Due to recent problems and complaints pertaining to PB pipe, some code jurisdictions may not allow PB to be used for interior water distribution. This situation is changing as time goes along, so check your local code requirements closely before using PB pipe and fittings.

Cross-linked polyethylene

Cross-linked polyethylene is a plastic tubing that is approved for multiple uses. This tubing can be used for water service and water distribution for hot and cold water. It is becoming common for PEX tubing to be used in place of PB tubing. PEX is inexpensive, easy to use, and especially useful in remodeling jobs. Since PEX is made of plastic, it

does not corrode and is resistant to most elements that may deteriorate other types of pipe and tubing. The ease of installation and the
affordable cost of PEX tubing make it a personal favorite of many
plumbers.

Polyethylene

PE is a black, or sometimes bluish, plastic pipe that is frequently
used for water services. It resists chemical reactions, as does PB, and
it is fairly flexible. This pipe is available in long coils, allowing it to be
rolled out for great distances, without joints. PB may be one of the
most common materials used for water services. It, however, is not
rated as a water-distribution pipe. PE pipe is subject to crimping in
tight turns, but it is a good pipe that will give years of satisfactory
service.

Polyvinyl chloride pipe

PVC pipe is well known as a drain and vent pipe, but the PVC used
for water services is not the same pipe. Both pipes are white, but the
PVC used for water services must be rated for use with potable water.
PVC water pipe is not acceptable as a water-distribution pipe. It is
not approved for hot-water usage. Remember, when we talk of water-
distribution pipes, it is assumed the building is supplied with hot and
cold water. If the building's only water distribution is cold water,
some of the pipes, like PVC and PE, can be used.

Water-Distribution Pipe

What is water-distribution pipe? Water-distribution pipe is the piping
located inside a building that delivers potable water to plumbing
fixtures. The water-distribution system normally comprises both hot
and cold water. Because of this fact, the materials approved for water-
distribution systems are more limited than those allowed for water-
service piping.

A determining factor in choosing a pipe for water distribution is the
pipe's ability to handle hot water. A water-distribution pipe must be
approved for conveying hot water. In zone two, this means a pipe
rated for a minimum working pressure of 100 psi, at a temperature of
180°F. Zone three requires the rating to be 80 psi, at 180°F. The reason the pressure rating is lower than that of a water-service pipe is
simple. If the water pressure coming into a building exceeds 80 psi, a
pressure-reducing valve must be installed at the water service to
reduce the pressure to no more than 80 psi.

Water-Distribution Materials

Many water-distribution materials are also approved for use for water service. However, the reverse is not true; not all water-service materials are acceptable as water-distribution materials. Again in alphabetical order, you will find below the types of piping approved for water-distribution pipes.

Brass pipe

Brass pipe is suitable for water distribution, but it is not normally used in modern applications. While once popular, brass pipe has been replaced, in preference, by many new types of materials. The newer materials are easier to work with and usually provide longer service, with less problems.

Copper pipe and copper tubing

Both copper pipe and copper tubing are acceptable choices for water distribution. Copper tubing, sometimes mistakenly called pipe, is by far the more common choice. Copper pipe has been around for many years and has proved itself to be a good water-distribution pipe. If water has an unusually high acidic content, copper can be subject to corrosion and pinhole leaks. If acidic water is suspected, a thick-wall copper or a plastic-type pipe should be considered over the use of a thin-wall copper.

Zones two and three allow types M, L, and K to be used above and below ground, but zone one does not allow the use of type M copper when the copper is installed underground, within a building.

Galvanized steel pipe

Galvanized steel pipe remains in the approved category, but it is hardly ever used in new plumbing systems. The characteristics of galvanized pipe remove it from the competition. It is difficult to work with and is subject to rust-related problems. The rust can cause leaks and reduced water pressure and volume.

Polybutylene

Polybutylene tubing got of to a rocky start when it was first used for water distribution. The fittings used with the pipes, when not installed properly, failed and flooded buildings. New types of connections and fittings were created. The new connection methods seemed to cure the problem. However, in recent times, PB tubing has once again fallen on hard times with the public. The tubing that seemed to be a rising star now seems to be on its way out. PB is still approved in many areas, but public distrust of PB tubing is a major factor in bringing PEX tubing to the plumbing industry.

Cross-linked polyethylene

Cross-linked polyethylene is a plastic tubing that is approved for multiple uses. This tubing can be used for water service and water distribution for hot and cold water. It is becoming common for PEX tubing to be used in place of PB tubing. PEX is inexpensive, easy to use, and especially useful in remodeling jobs. Since PEX is made of plastic, it does not corrode and is resistant to most elements that may deteriorate other types of pipe and tubing. The ease of installation and the affordable cost of PEX tubing make it a personal favorite of many plumbers.

Drain Waste and Vent Pipe

Pipe used for the DWV system is considerably different from its water pipe cousins. The most noticeable difference is size. DWV pipes typically range in size from 1½ to 4 in, in diameter; some are smaller and some are much larger. When first glancing at the names of DWV pipe materials, they may seem the same as the water pipe variety, but there are differences. Remember, for plastic water pipes to carry potable water, they must be tested and approved for potable water use.

There are a number of materials approved for DWV purposes, but in practice, only a few are commonly used in modern plumbing applications. Let's look, in alphabetical order, at the materials approved for DWV systems.

Acrylonitrile butadiene styrene

ABS pipe is black, or sometimes a dark gray color, and will be labeled as a DWV pipe when it is meant for DWV purposes. ABS is normally used as a DWV pipe instead of as a water pipe. The standard weight rating for common DWV pipe is schedule 40.

ABS pipe is easy to work with and may be used above or below ground. It cuts well with a hacksaw or regular handsaw. This material is joined with a solvent-weld cement and rarely leaks, even in less than desirable installation circumstances. ABS was very popular for a long time, but in many areas it is being pushed aside by PVC pipe, the white plastic DWV pipe. ABS is extremely durable and can take hard abuse without breaking or cracking.

In zone one, the use of ABS is restricted to certain types of structures. ABS may not be used in buildings that have more than three habitable floors. The building may have a buried basement, where at least one-half of the exterior wall sections are at ground level or below. The basement may not be used as habitable space. So, it is possible to use ABS in a four-story building so long as the first story is buried in the ground, as stated above, and not used as living space.

Aluminum tubing

Aluminum tubing is approved for above-ground use only. Aluminum tubing may not be allowed in zone one. Aluminum tubing is usually joined with mechanical joints and coated to prevent corrosive action. This material is, like most others, available in many sizes. The use of aluminum tubing has not become common for average plumbing installations in most regions.

Borosilicate glass

Borosilicate glass pipe may be used above or below ground for DWV purposes in zone two. Underground use requires a heavy schedule of pipe. Zones one and three do not recognize this pipe as an approved material. However, as with all regulations, local authorities have the power to amend regulations to suit local requirements.

Brass pipe

Brass pipe could be used as a DWV pipe, but it rarely is. The degree of difficulty in working with it is one reason it is not used more often. Zones two and three don't allow brass pipe to be used below grade for DWV purposes.

Cast-iron pipe

Cast-iron pipe has long been a favored DWV pipe. Cast iron has been used for many years and provides good service for extended periods of time. The pipe is available in a hub-and-spigot style, the type used years ago, and in a hubless version. The hubless version is newer and is joined with mechanical joints, resembling a rubber coupling (Fig. 3.2), surrounded and compressed by a stainless steel band.

The older, bell-and-spigot, or hub-and-spigot, type of cast iron is what is most often encountered during remodeling jobs. This type of cast iron was normally joined with oakum and molten lead. Today,

Figure 3.2 Hubless band for cast-iron connections.

however, there are rubber adapters available for creating joints with this type of pipe. These rubber adapters will also allow plastic pipe to be mated to the cast iron.

Cast-iron pipe is frequently referred to as soil pipe. This nickname separates DWV cast iron from cast iron designed for use as a potable water pipe. Cast iron is available as a service-weight pipe and as an extra-heavy pipe. Service-weight cast iron is the most commonly used. Even though the cost of labor and material for installing cast iron is more than it is for schedule 40 plastic, cast iron still sees frequent use, both above and below grade.

Cast iron is sometimes used in multifamily dwellings and custom homes to deaden the sound of drainage as it passes down the pipe in walls adjacent to living space. If chemical or heat concerns are present, cast iron is often chosen over plastic pipe.

Copper

Copper pipe is made in a DWV rating. This pipe is thin walled and identified with a yellow marking. The pipe is a good DWV pipe, but it is expensive and time-consuming to install. DWV copper is not normally used in new installations unless extreme temperatures, such as those from a commercial dishwasher, warrant the use of a nonplastic pipe. DWV copper is approved for use above and below ground in zones one and two. Zone three requires a minimum copper rating of type L for copper used underground for DWV purposes.

Galvanized steel pipe

Galvanized steel pipe keeps popping up as an approved material, but it is no longer a good choice for most plumbing jobs. As galvanized pipe ages and rusts, the rough surface, from the rust, is prone to catching debris and creating pipe blockages. Another disadvantage to galvanized DWV pipe is the time it takes to install the material.

Galvanized pipe is not allowed for underground use in DWV systems. When used for DWV purposes, galvanized pipe should not be installed closer than 6 in to the earth.

Lead pipe

Lead pipe is still an approved material, but like galvanized pipe, it has little place in modern plumbing applications. Zone two does not allow the use of lead for DWV installations. Zone three limits the use of lead to above-grade installations.

Polyvinyl chloride pipe

PVC is probably the leader in today's DWV pipe. This plastic pipe is white and is normally used in a rating of schedule 40. PVC pipe uses a solvent weld joint and should be cleaned and primed before being glued together. This pipe will become brittle in cold weather. If PVC is dropped on a hard surface while the pipe is cold, it is likely to crack or shatter. The cracks can go unnoticed until the pipe is installed and tested. Finding a cracked pipe moments before an inspector is to arrive is no fun, so be advised: handle cold PVC with care. PVC may be used above or below ground.

In zone one, the use of PVC is restricted to certain types of structures. PVC may not be used in buildings that have more than three habitable floors. The building may have a buried basement, where at least one-half of the exterior wall sections are at ground level or below. The basement may not be used as habitable space. So, it is possible to use PVC in a four-story building so long as the first story is buried in the ground, as stated above, and not used as living space.

If the underground piping will be used as a building sewer, one of the following types of pipes may be used:

- ABS
- Cast iron
- Vitrified clay

- PVC
- Concrete
- Asbestos cement

In zone three, bituminized-fiber pipe, type L copper pipe, and type K copper pipe may be used. Zone one tends to stick to the general guidelines, as given in the previous paragraphs.

If a building sewer will be installed in the same trench that contains a water service, some of the above pipes may not be used. Standard procedure for pipe selection under these conditions calls for the use of a pipe approved for use inside a building. These types of pipes could include ABS, PVC, and cast iron. Pipes that are more prone to breakage, like a clay pipe, are not allowed unless special installation precautions are taken.

Sewers installed in unstable ground are also subject to modified rulings. Normally, any pipe approved for use underground, inside a building, will be approved for use with unstable ground. But, the pipe must be well supported for its entire length.

Chemical wastes must be conveyed and vented with a system separated from the building's normal DWV system. The material requirements for chemical-waste piping must be obtained from the local code enforcement office.

Storm Drainage Materials

The materials used for interior and underground storm drainage may generally be the same materials used for sanitary drainage. Any approved DWV material is normally approved for use in storm drainage.

Inside storm drainage

Materials commonly approved in zone three for interior storm drainage include the following:

- ABS
- PVC
- Type DWV copper
- Type M copper
- Type L copper
- Type K copper
- Asbestos cement
- Bituminized fiber
- Cast iron
- Concrete
- Vitrified clay
- Aluminum
- Brass
- Lead
- Galvanized

Zone two's approved pipe materials for interior storm drainage include:

- ABS
- Type DWV copper
- Type L copper
- Asbestos cement
- Concrete
- Galvanized
- Brass

- PVC
- Type M copper
- Type K copper
- Cast iron
- Aluminum
- Black steel
- Lead

Zone one follows its standard pipe approvals for storm-water piping. Storm-drainage sewers are a little different. If you are installing a storm-drainage sewer, use one of the following types of pipes:

Zone one. Follow the basic guidelines for approved piping.

Zone two

- Cast iron
- Vitrified clay
- ABS
- Aluminum (coated to prevent corrosion)

- Asbestos cement
- Concrete
- PVC

Zone three

- Cast iron
- Vitrified clay
- ABS
- Bituminized fiber
- Type K copper
- Type DWV copper

- Asbestos cement
- Concrete
- PVC
- Type L copper
- Type M copper

Subsoil Drains

Subsoil drains are designed to collect and drain water entering the soil. They are frequently slotted pipes and could be made from any of the following materials:

- Asbestos cement
- Vitrified clay
- Cast iron

- Bituminized fiber
- PVC
- PE

Zone two does not allow bituminized fiber pipe to be used as a subsoil drain. It may not allow cast iron or some plastics. Zone one sticks to its normal pipe approvals. As always, check with local authorities before using any material.

Other Types of Materials

We have concluded our look at the various types of pipes approved for plumbing, but there are other types of materials to take into consideration. Valves, fittings, and nipples all fall under the watchful eye of the code enforcement office. In addition to these materials, there are still others to be discussed.

Fittings

Fittings that are made from cast iron, copper, plastic, steel, and other types of iron are all approved for use in their proper place. Generally speaking, fittings must either be made from the same material as the pipe they are being used with or be compatible with the pipe.

Valves

Valves have to meet some standards, but most of the decision for the use of valves will come from the local code enforcement office. Valves, like fittings, must be either of the same material as the pipe they are being used with or compatible with the pipe. Size and construction requirements will be stipulated by local jurisdictions.

Nipples

Manufactured pipe nipples are normally made from brass or steel. These nipples range in length from ⅛ to 12 in. Nipples must live up to certain standards, but they should be rated and approved before you are able to obtain them.

Flanges

Closet flanges (Fig. 3.3) made from plastic must have a thickness of ¼ in. Brass flanges may have a thickness of only ⅛ in. Flanges intended for caulking must have a thickness of ¼ in, with a caulking depth of 2 in. The screws or bolts used to secure flanges to a floor must be brass. All flanges must be approved for use by the local authorities.

Zone two prohibits the use of offset flanges (Fig. 3.4), without prior approval. Zone three requires hard-lead flanges to weigh at least 25 ounces (oz) and to be made from a lead ally with no less than a 7.75 percent antimony, by weight. Zone one requires flanges to have a

Figure 3.3 Closet flange.

Figure 3.4 Offset flange.

diameter of about 7 in. In zone one, the combination of the flange and the pipe receiving it must provide about 1½ in of space to accept the wax ring or sealing gasket.

Cleanout plugs

Cleanout plugs will be made of plastic or brass. Brass plugs are to be used only with metallic fittings. Unless they create a hazard, cleanout plugs shall have raised, square heads. If located where a hazard from the raised head may exist, countersunk heads may be used. Zone two requires borosilicate glass plugs to be used with cleanouts installed on borosilicate pipe.

Fixtures

Plumbing fixtures are regulated and must have smooth surfaces. These surfaces must be impervious. All fixtures must be in good working order and may not contain any hidden surfaces that may foul or become contaminated.

The rest of them

Lead bends and traps are not used much anymore, but if you decide to use these items, check with your code officer for guidance. These units must have a wall thickness of at least ⅛ in.

The days of using sheet lead and copper to form shower pans are all but gone, but the code does still offer minimum standards for these materials. Lead shower pans must not be rated at less than 4 pounds per square foot (lb/ft^2). If you need to use a lead pipe flashing, it should be rated at a minimum of 3 lb/ft^2. Copper shower pans should weigh in at 12 oz/ft^2, and copper flashings should have a minimum weight of 8 oz/ft^2.

When nonmetallic material is used for a shower pan, it must meet minimum standards. The material must be marked to indicate its approved qualities. Normally, membrane-type material is required to have a minimum thickness of 0.040 in. If the material must be joined together at seams, it must be joined in accordance with the manufacturer's recommended procedure. Paper-type shower pans are also allowed for use when they meet minimum construction requirements.

Soldering bushings, once used to adapt lead pipe to other materials, and caulking ferrules, used in the conversion of cast iron to other materials, are all but a thing of the past. Lead pipe is usually removed in today's plumbing and rubber adapters are used in place of caulking ferrules. If you have a nostalgic interest in these old-school items, you can find standards for them in your code book.

Connecting Your Materials

You now know what materials you can use for various purposes; next, you are going to learn how you may connect them. The connection methods approved are normally approved on a local level. But, there are some basics, and that's what follows.

Compatibility and performance

The main considerations in a good connection are compatibility and performance. A connection must be able to endure the pressure exerted on it in a normal testing procedure for the plumbing. For DWV connections, this usually amounts to a pressure of between 4 and 5 psi. On water pipes, the pressure for a test should be equal to the highest pressure expected to be placed on the system once it is in use. Zone two requires a pressure test with a pressure of at least 25 psi more than the highest working pressure.

Basic preparation

Before pipe is connected with fittings, the pipe should be properly prepared. This means cutting the pipe evenly and clearing it of any burrs or obstructions. If a connection will be made with a material of different construction than that of the pipe, the connector must be compatible with the pipe. For example, a rubber coupling, held in place with stainless steel clamps, can be used to join numerous types of different pipe styles.

ABS and PVC

Plastic pipes should be joined with solvent cements designed for the specific type of pipe being worked with. The plastic pipe and fittings should be clean, dry, and grease-free before a joint is made. When working with plastic pipes, a cleaner and primer are often recommended, prior to the application of solvent cement. Always follow the manufacturer's recommendations for joining pipe and fittings. For best results, once the pipe has been coated with cement and inserted in a fitting, it should be turned about a quarter of a turn. This helps to spread the glue and make a better joint.

ABS pipe and fittings tend to harden much quicker than PVC. ABS is also less sensitive to dirt and water in its joints. This is not to say that you can ignore proper procedures with ABS, but it is more forgiving than PVC.

Unusual pipes

Some types of materials, like asbestos cement pipe and bituminized-fiber pipe, only lend themselves to mechanical joints. Refer to manufacturer's recommendations, your code book, and the local code officer for the best methods in joining this type of pipe.

Cast-iron pipe

Cast-iron pipe offers a myriad of ways to join it with fittings. You can use hot lead, rubber doughnuts, rubber couplings, or special bands designed for use with hubless cast iron, just to name a few.

In the old days, cast iron was almost always joined with a caulked joint, using molten lead. This procedure is still used today. To make this kind of joint, oakum or hemp is placed into the hub of a cast-iron fitting, around the pipe being joined to the fitting. The oakum or hemp must be dry when it is installed if it is to make a good joint. Once the packing is in place, molten lead is poured into the hub. Once the lead is poured, a packing tool, basically a special chisel, is used with a hammer to drive the lead down deeper into the hub. Once the oakum becomes wet, it expands and seals the joint.

Without experience or an instructor, this type of joint can be extremely dangerous to make. The hot lead can take the skin right off your bones, and if the hot lead comes into contact with a wet surface, it can explode, causing personal injury. Proper clothing and safety gear are a necessity on this type of job, even for seasoned professionals.

If you are using lead to make joints in cast iron meant to carry potable water, you will not use oakum or hemp. Instead, you will use a rope packing. This type of packing has a higher density than oakum and is meant for use with potable water installations.

Rubber doughnuts, as they are called in the trade, offer an alternative to caulking with hot lead. These special rubber adapters are placed in the hub of a fitting and lubricated. The end of the pipe to be joined is inserted into the rubber gasket and driven or pulled into place. There are special tools used to join soil pipe in this manner (Fig. 3.5), but some plumbers use a block of wood and a sledge hammer to drive the pipe into the doughnut.

Figure 3.5 Soil pipe assembly tool.

If a mechanical joint is being used on cast iron meant for potable water, there must be an elastomeric gasket on the joint, held in place with an approved flange. One example of a mechanical joint with an elastomeric gasket and approved flange is the standard band used with hubless cast iron, but there are other types of mechanical joints available and approved for these types of joints.

Copper

When it comes to copper, your options for joints include compressions fittings (Fig. 3.6), soldered joints, screw joints, and other types of mechanical joints. Solder used to join copper for potable water systems must contain less than 0.02 percent lead. Threaded joints must be sealed with an acceptable pipe compound or tape. Welded and brazed joints offer two other ways to join your copper. Flared joints are still another way to mate copper to its fittings (Fig. 3.7). Unions (Figs. 3.8 and 3.9) are allowed for connecting copper and are usually required with the installation of water heaters.

Figure 3.6 Compression tee.

Figure 3.7 Flare joint.

Figure 3.8 Closed union.

Figure 3.9 Open union.

CPVC pipe

CPVC pipe is the homeowner's friend. Homeowners flock to this plastic water pipe because all they have to do is glue it together, or at least, that's what they think. CPVC is a finicky pipe. It must be primed with an approved primer before it is glued. If this step is ignored, the joint will not be as strong as it should be. CPVC also requires a long time for its joints to set up. This pipe must be clean, dry, and grease-free before you begin the connection process. Don't eliminate the priming process; it is essential for a good joint.

Figure 3.10 Rubber coupling.

Galvanized pipe

Galvanized pipe lends itself to threaded connections, but rubber couplings can also be used. Use the large rubber couplings, not the hubless cast-iron type (Fig. 3.10). Remember to apply the appropriate pipe dope or tape to the threads if you are making a screw connection. There are other types of approved ways to connect galvanized pipe, but these are the easiest.

Polybutylene

PB pipe is normally connected to insert fittings with special clamps (Fig. 3.11). These clamps are installed with a crimping tool designed for use with PB pipe (Fig. 3.12). Compression fittings are also allowed with PB pipe, but the ferrules should be nonmetallic to avoid cuts in the pipe. Flaring is possible with a special tool, but it is rarely needed. Another method for joining PB pipe is heat fusion. This process is not normally used with water-distribution piping; crimp rings are the most common method of joining PB pipe in a water-distribution system.

Cross-linked polyethylene

PEX tubing is connected with the same procedures described for PB tubing. It is essential that the proper fittings, couplings, and rings be used. It is not acceptable to mix PB connection applications with PEX materials.

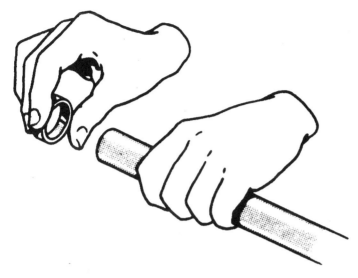

Slide a copper crimp ring over the pipe end.

Figure 3.11 PB crimp ring. (*Courtesy of Vanguard Plastics, Inc.*)

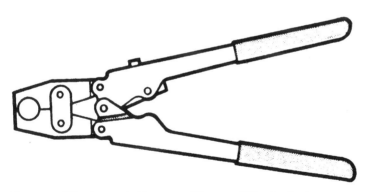

Figure 3.12 PB crimp tool. (*Courtesy of Vanguard Plastics, Inc.*)

Polyethylene

PE pipe is typically joined with inscrt fittings and stainless steel clamps (Fig. 3.13). The insert fitting is placed inside the pipe and a stainless steel clamp is applied outside of it over the insert fitting's shank. For best results, use two clamps.

Figure 3.13 Plastic insert tee in PE pipe.

Fixtures

Fixtures must be connected to their drains in an approved manner. For sinks and such, the typical connection is made with slip nuts and washers. The washers may be rubber or nylon. For toilets, the seal is normally made with a wax ring.

Pipe penetrations

Pipe penetrations must be sealed to protect against water infiltration, the spreading of fire, and rodent activity. When a pipe penetrates a wall or roof, it must be adequately sealed to prevent the above-mentioned problems. Many styles of roof flashings are available for sealing pipe holes (Figs. 3.14 and 3.15).

Reminder Notes

Zone one

1. Type M copper tubing may not be used below grade when used inside a building.

Figure 3.14 Metal roof flange.

Figure 3.15 Flexible roof flange.

2. ABS and PVC pipe for drainage is restricted to use in buildings having no more than three levels of habitable space. A fourth level is allowed under certain conditions. The conditions are that the first of the four levels must have no more than one-half of its exterior wall surface above grade, and this level may not be used as living space.

3. Aluminum tubing may not be approved for use as a DWV pipe. Check with local regulations.

4. Closet flanges should have a diameter of about 7 in. When combined with the pipe receiving it, the flange should offer an area of about 1½ in to accept a wax ring or sealing gasket.

Zone two

1. Pipe used to convey hot water must have a minimum working pressure of 100 psi, at 180°F.

2. Brass pipe may not be used for DWV purposes below ground level.

3. Borosilicate glass pipe is approved for above and below ground use in DWV systems.

4. Lead pipe may not be used in a new DWV system.

5. Copper pipe and bituminized-fiber pipe are not allowed for use as building sewers.

6. Zone two allows coated aluminum tubing to be used for underground storm-water sewer applications.

7. Zone two does not allow bituminized-fiber pipe for subsoil drains. It may not allow cast-iron pipe and some plastic pipes.

8. Offset closet flanges are prohibited without prior approval.

9. Borosilicate-glass cleanouts are required to have borosilicate glass plugs.

10. When testing water pipe, the test pressure must be at least 25 psi higher than the highest working pressure.

Zone three

1. Water-service pipe must not extend more than 5 ft inside a building before being converted to a water-distribution pipe unless the pipe used for the water service is also approved for water distribution.

2. Pipe used to convey hot water must have a minimum working pressure of 80 psi, at 180°F.

3. Brass pipe may not be used for DWV purposes below ground level.

4. When used for underground DWV purposes, copper must have a minimum rating of type L.

5. Lead pipe, when used as a DWV material, is limited to use above ground.

6. Zone three allows the use of vitrified-clay and bituminized-fiber pipe in interior storm-water applications.

7. Zone three allows the use of vitrified-clay, bituminized-fiber, and copper pipe in storm-water sewer applications.

8. Lead closet flanges must weigh at least 25 oz and be made from a lead alloy with no less than 7.75 percent antimony, by weight.

Drainage Systems

Drainage systems intimidate many people. When these people look at their code books, they see charts and math requirements that make them nervous. Their fear is largely unjustified. For the inexperienced, the fundamentals of building a suitable drainage system can appear formidable. But, with a basic understanding of plumbing, the process becomes much less complicated. This chapter is going to take you, step by step, through the procedures of making a working drainage system.

You are going to learn the criteria for sizing pipe. You will be shown which types of fittings can be used in various applications. During the process, you will be given instructions for the proper installation of a drainage system. Then, the focus will shift to indirect and special wastes. Storm drainage will complete the chapter. By the end of this chapter, you will be quite knowledgeable on the subject of drainage systems.

Pipe Sizing

Sizing pipe for a drainage system is not difficult. You must know a few benchmark numbers, but you don't have to memorize these numbers. Your code book will have charts and tables that provide the benchmarks. All you must know is how to interpret and use the information provided.

The size of a drainage pipe is determined by using various factors, the first of which is the drainage load. This refers to the volume of drainage the pipe will be responsible for carrying. When you refer to your code book, you will find ratings that assign a fixture-unit value to various plumbing fixtures. For example, a residential toilet has a fixture-unit value of 4. A bathtub's fixture-unit value is 2.

By using the ratings given in your code book, you can quickly assess the drainage load for the system you are designing. Since plumbing fixtures require traps, you must also determine what size traps are required for particular fixtures. Again, you don't need a math degree to accomplish this task. In fact, your code book will tell you what trap sizes are required for most common plumbing fixtures.

Your code book will provide trap-size requirements for specific fixtures. For example, by referring to the ratings in your code book, you will find that a bathtub requires a 1½-in trap. A lavatory may be trapped with a 1¼-in trap. The list will go on to describe the trap needs for all common plumbing fixtures. Trap sizes will not be provided for toilets, since toilets have integral traps.

When necessary, you can determine a fixture's drainage-unit value by the size of the fixture's trap. A 1¼-in trap, the smallest trap allowed, will carry a fixture-unit rating of 1. A 1½-in trap will have a fixture-unit rating of 2. A 2-in trap will have a rating of 3 fixture units. A 3-in trap will have a fixture-unit rating of 5, and a 4-in trap will have a fixture-unit rating of 6. This information can be found in your code book and may be applied for a fixture not specifically listed with a rating in your code book. Table 4.1 shows this information for zone two. Table 4.2 shows the same information for zone three. Zone one uses a different rating for the fixture-unit loads on traps. Refer to Table 4.3 for zone one's requirements.

TABLE 4.1   Zone Two's Fixture-Unit Requirements on Trap Sizes

Trap size (in)	No. of fixture units
1¼	1
1½	2
2	3
3	5
4	6

TABLE 4.2   Zone Three's Fixture-Unit Requirements on Trap Sizes

Trap size (in)	No. of fixture units
1¼	1
1½	2
2	3
3	5
4	6

TABLE 4.3   Zone One's Fixture-Unit Requirements on Trap Sizes

Trap size	No. of fixture units
1¼	1
1½	3
2	4
3	6
4	8

Determining the fixture-unit value of a pump does require a little math, but it's simple. By taking the flow rate, in gallons per minute (gpm), assign 2 fixture units for every gpm of flow. For example, a pump with a flow rate of 30 gpm would have a fixture-unit rating of 60. Zone three is more generous. In zone three, 1 fixture unit is assigned for every 7½ gpm. With the same pump, producing 30 gpm, zone three's fixture-unit rating would be 4. That's quite a difference from the ratings in zones one and two.

Other considerations when sizing drainage pipe is the type of drain you are sizing and the amount of fall that the pipe will have. For example, the sizing for a sewer will be done a little differently than the sizing for a vertical stack. A pipe with a ¼-in fall will be rated differently than the same pipe with a ⅛-in fall.

Sizing Building Drains and Sewers

Building drains and sewers use the same criteria in determining the proper pipe size. The two components you must know to size these types of pipes are the total number of drainage fixture units entering the pipe and the amount of fall placed on the pipe. The amount of fall is based on how much the pipe drops in each foot it travels. A normal grade is generally ¼ in/ft, but the fall could be more or less.

When you refer to your code book you will find information, probably a table, to aid you is sizing building drains and sewers. Let's take a look at how a building drain for a typical house would be sized in zone three.

Sizing—Example 1

Our sample house has two and one-half bathrooms, a kitchen, and a laundry room. To size the building drain for this house, we must determine the total fixture-unit load that may be placed on the building drain. To do this, we start by listing all of the plumbing fixtures producing a drainage load. In this house we have the following fixtures:

- One bathtub
- Three toilets
- One shower
- Three lavatories

- One kitchen sink
- One dishwasher
- One laundry tub
- One clothes washer

By using Table 4.4, we can determine the number of drainage fixture units assigned to each of these fixtures. When we add up all the fixture units, we have a total load of 28 fixture units. It is always best to allow a little extra in your fixture-unit load so your pipe will be in no danger of becoming overloaded. The next step is to look at Table 4.5 to determine the sizing of our building drain. Table 4.6 shows pitch allowances for zone two, and Table 4.7 shows standards for zone one.

TABLE 4.4   Fixture-Unit Ratings in Zone Three

Fixture	Rating
Bathtub	2
Shower	2
Residential toilet	4
Lavatory	1
Kitchen sink	2
Dishwasher	2
Clothes washer	3
Laundry tub	2

TABLE 4.5   Zone Three's Minimum Drainage-Pipe Pitch

Pipe diameter (in)	Pitch (in/ft)
Under 3	¼
3–6	⅛
8 or larger	1/16

TABLE 4.6   Zone Two's Minimum Drainage-Pipe Pitch

Pipe diameter (in)	Pitch (in/ft)
Under 3	¼
3 or larger	⅛

TABLE 4.7   Zone One's Minimum Drainage-Pipe Pitch

Pipe diameter (in)	Pitch (in/ft)
Under 4	¼
4 or larger	⅛

TABLE 4.8   **Building-Drain Sizing Table for Zone Three**

Pipe size (in)	Pipe grade (in/ft)	Maximum no. of fixture units
2	¼	21
3	¼	42*
4	¼	216

* No more than two toilets may be installed on a 3-in building drain.

Our building drain will be installed with a ¼-in fall. By looking at Table 4.8, we see that we can use a 3-in pipe for our building drain, based on the number of fixture units, but, notice the footnote below the chart. The note indicates that a 3-in pipe may not carry the discharge of more than two toilets, and our test house has three toilets. This means we will have to move up to a 4-in pipe.

Suppose our test house only had two toilets, what would the outcome be then? If we eliminate one of the toilets, our fixture load drops to 24. According to the table, we could use a 2½-in pipe, but we know our building drain must be at least a 3-inch pipe, to connect to the toilets. A fixture's drain may enter a pipe the same size as the fixture drain or a pipe that is larger, but it may never be reduced to a smaller size, except with a 4- by 3-in closet bend.

So, with two toilets, our sample house could have a building drain and sewer with a 3-in diameter. But, should we run a 3- or a 4-in pipe? In a highly competitive bidding situation, 3-in pipe would probably win the coin toss. It would be less expensive to install a 3-in drain, and you would be more likely to win the bid on the job. However, when feasible, it would be better to use a 4-in drain. This allows the homeowner to add another toilet at some time in the future. If you install a 3-in sewer, the homeowner would not be able to add a toilet without replacing the sewer with 4-in pipe.

Horizontal Branches

Horizontal branches are the pipes that branch off from a stack to accept the discharge from fixture drains. These horizontal branches normally leave the stack as a horizontal pipe, but they may turn to a vertical position, while retaining the name of horizontal branch. The procedure for sizing a horizontal branch is similar to the one used to size a building drain or sewer, but the ratings are different. Your code book will contain the benchmarks for your sizing efforts, but here are some examples.

The number of fixture units allowed on a horizontal branch is determined by pipe size and pitch. All of the following examples are based on a pitch of 1¼ in/ft. A 2-in pipe can accommodate up to 6 fix-

TABLE 4.9   Example of Horizontal-Branch Sizing Table in Zone Two*

Pipe size (in)	Maximum no. of fixture units on a horizontal branch
1¼	1
1½	2
2	6
3	20[†]
4	160
6	620\

* Table does not represent branches of the building drain, and other restrictions apply under battery-venting conditions.
[†] Not more than two toilets may be connected to a single 3-in horizontal branch. Any branch connecting with a toilet must have a minimum diameter of 3 in.

ture units, except in zone one, where it can have 8 fixture units. A 3-in pipe can handle 20 fixture units but not more than two toilets. In zone one, a 3-in pipe is allowed up to 35 units and up to three toilets. A 1½-in pipe will carry 3 fixture units, unless you are in zone one. Zone one only allows a 1½-in pipe to carry 2 fixture units, and they may not be from sinks, dishwashers, or urinals. A 4-in pipe will take up to 160 fixture units, except in zone one, where it will take up to 216 units. Table 4.9 gives you an example of how a table for sizing horizontal fixture units might be assembled in your code book.

Stack Sizing

Stack sizing is not too different from the other sizing exercises we have studied. When you size a stack, you must base your decision on the total number of fixture units carried by the stack and the amount of discharge into branch intervals. This may sound complicated, but it isn't.

Look at Tables 4.10 and 4.11 You will notice that there are three columns. The first is for pipe size, the second represents the discharge of a branch interval, and the last column shows the ratings for the total fixture-unit load on a stack. This table is based on a stack with no more than three branch intervals. See Fig. 4.1 for an example of what is meant by the limit of three branch intervals.

Sizing the stack requires you to first determine the fixture load entering the stack at each branch interval. Here is an example of how this type of sizing works. In our example we will size a stack that has two branch intervals. Fig. 4.2 shows you what the stack and branches look like. The lower branch has a half-bath and a kitchen on it. Using the ratings from zones two and three, the total fixture-unit count for this branch is 6. This is determined by the table providing fixture-unit ratings for various fixtures.

TABLE 4.10   Stack-Sizing Table for Zone Three

Pipe size (in)	Fixture-unit discharge on stack from a branch	Total fixture units allowed on stack
1½	2	4
2	6	10
3	20*	48*
4	90	240

* No more than two toilets may be placed on a 3-in branch, and no more than six toilets may be connected to a 3-in stack.

TABLE 4.11   Stack-Sizing Table for Zone Two

Pipe size (in)	Fixture-unit discharge on stack from a branch	Total fixture units allowed on stack
1½	3	4
2	6	10
3	20*	30*
4	160	240

* No more than two toilets may be placed on a 3-in branch, and no more than six toilets may be connected to a 3-in stack.

The second stack has a full bathroom group on it. The total fixture-unit count on this branch, using sizing from zone two and three, is six, if you use a bathroom group rating, or seven, if you count each fixture individually. It is better to use the larger of the two numbers, for a total of seven.

When you look at Table 4.10, look at the horizontal listings for a 3-in pipe. You know the stack must have a minimum size of 3 in to accommodate the toilets. As you look across the table, you will see that each 3-in branch may carry up to 20 fixture units. Well, your first branch has 6 fixture units and the second branch has 7 fixture units, so both branches are within their limits.

When you combine the total fixture units from both branches, you have a total of 13 fixture units. Continuing to look across the table you see that the stack can accommodate up to 48 fixture units in zone three and up to 30 fixture units in zone two. Obviously, a 3-in stack is adequate for your needs. If the fixture-unit loads had exceeded the numbers in either of the columns, the pipe size would have had to have been increased.

If you are sizing a stack that extends upward with more than three branch intervals, you must use different numbers. See Tables 4.12 and 4.13 for examples of how the fixture units on the stack are allowed to be increased with the taller stack. The numbers used for

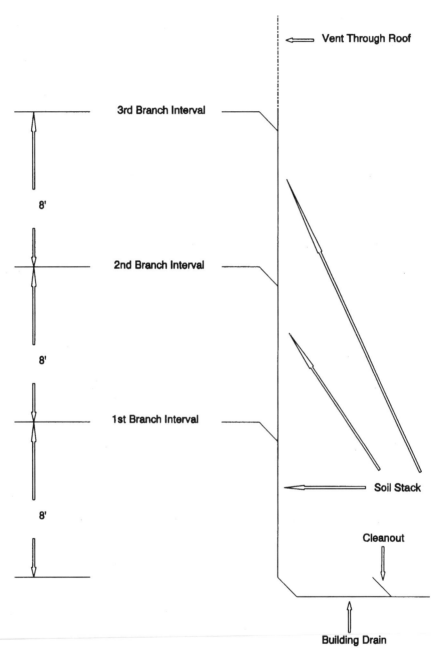

Figure 4.1 Branch-interval detail.

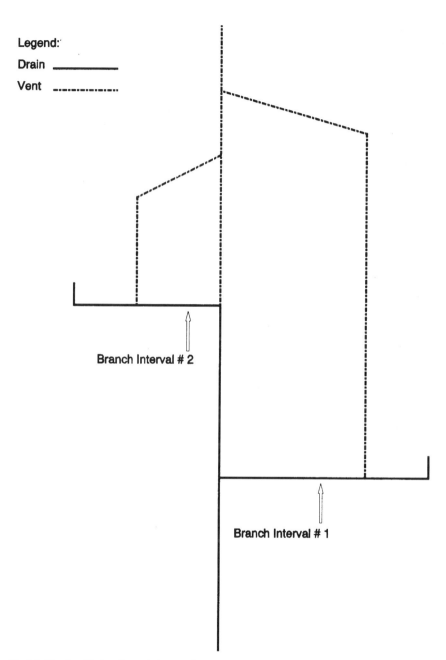

Legend:

Drain ──────────

Vent ·─··─··─··─··─··

Branch Interval # 2

Branch Interval # 1

Fig 4.2 Stack with two branch intervals.

TABLE 4.12   Stack-Sizing Tall Stacks in Zone Two (Stacks with More Than Three Branch Intervals)

Pipe size (in)	Fixture-unit discharge on stack from a branch	Total fixture units allowed on stack
1½	2	8
2	6	24
3	16*	60*
4	90	500

* No more than two toilets may be placed on a 3-in branch, and no more than six toilets may be connected to a 3-in stack.

TABLE 4.13   Stack-Sizing Tall Stacks in Zone Three (Stacks with More Than Three Branch Intervals)

Pipe size (in)	Fixture-unit discharge on stack from a branch	Total fixture units allowed on stack
1½	2	8
2	6	24
3	20*	72*
4	90	500

* No more than two toilets may be placed on a 3-in branch, and no more than six toilets may be connected to a 3-in stack.

rating the load on branch intervals will remain the same, but the number of total fixture units allowed on the taller stack is increased.

When sizing a stack, it is possible that the developed length of the stack will comprise different sizes of pipe. For example, at the top of the stack the pipe size may be 3 in, but at the bottom of the stack the pipe size may be 4 in. This is because as you get to the lower portion of the stack, the total fixture units placed on the stack are greater.

Pipe Installations

Once the pipe is properly sized, it is ready for installation. There are a few regulations pertaining to pipe installation that you need to be aware of.

Grading your pipe

When you install horizontal drainage piping, you must install it so that it falls toward the waste-disposal site. A typical grade for drainage pipe is ¼ in of fall per foot. This means the lower end of a 20-ft piece of pipe would be 5 in lower than the upper end, when properly installed.

While the ¼-in/ft grade is typical, it is not the only acceptable grade for all pipes.

If you are working with pipe that has a diameter of 2½ in, or less, the minimum grade for the pipe is ¼ in/ft. Pipes with diameters between 3 and 6 in are allowed a minimum grade of ⅛ in/ft. Zone one requires special permission to be granted prior to installing pipe with a ⅛-in/ft grade. In zone three, pipes with diameters of 8 in or more, an acceptable grade is ¹⁄₁₆ in/ft.

Supporting Your Pipe

How you support your pipes is also regulated by the plumbing code. There are requirements for the type of materials you may use and how they may be used. Let's see what they are.

One concern with the type of hangers used is their compatibility with the pipe they are supporting. You must use a hanger that will not have a detrimental effect on your piping. For example, you may not use galvanized strap hanger to support copper pipe. As a rule of thumb, the hangers used to support a pipe should be made from the same material as the pipe being supported. For example, copper pipe should be hung with copper hangers. This eliminates the risk of a corrosive action between two different types of materials. If you are using a plastic or plastic-coated hanger, you may use it with all types of pipe. The exception to this rule might be when the piping is carrying a liquid with a temperature that might affect or melt the plastic hanger.

The hangers used to support pipe must be capable of supporting the pipe at all times. The hanger must be attached to the pipe and to the member holding the hanger in a satisfactory manner. For example, it would not be acceptable to wrap a piece of wire around a pipe and then wrap the wire around the bridging between two floor joists. Hangers should be securely attached to the member supporting it. For example, a hanger should be attached to the pipe and then nailed to a floor joist. The nails used to hold a hanger in place should be made of the same material as the hanger if corrosive action is a possibility.

Both horizontal and vertical pipes require support. The intervals between supports will vary, depending upon the type of pipe being used and whether it is installed vertically or horizontally. The following examples will show you how often you must support the various types of pipes when they are hung horizontally. These examples are the maximum distances allowed between supports for zone three:

- ABS—every 4 in
- Galvanized—every 12 in
- DWV copper—every 10 in

- Cast iron—every 5 in
- PVC—every 4 in

When these same types of pipes are installed vertically, in zone three, they must be supported at no less than the following intervals:

- ABS—every 4 in
- Galvanized—every 15 in
- DWV copper—every 10 in
- Cast iron—every 15 in
- PVC—every 4 in

Table 4.14 for horizontal pipe support intervals required in zone one and Table 4.15 for vertical support intervals. For zone two, see Table 4.16 for horizontal support requirements and Table 4.17 for vertical support intervals.

When installing cast-iron stacks, the base of each stack must be supported because of the weight of cast-iron pipe. See Figure 4.3 for an example of how to support the base of a cast-iron stack.

TABLE 4.14   Horizontal Pipe-Support Intervals in Zone One

Support material	Maximum distance of supports (ft)
ABS	4
Cast iron	At each pipe joint*
Galvanized (1 in and larger)	12
Galvanized (¾ in and smaller)	10
PVC	4
Copper (2 in and larger)	10
Copper (1½ in and smaller)	6

* Cast-iron pipe must be supported at each joint, but supports may not be more than 10 ft apart.

TABLE 4.15   Vertical Pipe-Support Intervals in Zone One

Type of drainage pipe	Maximum distance of supports*
Lead pipe	4 ft
Cast iron	At each story
Galvanized	At least every other story
Copper	At each story†
PVC	Not mentioned
ABS	Not mentioned

* All stacks must be supported at their bases.
† Support intervals may not exceed 10 ft.

TABLE 4.16   Horizontal Pipe-Support Intervals in Zone Two

Type of drainage pipe	Maximum distance of supports (ft)
ABS	4
Cast iron	At each pipe joint
Galvanized (1 in and larger)	12
PVC	4
Copper (2 in and larger)	10
Copper (1½ in and smaller	6

TABLE 4.17   Vertical Pipe-Support Intervals in Zone Two

Type of drainage pipe	Maximum distance of supports (ft)*
Lead pipe	4
Cast iron	At each story[†]
Galvanized	At each story[‡]
Copper (1¼ in and smaller)	4
Copper (1½ in and larger)	At each story
PVC (1½ in and smaller)	4
PVC (2 in and larger)	At each story
ABS (1½ in and smaller)	4
ABS (2 in and larger)	At each story

* All stacks must be supported at their bases.
[†] Support intervals may not exceed 15 ft.
[‡] Support intervals may not exceed 30 ft.

When installing pipe with flexible couplings, bands, or unions, the pipe must be installed and supported to prevent these flexible connections from moving. In larger pipes, pipes larger than 4 in, all flexible couplings must be supported to prevent the force of the pipe's flow from loosening the connection at changes in direction.

Pipe-size reduction

As mentioned earlier, you may not reduce the size of a drainage pipe as it heads for the waste-disposal site. The pipe size may be enlarged, but it may not be reduced. There is one exception to this rule. Reducing closet bends, such as a 4-by-3 closet bend, are allowed.

Other facts to remember

A drainage pipe installed underground must have a minimum diameter of 2 in. When you are installing a horizontal branch fitting near the base of a stack, keep the branch fitting away from the point where the vertical stack turns to a horizontal run. The branch fitting should be installed at least 30 in back on a 3-in pipe and 40 in back on a 4-in

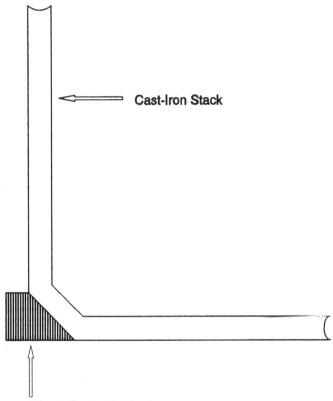

Figure 4.3 Support of a cast-iron stack.

pipe. By multiplying the size of the pipe by a factor of 10, you can determine how far back the branch fitting should be installed. See Fig. 4.4 for an example of this type of installation.

All drainage piping must be protected from the effects of flooding. When leaving a stub of pipe to connect with fixtures planned for the future, the stub must not be more than 2 ft in length and it must be capped. Some exceptions are possible on the prescribed length of a pipe stub. If you have a need for a longer stub, consult your local code officer. Cleanout extensions are not affected by the 2-ft rule.

Fittings

Fittings are also a part of the drainage system. Knowing when, where, and how to use the proper fittings is mandatory to the installation of a

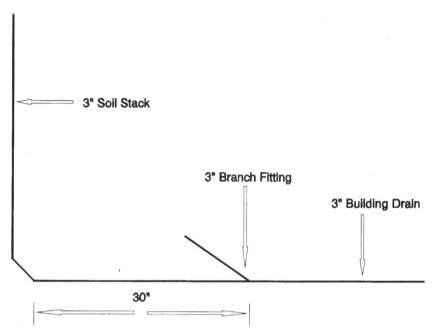

Figure 4.4 Example of the branch-fitting rule.

drainage system. Fittings are used to make branches and to change direction. The use of fittings to change direction is where we will start. When you wish to change direction with a pipe, you may have it change from a horizontal run to a vertical rise. You may be going from a vertical position to a horizontal one, or you might only want to offset the pipe in a horizontal run. Each of these three categories requires the use of different fittings. Let's take each circumstance and examine the fittings allowed.

Offsets in horizontal piping

When you want to change the direction of a horizontal pipe, you must use fittings approved for that purpose. You have six choices to choose from in zone three. Those choices are:

- Sixteenth bend (Fig. 4.5)
- Eighth bend (Fig. 4.6)
- Sixth bend (Fig. 4.7)
- Long-sweep fittings (Fig. 4.8)
- Combination wye and eighth bend (Fig. 4.9)
- Wye (Fig. 4.10)

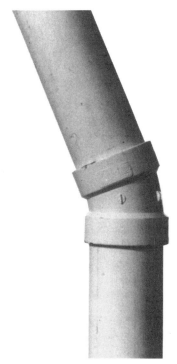

Figure 4.5 Sixteenth bend.

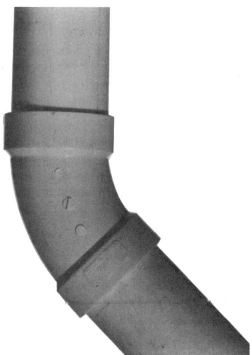

Figure 4.6 Eighth bend.

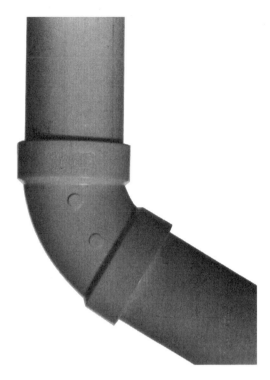

Figure 4.7 Sixth bend.

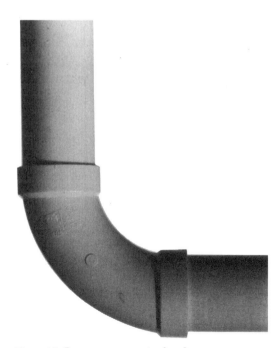

Figure 4.8 Long-sweep quarter bend.

Figure 4.9 Combination wye and eighth bend.

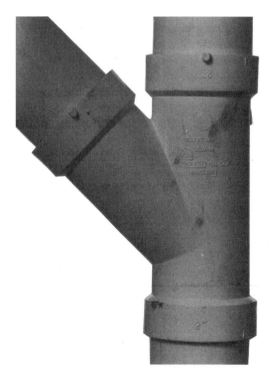

Figure 4.10 Wye.

TABLE 4.18   Fittings Approved for Horizontal Changes in Zone One*

45° wye
Combination wye and eighth bend

  * Other fittings with similar sweeps may also be approved.

Any of these fittings are generally approved for changing direction with horizontal piping, but as always, it is best to check with your local code officer for current regulations. For zone one's requirements of fittings used to change direction with horizontal piping, refer to Table 4.18.

Going from horizontal to vertical

You have a wider range of choice in selecting a fitting for going from a horizontal to a vertical position. There are nine possible candidates available for this type of change in direction, when working in zone three. The choices are:

- Sixteenth bend
- Eighth bend
- Sixth bend
- Long-sweep fittings
- Combination wye and eighth bend
- Quarter bend (Fig. 4.11)
- Wye
- Short-sweep fittings
- Sanitary tee (Fig. 4.12)

You may not use a double sanitary tee in a back-to-back situation if

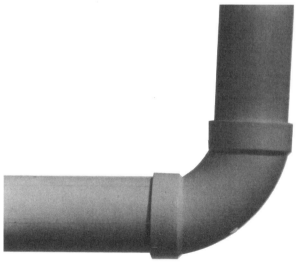

Figure 4.11 Quarter bend.

Figure 4.12 Sanitary tee.

the fixtures being served are of a blow-out or pump type. For example, you could not use a double sanitary tee to receive the discharge of two washing machines if the machines were positioned back to back. The sanitary tee's throat is not deep enough to keep drainage from feeding back and forth between the fittings. In a case like this, use a double combination wye and eighth bend. The combination fitting (Fig. 4.13) has a much longer throat and will prohibit waste water from transferring across the fitting to the other fixture.

Vertical-to-horizontal changes in direction

There are seven fittings allowed to change direction from vertical to horizontal. These fittings are:

- Sixteenth bend
- Sixth bend
- Eighth bend
- Long-sweep fittings
- Wye
- Combination wye and eighth bend
- Short-sweep fittings that are 3-in or larger

For fittings allowed to change direction from horizontal to vertical in zone one, refer to Table 4.19. Table 4.20 shows the fittings allowed in zone one for changing direction from vertical to horizontal.

Figure 4.13 Double combination wye and eighth bend.

TABLE 4.19   Fittings Approved for Horizontal-to-Vertical Changes in Zone One*

45° wye
60° wye
Combination wye and eighth bend
Sanitary tee
Sanitary tapped tee branches

* Cross-fittings, like double sanitary tees, cannot be used when they are of a short-sweep pattern; however, double sanitary tees can be used if the barrel of the tee is at least two pipe sizes larger than the largest inlet.

TABLE 4.20   Fittings Approved for Vertical-to-Horizontal Changes in Zone One

45° branches
60° branches and offsets if they are installed in a true vertical position

Zone one prohibits a fixture outlet connection within 8 ft of a vertical-to-horizontal change in direction of a stack, if the stack serves a suds-producing fixture. A suds-producing fixture could be a laundry fixture, a dishwasher, a bathing unit, or a kitchen sink. This rule does not apply to single-family homes and stacks in buildings with less than three stories.

Indirect Wastes

Indirect-waste requirements can pertain to a number of types of plumbing fixtures and equipment. These might include a clothes washer drain, a condensate line, a sink drain, or the blow-off pipe from a relief valve, just to name a few. These indirect wastes are piped in this manner to prevent the possibility of contaminated matter backing up the drain into a potable water or food source, among other things.

Most indirect-waste receptors are trapped. If the drain from the fixture is more than 2 ft long, the indirect-waste receptor must be trapped. However, this trap rule applies to fixtures like sinks, not to an item such as a blow-off pipe from a relief valve. The rule is different in zone one where a drain more than 4 ft long must be trapped.

The safest method for indirect waste is accomplished by using an air gap. When an air gap is used, the drain from the fixture terminates above the indirect-waste receptor, with open-air space between the waste receptor and the drain. This prevents any back-up or back-siphonage.

Some fixtures, depending on local code requirements, may be piped with an air break, rather than an air gap. With an air break, the drain may extend below the flood-level rim and terminate just above the trap's seal. The risk to an air break is the possibility of a back-up. Since the drain is run below the flood-level rim of the waste receptor, it is possible that the waste receptor could overflow and back up into the drain. This could create contamination, but in cases where contamination is likely, an air gap will be required. Check with your local codes office before using an air break.

Standpipes, like those used for washing machines, are a form of indirect-waste receptors. A standpipe used for this purpose in zones one and three must extend at least 18 in above the trap's seal, but they may not extend more than 30 in above the trap seal. If a clear-water waste receptor is located in a floor, zone three requires the lip of the receptor to extend at least 2 in above the floor. This eliminates the waste receptor from being used as a floor drain.

Choosing the proper size for a waste receptor is generally based on the receptor's ability to handle the discharge from a drain without excessive splashing. If you are concerned with sizing a particular waste receptor, discuss it with your local code officer in order to obtain a ruling.

Buildings used for food preparation, storage, and similar activities are required to have their fixtures and equipment discharge drainage through an air gap. Zone three provides an exception to this rule. In zone three, dishwashers and open culinary sinks are excepted. Zone two requires the discharge pipe to terminate at least 2 in above the receptor. Zone one requires the distance to be a minimum of 1 in. Zones two and three require the air-gap distance to be a minimum of twice the size of the pipe discharging the waste. For example, a ½-in discharge pipe would require a 1-in air gap.

Zones two and three prohibit the installation of an indirect-waste receptor in any room containing toilet facilities. Zone one goes along with this ruling but allows one exception: the installation of a receptor for a clothes washer when the clothes washer is installed in the same room. Indirect-waste receptors are not allowed to be installed in closets and other unvented areas. Indirect-waste receptors must be accessible. Zone two requires all receptors to be equipped with a means of preventing solids with diameters of ½ in or larger from entering the drainage system. These straining devices must be removable to allow for cleaning.

When you are dealing with extreme water temperatures in waste water, such as with a commercial dishwasher, the dishwasher drain must be piped to an indirect waste. The indirect waste will be connected to the sanitary plumbing system, but the dishwasher drain may not connect to the sanitary system directly if the waste water temperature exceeds 140°F. Steam pipes may not be connected directly to a sanitary drainage system. Local regulations may require the use of special piping, sumps, or condensers to accept high-temperature water. Zone one prohibits the direct connection of any dishwasher to the sanitary drainage system.

Clear-water waste, from a potable source, must be piped to an indirect waste, with the use of an air gap. Sterilizers and swimming pools might provide two examples of when this rule would be used. Clear water from nonpotable sources, such as a drip from a piece of equipment, must be piped to an indirect waste. In zone three, an air break is allowed in place of an air gap. Zone two requires any waste entering the sanitary drainage system from an air conditioner to do so through an indirect waste.

Special Wastes

Special wastes are those wastes that may have a harming effect on a plumbing or waste-disposal system. Possible locations for special waste piping might include photographic labs, hospitals, or buildings where chemicals or other potentially dangerous wastes are dispersed. Small, personal-type photo darkrooms do not generally fall under the scrutiny of these regulations. Buildings that are considered to have a need for special waste plumbing are often required to have two plumbing systems, one system for normal sanitary discharge and a separate system for the special wastes. Before many special wastes are allowed to enter a sanitary drainage system, the wastes must be neutralized, diluted, or otherwise treated.

Depending upon the nature of the special wastes, special materials may be required. When you venture into the plumbing of special wastes, it is always best to consult the local code officer before proceeding with your work.

Sewer Pumps and Pits

There will be times when conditions will not allow a plumbing system to flow in the desired direction by gravity. When this is the case, sewer pumps and pits become involved. It is also possible that sump pumps and sumps will be used to remove water collected below the level of the building drain or sewer. When you plan to install a pump or sump pit, you must abide by certain regulations.

All sump pits must have a sealed cover that will not allow the escape of sewer gas. The pit size will be determined by the size and performance of the pump being housed in the sump. But, the sump generally must have a minimum diameter of 18 in and a minimum depth of 24 in. If a sewer pump is installed in a pit, the pump must be capable of lifting solids, with a diameter of 2 in, up into a gravity drain or sewer. The discharge pipe from these sumps must have a minimum diameter of 2 in. Zone one requires the sizing of the drain receiving the discharge from a sewer sump to be sized with a rating of 2 fixtures units for every gallon per minute the pump is capable of producing.

If the sump pit will not receive any discharge from toilets, the pump is not required to lift the 2-in solids and may be smaller. A standard procedure is to install a pump capable of lifting ½-in solids to a gravity drain if no toilets discharge into the sump.

It is a good idea to install two pumps in the sump. The pumps may be installed in a manner to take turns with the pumping chores, but most importantly, if one pump fails, the other pump can continue to operate. Zone two requires the installation of this type of two-pump system when six or more water closets discharge into the sump. Zone one requires any public installation of a sewer sump to be equipped with a two-pump system. Alarm systems are often installed on sewer pump systems. These alarms warn building occupants if the water level in the sewer pit rises to an unusually high level. Zones one and two require the effluent level to remain at least 2 in below the inlet of the sump.

All sewer sumps should be equipped with a vent. Ideally, the vent should extend upward to open air space without tying into another vent. Most sump vents are 2 in in diameter, but in no case shall they have a diameter of less than 1½ in.

There should be a check valve and a gate valve installed on the discharge piping from the pump. These devices prevent water from running back into the sump and allow the pump to be worked on with relative ease.

Most sewer pumps are equipped with a 2-in discharge outlet. An ejector pump with a 2-in outlet should be able to pump 21 gpm. If the discharge outlet is 3 in in diameter, the pump should have a flow rate of 46 gpm.

Storm-Water Drainage Piping

Storm-water drainage piping is a piping system designed to control and convey excess groundwater to a suitable location, which might be a catch basin, storm-sewer, or a pond. But, storm-water drainage may never be piped into a sanitary sewer or plumbing system. Up until now, our math requirements have been fairly simple, but that is about to change. Not that this section is a brain-buster, but it will require a little extra effort. The following sizing examples are based on zone three's requirements.

When you wish to size a storm-water drainage system, you must have some benchmark information to work with. One consideration is the amount of pitch a horizontal pipe will have on it. Another piece of the puzzle is the number of square feet of surface area your system will be required to drain. You will also need data on the rainfall rates in your area.

When you use your code book to size a storm-water system, you should have access to all the key elements required to size the job, except for the possibility of the local rainfall amounts. You should be able to obtain rainfall figures from your state or county offices. Your code book should provide you with a table to use in making your sizing calculations. Now, let's get into the computations needed to size a horizontal storm drain or sewer.

Sizing a horizontal storm drain or sewer

The first step to take when sizing a storm drain or sewer is to establish your known criteria. How much pitch will your pipe have on it? In the following example, the pipe will have a ¼-in/ft pitch. Knowing the pitch gives us a starting point and begins to take the edge off of an intimidating task.

Table 4.21 shows an example of a sizing table. To keep this process as simple as possible, it only includes a column in the table for our

TABLE 4.21   **Example of a Horizontal Storm-Water Sizing Table***

Pipe grade (in/ft)	Pipe size (in)	Gallons per minute	Number of square feet of surface area
¼	3	48	4,640
¼	4	110	10,600
¼	6	314	18,880
¼	8	677	65,200

* These figures are based on a rainfall with a maximum rate of 1 in of rain per hour, for a full hour, and occurring once every 100 years.

known grade of ¼ in/ft. In your code book you should be offered a few other choices, but since you are only going to use one pitch, once you know what it is, that is the only column that you need to pay attention to, as far as pitch is concerned.

What else do we know? Well, we know that the subject system is going to be located in Portland, Maine. Portland's rainfall is rated at 2.4 in/h. This rating assumes a 1-h storm that is only likely to occur once every 100 years. Now we have two of the factors needed to size our system.

We also know that the surface area that the system will be required to drain is 15,000 ft²; this includes the roof and parking area. We're getting close to home now. We've got three of the elements needed to get this job done. But, how do we use the numbers in Table 4.21 to make any sense of this? Well, there are a couple of ways to ease the burden. When you are working with a standard table, like the ones found in most code books, you must convert the information to suit your local conditions. For example, if a standardized table is based on 1 in of rainfall an hour and your location has 2.4 in of rainfall per hour, you must convert the table, but this is not difficult.

When we want to convert a table based on a 1-in rainfall to meet local needs, all we have to do is divide the drainage area in the table by our rainfall amount. For example, if the standard chart shows an area of 10,000 ft² requiring a 4-in pipe, we can change the table by dividing our rainfall amount, 2.4, into the surface area of 10,000 ft².

If we divide 10,000 by 2.4, we get 4167. All of a sudden, we have solved the mystery of computing storm-water piping needs. With this simple conversion, we know that if our surface area was 4167 ft², we would need a 4-in pipe. But, our surface area is 15,000 ft², so, what size pipe do we need? Well, we know it will have to be larger than 4 in. So, we look down the conversion chart and find the appropriate surface area. Our 15,000 ft² of surface area will require a storm-water drain with a diameter of 8 in. We found this by dividing the surface areas of the numbers in Table 4.21 by 2.4 until we reached a number equal to, or greater than, our surface area. We could almost get by with a 6-in pipe, but not quite.

Now, let's recap this exercise. To size a horizontal storm drain or sewer, decide what pitch you will put on the pipe. Next, determine what your area's rainfall is for a 1-h storm, occurring each 100 years. If you live in a city, your city may be listed, with its rainfall amount, in your code book. Using a standardized chart, rated for 1 in of rainfall per hour, divide the surface area by a factor equal to your rainfall index; in the example it was 2.4. This division process converts a generic table into a customized table, just for your area.

Once the math is done, look down the table for the surface area that most closely matches the area you have to drain. To be safe, go with a number slightly higher than your projected number. It is better to have a pipe sized one size too large than one size too small. When you have found the appropriate surface area, look across the table to see what size pipe you need. See how easy that was. Well, maybe it's not easy, but it is a chore you can handle.

Sizing rain leaders and gutters

When you are required to size rain leaders or downspouts, you use the same procedure described above, with one exception. You use a table, supplied in your code book, to size the vertical piping. Determine the amount of surface area your leader will drain and use the appropriate table to establish your pipe size. The conversion factors are the same.

Sizing gutters is essentially the same as sizing horizontal storm drains. You will use a different table, provided in your code book, but the mechanics are the same.

Roof drains

Roof drains are often the starting point of a storm-water drainage system. As the name implies, roof drains are located on roofs. On most roofs, the roof drains are equipped with strainers that protrude upward, at least 4 in, to catch leaves and other debris. Roof drains should be at least twice the size of the piping connected to them. All roofs that do not drain to hanging gutters are required to have roof drains. A minimum of two roof drains should be installed on roofs with a surface area of 10,000 ft^2, or less. If the surface area exceeds 10,000 ft^2, a minimum of four roof drains should be installed.

When a roof is used for purposes other than just shelter, the roof drains may have a strainer that is flush with the roof's surface. Roof drains should obviously be sealed to prevent water from leaking around them. The size of the roof drain can be instrumental in the flow rates designed into a storm-water system. When a controlled flow from roof drains is wanted, the roof structure must be designed to accommodate the controlled flow.

More sizing information

If a combined storm-drain and sewer arrangement is approved, it must be sized properly. This requires converting fixture-unit loads into drainage surface area. For example, 256 fixture units will be treated as 1000 ft^2 of surface area. Each additional fixture unit, in excess of 256, will be assigned a value of 3$\frac{9}{10}$ ft^2. In the case of sizing for continuous flow, each gallon per minute is rated as 96 ft^2 of drainage area.

Some facts about storm-water piping

Storm-water piping requires the same amount of cleanouts, with the same frequency, as a sanitary system. Just as regular plumbing pipes must be protected, so shall storm-water piping. For example, if a downspout is in danger of being crushed by automobiles, you must install a guard to protect the downspout.

As stated earlier, storm-water systems and sanitary systems should not be combined. There may be some cities where the two are combined, but they are the exception rather than the rule. Area-way, or floor, drains must be trapped. When rain leaders and storm drains are allowed to connect to a sanitary sewer, they are required to be trapped. The trap must be equal in size to the drain it serves. Traps must be accessible for cleaning the drainage piping. Storm-water piping may not be used for conveying sanitary drainage.

Sump pumps

Sump pumps are used to remove water collected in building subdrains. These pumps must be placed in a sump, but the sump need not be covered with a gastight lid or vented. Many people are not sure what to do with the water pumped out of their basement by a sump pump. Do you pump it into your sewer? No, the discharge from a sump pump should not be pumped into a sanitary sewer. The water from the pump should be pumped to a storm-water drain or, in some cases, to a point on the property where it will not cause a problem.

All sump-pump discharge pipes should be equipped with a check valve. The check valve prevents previously pumped water from running down the discharge pipe and refilling the sump, forcing the pump to pull double duty. Here sump pumps are pumps removing groundwater, not waste or sewage.

Zone one's differences

Zone one has some additional requirements. See Tables 4.22 and 4.23 for zone one's requirements for storm-water materials. Note that once storm-water piping extends at least 2 ft from a building, any approved material may be used.

With zone one, the inlet area of a roof drain is generally only required to be 1½ times the size of the piping connected to the roof drain. However, when positioned on roofs used for purposes other than weather protection, roof drain openings must be sized to be twice as large as the drain connecting to them.

Zone one also provides tables for sizing purposes. When computing the drainage area, you must take into account the effect vertical walls will have on the drainage area. For example, a vertical wall that reflects water onto the drainage area must be allowed for in your surface-area computations. In the case of a single vertical wall, add one-half of the wall's total square footage to the surface area.

Two vertical walls that are adjacent to each other require you to add 35 percent of the combined wall square footage to your surface area.

If you have two walls of the same height that are opposite of each other, no added space is needed. In this case, each wall protects the other one and does not allow extra water to collect on the roof area.

When you have two opposing walls with different heights, you must make a surface-area adjustment. Take the square footage of the high-

TABLE 4.22   Approved Materials for Storm-Water Drainage in Zone One for Use Inside Buildings, Above Ground

Galvanized
Wrought iron
Brass
Copper
Cast iron
ABS*
PVC*
Lead

* ABS and PVC may not be used in buildings that have more than three floors above grade.

TABLE 4.23   Approve Materials for Storm-Water Drainage in Zone One for Use Inside Buildings, Below Ground

Service-weight cast iron
DWV copper
ABS
PVC
Extra-strength vitrified clay

est wall, as it extends above the other wall, and add half of the square footage to your surface area.

When you encounter three walls, you use a combination of the above instructions to reach your goal. Four walls of equal height do not require an adjustment. If the walls are not of equal height, use the procedures listed above to compute your surface area.

Zone two's differences

It would be nice if all plumbing codes were the same, but they are not. The following information provides insight into how zone two varies from zones one and three. See Tables 4.24, 4.25, and 4.26 for material requirements in zone two.

Sump pits are required to have a minimum diameter of 18 in. Floor drains may not connect to drains intended solely for storm water. When computing surface area to be drained for vertical walls, such as walls enclosing a roof-top stairway, use one-half of the total square footage from the vertical wall surface that reflects water onto the drainage surface.

Some roof designs require a back-up drainage system for emergen-

---

TABLE 4.24   Approved Materials for Storm-Water Drainage in Zone Two for Underground Use

Cast iron
Coated aluminum
ABS*
PVC*
Copper*
Concrete*
Asbestos-cement*
Vitrified clay*

* These materials may be allowed for use, subject to local code authorities.

TABLE 4.25   Approved Materials for Storm-Water Drainage in Zone Two for Building Storm Sewers

Cast iron
Aluminum*
ABS
PVC
Vitrified clay
Concrete
Asbestos-cement

* Buried aluminum must be coated.

TABLE 4.26   Approved Materials for Storm-Water Drainage in Zone Two for Above-Ground Use

Galvanized
Black steel
Brass
DWV thicker types of copper
Cast iron
ABS
PVC
Aluminum
Lead

cies. These roofs are generally roofs that are surrounded by vertical sections. If these vertical sections are capable of retaining water on the roof if the primary drainage system fails, a secondary drainage system is required. In these cases, the secondary system must have independent piping and discharge locations. These special systems are sized with the use of different rainfall rates. The ratings are based on a 15-minute (min) rainfall. Otherwise, the 100-year conditions still apply.

Zone two's requirements for sizing a continuous flow require a rating of 24 ft^2 of surface area to be given for every gallon per minute generated. For regular sizing, based on 4 in of rain per hour, 256 fixture units equal 1000 ft^2 of surface area. Each additional fixture unit is rated at 3%$_0$ in. If the rainfall rate varies, a conversion must be done.

To convert the fixture-unit ratings to a higher or lower rainfall, you must do some math. Take the square foot rating assigned to fixture units and multiply it by 4. For example, 256 fixture units equal 1000 ft^2. Multiply 1000 by 4, and get 4000. Now, divide the 4000 by the rate of rainfall for 1 h. Say for example that the hourly rainfall was 2 in; the converted surface area would be 2000.

Well, you have made it past a section of code regulations that gives professional plumbers the most trouble. Storm-water drains are despised by some plumbers because they have little knowledge of how to compute them. With the aid of this chapter, you should be able to design a suitable system with minimal effort.

Reminder Notes

Zone one

1. Fixture-unit loads on traps are different from those found in zones two and three. Refer to Table 4.3 for zone one's requirements.

2. A 2-in horizontal drain may carry up to eight fixture units, with exceptions.

3. A 3-in horizontal drain may carry up to 35 fixture units and of these units as many as three water closets may be installed.

4. A 1½-in horizontal drain may carry up to two fixture units, except it may not carry the drainage of sinks, urinals, and dishwashers.

5. A 4-in horizontal drain is rated for up to 216 fixture units.

6. Special permission is required to install pipe with a ⅛-in/ft grade.

7. See Table 4.14 for pipe support intervals of horizontal piping.

8. See Table 4.15 for pipe support intervals of vertical piping.

9. See Table 4.18 for fittings allowed to change the direction of horizontal piping.

10. See Table 4.20 for fittings allowed to change from a vertical to a horizontal direction.

11. See Table 4.19 for fittings allowed to change from a horizontal to a vertical direction.

12. In buildings, other than single-family homes and buildings with less than three stories, a fixture outlet may not connect to a stack within 8 ft of a vertical to horizontal change of direction, when the stack is receiving the discharge of a suds-producing fixture.

13. An indirect-waste drain that is 5 ft long, or longer, must be trapped.

14. Indirect-waste receptors may not be installed in rooms with toilet facilities, except for washing-machine receptors, when the clothes washer is installed in the same room.

15. No dishwasher is allowed to be connected directly to a sanitary drainage system.

16. Indirect-waste pipes in buildings used for food preparation, and similar activities, must terminate at least 2 in above the indirect-waste receptor.

17. Refrigerators that are used for the storage of prepackaged goods, like bottles of soda, are excepted from the indirect-waste rules.

18. Air conditioning equipment may be piped with an air break, but food-related equipment and fixtures must be piped with air gaps.

19. A vent from an indirect waste may not tie into a vent that is connected to a sewer.

20. Fixtures that produce waste under pressure are required to be piped to indirect-waste receptors.

21. Condensate drains from air conditioning coils may be connected directly to a lavatory tailpiece or tub waste, under special conditions. The connection point must be accessible and in a place controlled by the same person controlling the air conditioner.

22. Pure condensate, from a fuel-burning appliance, being discharged into a drainage system must only discharge into drainage systems constructed of materials approved for this purpose.

23. It is permissible to discharge the drainage from a watercooler to an indirect waste.

24. Owners of buildings containing chemical-waste piping must make and keep a detailed record of the location of all piping in these systems.

25. Chemical-waste pipes should be installed in a way to make them as readily accessible as reasonably possible.

26. Drains accepting the waste for sewer sumps must be sized with a rating of 2 fixture units for every gallon per minute the pump is capable of producing.

27. Any installation of a sewer sump that will serve the public requires the installation of a two-pump system.

28. Effluent levels in sewer pump sumps must not rise to a level closer than 2 in to the sump inlet.

29. Sewer sump vents when serving an air-operated sewer ejector pump may not connect to other vents.

30. Water-operated sewer ejectors are not approved for use.

Zone two

1. See Table 4.16 for pipe support intervals of horizontal piping.

2. See Table 4.17 for pipe support intervals of vertical piping.

3. Indirect-waste pipes in buildings used for food preparation and similar activities, must terminate at least 2 in above the indirect-waste receptor.

4. Any waste from an air conditioner that discharges into the sanitary drainage system must do so through an indirect waste.

5. Indirect-waste receptors must be equipped with a means of preventing solids with diameters of ½ in or larger from entering the drainage system. The straining devices must be able to be removed for cleaning.

6. It is permissible to discharge the drainage from a watercooler to an indirect waste.

7. Sewer ejector sumps are required to be equipped with a duplex pumping system when six or more water closets discharge into the sump.

8. Effluent levels in sewer pump sumps must not rise to a level closer than 2 in to the sump inlet.

Zone three

1. Fixture-unit loads for a continuous flow are assigned at a rate of 1 fixture unit for every 7½ gpm.

2. Pipes with diameters in excess of 8 in can be installed with a grade of ⅟₁₆ in/ft.

3. Indirect-waste receptors for clear-water wastes that are in a floor must extend at least 2 in above the floor level.

4. Zone three does not require dishwashers and open culinary sinks to be piped to an indirect waste.

5. Clear-water wastes from nonpotable sources may be piped to an indirect waste by using an air break.

5

Using Combination Waste and Vent Systems

Using combination waste and vent systems can be profitable. Not all jurisdictions favor the use of combination drain-waste-and-vent (DWV) systems, but most code offices do recognize alternative forms of plumbing under special conditions. If you learn to use all of the options available to you under the plumbing code, you can make your job easier and your bank account fatter.

What's the big advantage to a combination waste and vent system? Actually, there are many advantages. When you are allowed to use a combination system, you can reduce the number of vent pipes that must be installed. This saves time and material, and should result in an opportunity to do more jobs. It can also make you a more competitive bidder. Think about it. If I bid a job with a full set of vents for each fixture and you bid the same job with combination vents, your price should be lower. If your price is not lower, your profits will be higher.

Combination waste and vent systems can work very well. Some don't work as well as customers would like them to, however. When installed properly, a combination waste and vent system will work fine. But, some systems are installed inadequately and don't provide substantial airflow for the drains to run quickly. This is often the case at many of the jobs I visit in Maine. Two major causes of poor drainage in a combination system are undersized pipe and vents that are too far from the drains being served by the vent.

The state of Maine has a very liberal code for alternative plumbing systems. It is used as a primary code for most residential plumbing. Compared to other parts of the country where I have worked, this is unusual. My experience in other states has shown that combination systems are not normally used on a routine basis. The reason for a lack of use is usually a result of code requirements, but it can be a matter of plumbers not knowing how to take advantage of alternative plumbing procedures.

All of the major plumbing codes share similarities, and most of them are very similar. This is not the case when it comes to combination waste and vent systems. The differences between local codes on the issue of alternative plumbing methods can be very different. To illustrate this point, let's take a quick look at some of the differences between various codes.

Since I have worked with the code serving zone three more than with any other code, I will start with it. A combination system in zone three can serve floor drains, standpipes, sinks, and lavatories, but not toilets. The only vertical pipe allowed is the connection between a fixture branch of a sink, lavatory, or standpipe and the horizontal combination drain and vent pipe. What is the maximum vertical distance allowed for the pipe? Eight feet is the maximum distance allowed.

When a combo system is used in zone three, the combo pipe must connect to either a horizontal drain that is vented or a vent must be installed on the combo system. Any vent must extend at least 6 in above the flood level rim of the highest fixture before being offset.

The sizing of a vent for a combination system must be determined by the total load of fixture units. Fixture branches and fixture drains must connect to the combination system with a distance specified within the local code, and the size of the combination system must also comply with regulations set forth in the code. To expand on this, let me give you an overview of the type of table that you would have to comply with in zone three when sizing a combination waste and vent system.

Pipe size (in)	Horizontal connection (fixture units)	Building drain connection (fixture units)
2	3	4
3	6	26
4	20	50
6	360	575

Now that we've examined the high points of combination waste and vent systems in zone three, let's compare the same type of information as it is presented in the code for zone two.

What fixtures are allowed on a combination waste and vent system in zone two? Sinks, floor sinks, indirect waste receptors, floor drains, dishwashers, and similar fixtures. The fixtures must not be adjacent to walls or partitions. High-output fixtures, such as pumps, must not be installed in a way that will overload a combination waste and vent system.

Any combination waste and vent system used in zone two must be presented to the local plumbing code enforcement office and approved. All waste pipes in a combination system must be at least two pipe sizes larger than typical pipe-sizing requirements, in order to provide adequate airflow. Waste pipes must also be at least two pipe sizes larger than any fixture tailpiece or connection, except for when a P-trap is installed above floor level.

When a vertical waste pipe is installed, it must be two pipe sizes larger and must extend above floor level, to a normal rough-in level. A cleanout must be installed in the top of the connecting waste tee. When a floor sink is connected to a combination system, it must be connected via a running trap that is two pipe sizes larger than the fixture outlet. Minimum pipe size in a combo system is 2 in when installed underground.

Vents in a combo system must be provided at the upstream end of each branch, and the vent must be washed over or under by the last fixture on the branch. When a vent leaves a horizontal waste branch it must leave at an angle of less than 45°, unless the vent is washed by a fixture. When branches intersect, they must be vented. A vent must be located downstream from all fixtures in the system.

Zone one does not allow the use of combination waste and vent systems unless structural conditions prevent the use of a normal installation. As is common, every proposal for a combination waste and vent system must be presented and approved before being installed. A simple rule in zone one is that every combo system must be provided with a vent that is adequate to assure free circulation of air.

Branches with developed lengths in excess of 15 ft must be vented separately and in an approved manner. Vents must be downstream of the uppermost fixture. And the minimum area of any vent installed for a combo system must be at least one-half the inside cross-sectional area of the drain being served.

Waste pipes and traps in combo systems must be at least two pipe sizes larger than normal regulations. This holds true for tailpieces and fixture connections in that the pipe serving these elements must be two pipe sizes larger that what would normally be required.

With few exceptions, the only vertical waste pipe allowed in a combo system is a tailpiece or connection between the outlet of a plumbing fixture and trap. Tailpieces and trap connections that are vertical may not be taller than 24 in, and they should be kept as short as possible.

A wet vented branch that serves a single trap and that is not less than 2 in in diameter may not be required to be fitted with a cleanout. One condition of this rule is that the vent be readily accessible for cleaning through a trap. Accessible cleanouts are required for each vent dedicated to a combo system.

There are a number of other differences within local codes for combination waste and vent systems. For example, in Maine, the alternative plumbing code is very liberal in terms of what fixtures can be connected to a combination waste and vent system. Toilets and bathing units can be connected to combo systems in Maine, so long as the job meets particular criteria. If the basement of a home contains sinks or bathrooms, the alternative code is not allowed. Washing machines and floor drains are allowed in basements when the alternative code is used.

What's the bottom line of all of this? Each plumbing code treats combination waste and vent systems differently. Local code officers have the authority to interpret and set the rules and regulations that plumbers must work with. You must verify your local code requirements before you attempt to use alternative methods to basic plumbing procedures.

Bidding Advantages

There are bidding advantages to using combo systems. Using a combination waste and vent system allows you to use less material and to finish jobs faster. In a flat-rate bid, this can mean putting more money into your bank account. I fight this battle often in Maine. My personal feelings support the use of conventional plumbing, under most circumstances. A majority of my competitors in Maine use the alternative code that allows extensive use of combo systems. If I bid a job with individual vents for all fixtures and offer to install a system that meets standard, national plumbing criteria, I'm at a disadvantage in a bidding war. A plumber who uses a complete combination waste and vent system can get by with a lot less material and time. For me to convince customers to pay more for my preferred method is not easy. I must spend time talking with customers to make them understand the advantages of a fully vented system.

There is certainly nothing wrong with using a combination waste and vent system. I use the systems from time to time, but I do prefer a fully vented system. Would I use a combination system to win a big bid? Yes, as long as I made it clear to the customer what I was providing a price for. You can use the advantages of a combo system to reduce costs, win more bids, and to make more money. Let's spend a little time in looking at how this is true.

In Practice

In practice, the use of a combination of waste and vent system is ideal from a financial point of view. Used properly, this type of system can provide satisfactory drainage for a plumbing system. As a plumber, you know the importance of venting and of air flow in the easy draining of fixtures. Do you understand why the pipe size of a combination system must be two sizes larger than normal? The reason is simple; by having the pipe larger in diameter, air can flow over the discharge of liquids, which allows a fixture to drain properly.

A combination system utilizes oversized drains to compensate for fewer vents and vents that are farther from trap connections than traditional plumbing installations. In Maine, the vent can be up to 10 ft away from the fixture being served. Zone three requires a maximum distance of 8 ft. I'm not an engineer, but as a master plumber, these distances seem excessive to me. Do combination systems work? Yes, they do. Are combination systems able to work as well as more traditional systems? Not in my opinion. Even as a firm supporter of fully vented systems, I must say that combo systems do have their place, and they can make you a more competitive plumbing contractor.

Your opportunities in working with combination systems are limited by your local code requirements. It may be that your only occasion to use such a system is to circumvent structural problems that prevent you from performing routine plumbing. If this is the case, you cannot use the alternative methods to capitalize on winning more bids. If you work in zone two or zone three, you may be able to take greater advantage of combination systems. Plumbers who are fortunate enough to work with liberal codes, such as the one in Maine, can use the code to their advantage. I don't know where you live or work, but I can give you some examples of how a combination waste and vent system can make your work easier and possibly more profitable.

Saving Time

Saving time is one way to make more money. There are two ways to look at this. If you bid a job at a flat rate and get the job done for less money than you would without the use of a combo system, you make more money. Now, let's say that you bid the job low to get it, since you will be using a combo system. Since your price is lower, you won't make as much money. Even so, you can still make more money. How is this possible? Since you finish the job faster, you are ready to move onto another job. By increasing your volume of business, you have a good probability of making more money. Either way you look at it, you have a good chance of enjoying a more profitable business.

Reducing Material Needs

Reducing material needs is always a good reason to consider using a combination waste and vent system. By using less material, you are saving money, extending the natural resources available to all of us, and offering your customers an opportunity for lower-priced jobs. Material suppliers would probably discourage the use of combo systems, but the fact that less material is needed can't be ignored.

How much material can be saved when a combination system is used? it depends on the layout of the job, but savings can be quite substantial. The amount of pipe and fittings eliminated when a full combination system is used can be a lot more than you might imagine. Honestly, I didn't know how much material could be saved until I moved to Maine. Working in Virginia for most of my career, I was used to full-vented systems. When I started working in Maine, I saw jobs that I thought were far too limited in venting. Frankly, some of the procedures used in Maine do result in what many master plumbers would consider unacceptable drainage. However, the jobs are within code compliance, they use little material, and they are the norm rather than the exception. When in Rome....

I have not worked in every state, county, or city in this country. Who has? However, I have worked in several states, dozens of counties, and numerous cities. My experience with the various major and local codes is extensive. Of all the codes I've worked with, the code in Maine is the most liberal in terms of combination waste and vent systems, so I'd like to use Maine's alternative code as an example for you. The standard code in Maine is similar to the major codes, but the alternative code is, in my experience, unique. I don't know of another code where so many types of fixtures, including water closets, are allowed to be installed on combo systems.

How does the time and material required to install a full-vented system compare with the same requirements of a combination waste and vent system? Well, let's find out. I'll give you a general idea of the types of differences that you might expect between the two types of systems.

Setting Up a Full-Vented System

Setting up a full-vented system is standard procedure. This is the type of system that most plumbers use. Is it the best system? I think it is, in most conditions, but there are clearly times when a combo system is warranted. To install a full-vented system, every fixture within the system must be vented. This may involve any number of vent types, ranging from wet vents, to individual vents, to battery vents, and others. In any case, pipe and fittings are needed to provide venting for each fixture. This is not the case in a combo system.

Running vent pipes from each and every vent is time-consuming, uses material, and costs more money than using a combo system. Why run the vents if you don't have to? Better drainage is one reason, but we are concentrating on how to save time and money with combo systems, so let's stick with the program.

In addition to the materials required for a full-vent system and the time spent installing pipe and fittings, there is more effort required. You probably have to drill a lot of holes to accommodate the pipe being installed. More pipe hangers are needed when you have to install vent pipes. Getting up on roofs to install flashings will take more time than you would spend if the vents didn't have to penetrate the roofs. The list of time and materials that you have to add to your job cost can be substantial.

If you are a licensed, working plumber, as I suspect you are, you are familiar with traditional methods of plumbing. I don't need to give you a lecture on how full-vented jobs have to be installed. So, let's cut to the chase and get right into the big differences that you would discover if you were to start plumbing under the alternative code in Maine.

Using Maine's Alternative Plumbing Code

Using Maine's alternative plumbing code is a real experience for plumbers who move to the state from other areas, such as Virginia. When I moved here, I suffered from major cultural shock. In all my years of plumbing, I have never seen plumbing done the way it is done on a daily basis in Maine. It is difficult for me, as a master plumber, to accept the type of work that is installed in many parts of the state of Maine. It's not that I have a higher standard, but I was taught a different type of plumbing. It's been said that old dogs can't learn new tricks, but I don't go along with this line of thinking. I've not only learned the alternative plumbing code in Maine, I've taught it at Central Maine Technical College and used it on a regular basis in the field. The learning experience was not difficult, but accepting the standards was a little tough for me to take. Even so, it's important that you get a grasp of how much difference such a system can make in your career and business, so let me give you some hard-line examples of how such a system works.

Assume that you have recently moved to Maine and that you and I are about to plumb a single-family home. The house is a two-story home with a full basement. The only plumbing fixture in the basement is a washing machine hook-up. This makes it possible to use the alternative plumbing code. When we do this, we are going to save a lot of time and material.

The sewer for this home comes in through the foundation wall. It is low enough for us to install the washer hook-up directly to the building drain. In Maine, this fixture hook-up is not required to be vented. We have a 4-in building drain to work with, one kitchen sink at the opposite end of the house, and two back-to-back bathrooms in the middle of the home.

When we run our drain to the kitchen sink we will have to run an individual vent through the roof. This would be the case in a full-vent system. But, things change a lot when we get to the bathroom group. A 3-in vent is required for the home, as it would be with any full-vented system. However, in this case, the main stack will serve as the only vent for both bathrooms.

In a regular job, the bathing units might be wet-vented through the lavatory vents. The water closets would probably be vented with the main stack. Dry vents from the lavatories would probably tie into the main stack. This, in and of itself, wouldn't amount to a lot of extra work or material. But, suppose the bathrooms were not so close together. You could be running horizontal vents a long way to make the tie-in. This is not necessary when the combo system is used. As long as the fixtures are within 10 ft of their vent, no additional vents are required.

I want you to think a moment about what you've just learned. If we were plumbing the bathroom described above, the only vent installed for two bathing units, two lavatories, and two water closets would be one 3-in pipe. Can you even imagine this? Well, I couldn't when I first encountered the Maine code for alternative plumbing, but it's real, and it's used daily.

You know from what I've told you previously that most codes are not as lenient on the use of combination systems as Maine is. But, there are opportunities waiting for enterprising plumbers who have a full knowledge of their local codes. When you use a combination system, you may have to install relief vents to make up for excessive distance between a fixture and a vent, but the chance to use less material and to spend less time on your installations does exist. Talk to your local plumbing inspectors and find out what you may be overlooking with combination waste and vent systems. If you are lucky enough to find that you have the freedom that I have in Maine, it can make a major difference in how you conduct your business.

6

Private Sewage Systems

Private sewage systems don't always, or even often, involve plumbers during the planning and installation stages. Many plumbers simply connect to an existing sewer pipe that has been provided by a septic installer. However, plumbers do work with septic pumps, sewers, and other elements of septic systems. A plumber may not be called to install a septic system, but when there is a problem with a private waste disposal system, plumbers are often called. The more you understand about septic systems, the better qualified you will be to serve your customers.

Design criteria for septic systems can vary from one jurisdiction to the next. Trench and bed systems are both types of drain fields that do not require the use of chambers. Bed systems are the more common of the two. The design criteria for either of these systems are different from what would be used to lay out a chamber system.

The landscape position which is normally considered suitable for a trench system should not have a slope of more than 25 percent. A slope greater than this can impair the use of equipment needed to install a system. Bed systems may be limited to a slope of no more than 5 percent. Keep in mind, the numbers I'm giving you are only suggestions. They do not necessarily represent the requirements in your area.

These systems can be installed on land that is level and well-drained and on the crests of slopes. Convex slopes are considered the best location. Areas where these systems should not be installed include depressions and the bases of slopes where suitable surface drainage is not available.

In terms of texture, a sandy or loamy soil is best-suited to trench and bed systems. Gravelly and cobbley soils are not as desirable. Clay soil is the least desirable.

Drain Pipes Bedded In Crushed Stone (end view)

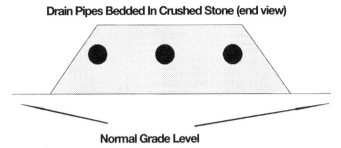

Normal Grade Level

Figure 6.1   Mound-type septic system.

When it comes to structure, a strong, granular, blocky or prismatic structure is best. Platy or unstructured massive soils are the least desirable.

When you are looking at the color of soil with the intent of installing a trench or bed system, look for bright, uniform colors. This indicates a well-drained, well-aerated soil. Ground with a dull, gray, or mottled appearance will usually be a sign of seasonal saturation. This makes soil unsuitable for a trench or bed system.

If you find soil that is layered with distinct textural or structural differences, be careful. This may indicate that water movement will be hindered, and this is not good.

Ideally, there should be between 2 and 4 ft of unsaturated soil between the bottom of a drain system and the top of a seasonally high water table or bedrock. Check with your local authorities to confirm the information I've given you here. You can also ask the authorities to give you acceptable ratings for percolation tests.

Mound and chamber systems (Figs. 6.1 and 6.2) can have different design criteria from those discussed for bed and trench systems. I chose bed and trench systems to use as our example since they are the types of systems most often used. Your local code office or county extension office can provide you with more detailed criteria for all types of systems.

Soil Types

What soil types are suitable for an absorption-based septic system? There are a great many types of soils which can accommodate a standard septic system. Naturally, some are better than others. Let's take a few moments to discuss briefly what you should look for in terms of soil types.

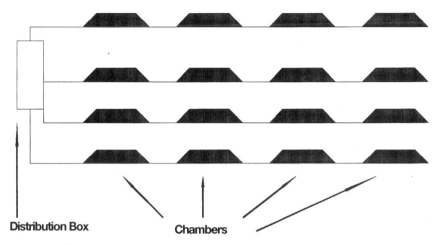

Distribution Box   **Chambers**

Figure 6.2   Chamber-type system.

The best

What is the best type of soil to have when you want to install a normal septic system? There is not necessarily one particular type of soil that is best. However, there are several types of soil which would fall into a category of being very desirable. Gravels and gravel-sand mixtures are some of the best soils to work with. Sandy soil is also very good. Soil that is made up of silty gravel or a combination of gravel, sand, and silt can be considered good to work with. Even silty sand and sand-silt combinations rate a good report card. In all of these soil types, it is best to avoid what is known as *fines.*

Pretty good

Just as there are a number of good soils, there are a good many soil types that are pretty good to install a septic system in. Gravel that has clay mixed with it is fairly good to deal with, and so is a gravel-sand-clay mixture. The same can be said for sand-clay mixtures. Moving down the list of acceptable soil types, you can find inorganic silts, fine sands, and silty or clayey fine sands.

Not so good

Inorganic clay, fat clay, and inorganic silt are not so good when it comes to drainage values. This is also true of micaceous or diatomaceous fine sandy or silty soils. These types of soils can be used in conjunction with absorption-based septic systems, but the systems will have to be designed to make up for the poor drainage characteristics of the soils.

Just won't do

Some types of soils just won't do when it comes to installing a standard septic system. Of these types of soils, organic silts and clays are included. So are peat and other soils that have a high organic rating.

Pipe and Gravel Septic Systems

Let's talk about the basic components of a pipe-and-gravel septic system (Fig. 6.3). Starting near the foundation of a building, there is a sewer. The sewer pipe should be made of solid pipe, not perforated pipe. The sewer pipe runs to the septic tank. An average size tank holds about 1000 gal. The connection between the sewer and the septic tank should be watertight.

The discharge pipe from the septic tank should be made of solid pipe, just like the sewer pipe. This pipe runs from the septic tank to a distribution box. Once the discharge pipe reaches the distribution box, the type of materials used changes.

The drain field is constructed according to an approved septic design. In basic terms, the excavated area for the septic bed is lined with crushed stone (Fig. 6.4). Perforated plastic pipe is installed in rows. The distance between the drain pipes and the number of drain pipes is controlled by the septic design. All of the drain-field pipes connect to the distribution box. The septic field is then covered with material specified in the septic design.

Chamber Systems

Chamber septic systems are used most often when the perk rate on ground is low. Soil with a rapid absorption rate can support a standard, pipe-and-gravel septic system. Clay and other types of soil may not. When bedrock is close to the ground surface, as is the case in much of Maine, chambers are often used.

What is a chamber system? A chamber system is installed very much like a pipe-and-gravel system, except for the use of chambers. The chambers might be made of concrete or plastic. Concrete chambers are naturally more expensive to install. Plastic chambers are shipped in halves and put together in the field. Since plastic is a very durable material, and it's relatively cheap, plastic chambers are more popular than concrete chambers.

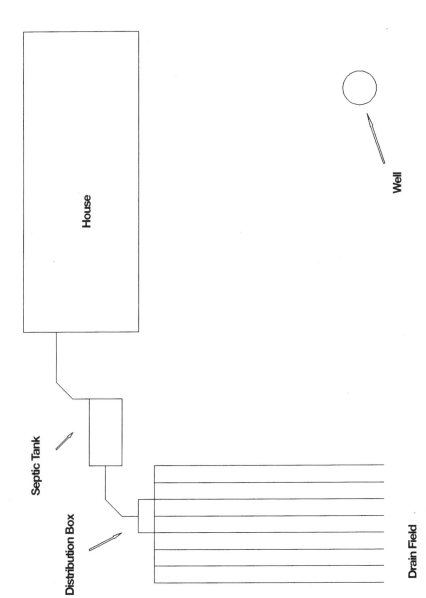

Figure 6.3 Typical septic system.

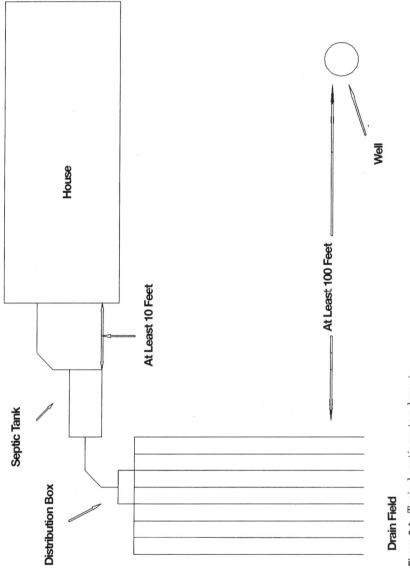

Figure 6.4 Typical septic system layout.

When a chamber system is called for, there are typically many chambers involved. These chambers are installed in the leach field, between sections of pipe. As effluent is released from a septic tank, it is sent into the chambers. The chambers collect and hold the effluent for a period of time. Gradually, the liquid is released into the leach field and absorbed by the earth. The primary role of the chambers is to retard the distribution rate of the effluent.

A chamber system is simple enough in its design. Liquid leaves a septic tank and enters the first chamber. As more liquid is released from the septic tank, it is transferred into additional chambers that are farther downstream. This process continues with the chambers releasing a predetermined amount of liquid into the soil as time goes on. The process allows more time for bacterial action to attack raw sewage, and it controls the flow of liquid into the ground.

If a perforated-pipe system is used in ground where a chamber system is recommended, the result could be a flooded leach field. This might create health risks. It would most likely produce unpleasant odors, and it might even shorten the life of the septic field.

Chambers are installed between sections of pipe within the drain field. The chambers are then covered with soil. The finished system is not visible above ground. All of the action takes place below grade. The only real downside to a chamber system is the cost.

Trench Systems

Trench systems are the least expensive version of special septic systems. They are comparable in many ways to a standard pipe-and-gravel bed system. The main difference between a trench system and a bed system is that the drain lines in a trench system are separated by a physical barrier. Bed systems consist of drain pipes situated in a rock bed. All of the pipes are in one large bed. Trench fields depend on separation to work properly. To expand on this, let me give you some technical information.

A typical trench system is set into trenches that are between 1 to 5 ft deep. The width of the trench tends to run from 1 to 3 ft. Perforated pipe is placed in these trenches on a 6-in bed of crushed stone. A second layer of stone is placed on top of the drain pipe. This rock is covered with a barrier of some type to protect it from the backfilling process. The type of barrier used will be specified in a septic design.

When a trench system is used, both the sides of the trench and the bottom of the excavation are an outlet for liquid. Only one pipe is placed in each trench. These two factors are what distinguishes a trench system from a standard bed system. Bed systems have all of the drain pipes in one large excavation. In a bed system, the bottom of the bed is the only significant infiltrative surface. Since trench systems use both the bottoms and sides of trenches as infiltrative surfaces, more absorption is potentially possible.

Neither bed or trench systems should be used in soils where the percolation rate is either very fast or slow. For example, if the soil will accept 1 in of liquid per minute, it is too fast for a standard absorption system. This can be overcome by lining the infiltrative surface with a thick layer (about 2 ft or more) of sandy loam soil. Conversely, land that drains at a rate of 1 in an hour is too slow for a bed or trench system. This is a situation where a chamber system might be recommended as an alternative.

More land area

Because of their design, trench systems require more land area than bed systems do. This can be a problem on small building lots. It can also add to the expense of clearing land for a septic field. However, trench systems are normally considered to be better than bed systems. There are many reasons for this.

Trench systems are said to offer up to 5 times more side area for infiltration to take place. This is based on a trench system with a bottom area identical to a bed system. The difference is in the depth and separation of the trenches. Experts like trench systems because digging equipment can straddle the trench locations during excavation. This reduces damage to the bottom soil and improves performance. In a bed system, equipment must operate within the bed, compacting soil and reducing efficiency.

If you are faced with hilly land to work with, a trench system is ideal. The trenches can be dug to follow the contour of the land. This gives you maximum utilization of the sloping ground. Infiltrative surfaces are maintained while excessive excavation is eliminated.

The advantages of a trench system are numerous. For example, trenches can be run between trees. This reduces clearing costs and allows trees to remain for shade and aesthetic purposes. However, roots may still be a consideration. Most people agree that a trench system performs better than a bed system. When you combine performance with the many other advantages of a trench system, you may want to consider trenching your next septic system. It costs more to dig individual trenches than it does to create a group bed, but the benefits may outweigh the costs.

Mound Systems

Mound systems, as you might suspect, are septic systems which are constructed in mounds that rise above the natural topography. This is done to compensate for high water tables and soils with slow absorption rates. Because of the large amount of fill material required to create a mound, the cost is naturally higher than it would be for a bed system.

Coarse gravel is normally used to build a septic mound. The stone is piled on top of the existing ground. However, top soil is removed before the stone is installed. When a mound is built, it contains suitable fill material, an absorption area, a distribution network, a cap, and top soil. Because of the raised height, a mound system depends on either pumping or siphonic action to work properly. Essentially, effluent is either pumped or siphoned into the distribution network.

As the effluent is passing through the coarse gravel and infiltrating the fill material, treatment of the wastewater occurs. This continues as the liquid passes through the unsaturated zone of the natural soil.

The purpose of the cap is to retard frost action, deflect precipitation, and to retain moisture that will stimulate the growth of ground cover. Without adequate ground cover, erosion can be a problem. There are a multitude of choices available as acceptable ground covers. Grass is the most common choice.

Mounds should be used only in areas that drain well. The topography can be level or slightly sloping. The amount of slope allowable depends on the perk rate. For example, soil that perks at a rate of 1 in every 60 min or less should not have a slope of more than 6 percent if a mound system is to be installed. If the soil absorbs water from a perk test faster than 1 in in 1 h, the slope could be increased to 12 percent. These numbers are only examples. A professional who designs mound systems will set the true criteria for slope values.

Ideally, about 2 ft of unsaturated soil should exist between the original soil surface and the seasonally saturated top soil. There should be 3 to 5 ft of depth to the impermeable barrier. The overall range of perk rate could go as high as 1 in in 2 h, but this, of course, is subject to local approval. Perk tests for this type of system are best when done at a depth of about 20 in. However, they can be performed at shallow depths of only 12 in. Again, you must consult and follow local requirements. The design and construction of mound systems can get quite complicated.

Sewage Pumps

Sewage pumps, also known as *black-water sumps,* work on a principle similar to that of gray-water sumps. However, black-water sumps may receive the discharge of toilets and other fixtures of a similar nature. The pumps used in black-water sumps are of a different type than those used in gray-water setups. Black-water sumps are normally installed below a finished floor level. They are quite often buried in concrete floors. It is possible to install such a sump in a crawl space by simply creating a stable base for it on the ground.

Black-water basins or sumps are available in different sizes. A typical residential sump may be 30 in deep with an 18-in diameter at the lid. Basin packages can be purchased that include all parts necessary to set up the basin.

A 2-in vent pipe should extend from the basin cover to open air, outside of a building. The use of a check valve is required in the vertical discharge line. A 4-in inlet opening is molded into the basin to accept the waste of all types of residential plumbing fixtures. Some type of float system is used to activate the pump that is housed in the basin. The exact type of float system depends on the type of pump being used.

Pumps used for a typical, in-house, black-water sumps are known as effluent pumps and sewage ejectors. The discharge pipe from the pump normally has a diameter of 2 in, although a 3-in discharge flange is available. Even if the waste of a toilet is being pumped, a 2-in discharge pipe is sufficient.

The cost of a complete black-water sump system will run into several hundred dollars, but this is much less than what a whole-house pump station would cost. If you have a basement bathroom that requires pumping, this type of system is ideal. I've installed dozens, if not hundreds of them, and I've never been called back for a failure. During my plumbing career, I have responded to failures in similar systems. The most common problem is a float that has become wedged against the side of the sump. This won't happen if the system is installed properly.

If you have a house where most of the plumbing can be drained by gravity, and only a few fixtures need to be pumped, as in the case of a basement bathroom, an in-house, black-water sump is the way to go.

Whole-House Pump Stations

Whole-house pump stations pose the potential threat of a house where none of the plumbing can be used if the pump fails. As bad as they are, pump stations are sometimes a blessing. Whole-house pump stations are normally located outside the foundation walls. This is not always the case, but it usually is. Since all of the plumbing in a house with a pump station is dependent on a pump's operation, you must take some special precautions when installing such a system. Code requirements are often very stringent on this issue, although they vary from place to place.

It is standard procedure to install a whole-house pump station in a location outside of the home being served. A sewer pipe runs from the house or building to the storage sump. The size and capacity of this sump is determined by local code requirements and anticipated use. A sump can be made of many types of material, such as concrete or fiberglass.

Fiberglass sumps are not uncommon. One fiberglass sump that I know of is 3 ft deep and has a 30-in diameter. This particular sump is designed for interior use with a grinder pump. It has a 4-in inlet and an outlet for a 2-in vent. The basin kit comes complete with a control panel, a hands-off automatic switch, a terminal strip, and an audible alarm. There are also three encapsulated mercury switches that provide positions for on, off, and alarm. This is only one example of a whole-house pumping system; there are many others available to choose from.

You may never deal with the service or installation of sewage pumps and ejectors. If you are involved in the installation of such a system, you must consult your local code for exact requirements. It may be a requirement for you to install two pumps in a whole-house pumping station, so that one pump will function if the other fails. Alarms, lights, and other safety precautions may also be required. The code requirements on black-water pump stations can be quite strict, so check carefully with your local code office before making an installation of this type.

7

Traps, Cleanouts, Interceptors, and More

We have covered most of the regulations you will need to know about drains and vents. This chapter will round out your knowledge. Here you will learn about traps. Traps have been mentioned before, and you have learned the importance of vents to trap seals, but here you will learn more about traps.

Cleanouts are a necessary part of the drainage system. This chapter will tell you what types of cleanouts you can use and when and where they must be used. Along with cleanouts, back-water valves will be explained. Grease receptors, or grease traps as they are often called, will be explored. By the end of this chapter you should be prepared to tackle just about any drain waste and vent (DWV) job.

Cleanouts

What are cleanouts, and why are they needed? Cleanouts are a means of access to the interior of drainage pipes. They are needed so that blockages in drains may be cleared. Without cleanouts, it is much more difficult to snake a drain. In general, the more cleanouts you have, the better. Plumbing codes establish minimums for the number of cleanouts required and their placement. Let's look at how these regulations apply to you.

Where Are Cleanouts Required?

There are many places in a plumbing system where cleanouts are required. Let's start with sewers. All sewers must have cleanouts. The distances between these cleanouts vary from region to region.

Generally, cleanouts will be required where the building drain meets the building sewer. The cleanouts may be installed inside the foundation or outside, but the cleanout opening must extend upward to the finished floor level or the finished grade outside.

Zone two prefers that the cleanouts at the junction of building drains and sewers be located outside. If the cleanout is installed inside, within zone two, it must extend above the flood level rim of the fixtures served by the horizontal drain. When this is not feasible, allowances may be made. Zone three will waive the requirement for a junction cleanout if there is a cleanout of at least a 3-in diameter within 10 ft of the junction.

Once the sewer is begun, cleanouts should be installed every 100 ft. In zone two, the interval distance is 75 ft for 4-in and larger pipe, and 55 ft for pipe smaller than 4-in. Cleanouts are also required in sewers when the pipe takes a change in direction. In zone three, a cleanout is required every time the sewer turns more than 45°. In zone one, a cleanout is required whenever the change in direction is more than 135°.

The cleanouts installed in a sewer must be accessible. This generally means that a standpipe will rise from the sewer to just below ground level. At that point, a cleanout fitting and plug are installed on the standpipe. This allows the sewer to be snaked out from ground level, with little to no digging required.

For building drains and horizontal branches, the cleanout location will depend upon pipe size, but cleanouts are normally required every 50 ft. For pipes with diameters of 4 in, or less, cleanouts must be installed every 50 ft. Larger drains may have their cleanouts spaced at 100-ft intervals. Cleanouts are also required on these pipes with a change in direction. For zone three, the degree of change is anything in excess of 45°. Cleanouts must be installed at the end of all horizontal drain runs. Zone one does not require cleanouts at 50-ft intervals, only at 100-ft intervals.

As with most rules, there are some exceptions to these. Zone one offers some exceptions to the cleanout requirements for horizontal drains. The following exceptions apply only to zone one. If a drain is less than 5 ft long and is not draining sinks or urinals, a cleanout is not required. A change in direction from a vertical drain with a fifth bend does not require a cleanout. Cleanouts are not required on pipes, other than building drains and their horizontal branches, that are above the first-floor level.

P-traps and water closets are often allowed to act as cleanouts. When these devices are approved for cleanout purposes, the normally required cleanout fitting and plug at the end of a horizontal pipe run

Figure 7.1 P-trap.

may be eliminated. Not all jurisdictions will accept P-traps (Fig. 7.1) and toilets as cleanouts; check your local requirements before omitting standard cleanouts.

Cleanouts must be installed in a way that the cleanout opening is accessible and allows adequate room for drain cleaning. The cleanout must be installed to go with the flow. This means that when the cleanout plug is removed, a drain-cleaning device should be able to enter the fitting and the flow of the drainage pipe without trouble.

When you are installing your plumbing in zone three, you must include a cleanout at the base of every stack (Fig. 7.2). This is good procedure at any time, but it is not required by all codes. The height of this cleanout should not exceed 4 ft. Many plumbers install test tees at these locations to plug their stacks for pressure testing (Fig. 7.3). The test tee doubles as a cleanout.

When the pipes holding cleanouts will be concealed, the cleanout must be made accessible. For example, if a stack will be concealed by

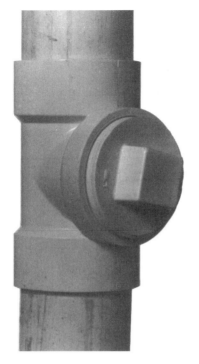

Figure 7.2 Test tee.

Figure 7.3 Test tee with test ball installed.

a finished wall, provisions must be made for access to the cleanout. This access could take the form of an access door, or the cleanout could simply extend past the finished wall covering. If the cleanout is serving a pipe concealed by a floor, the cleanout must be brought up to floor level and made accessible. This ruling applies not only to cleanouts installed beneath concrete floors but also to cleanouts installed in crawl spaces, with very little room to work.

What Else Do I Need to Know about Cleanouts?

There is still more to learn about cleanouts. Size is one of the lessons to be learned. Cleanouts are required to be the same size as the pipe they are serving, unless the pipe is larger than 4 in. If you are installing a 2-in pipe, you must install 2-in cleanouts. However, when a P-trap is allowed for a cleanout, it may be smaller than the drain (Fig. 7.4). An example would be a 1¼-in trap on a 1½-in drain. Remember though, not all code enforcement officers will allow P-traps as cleanouts, and they may require the P-trap to be the same size as

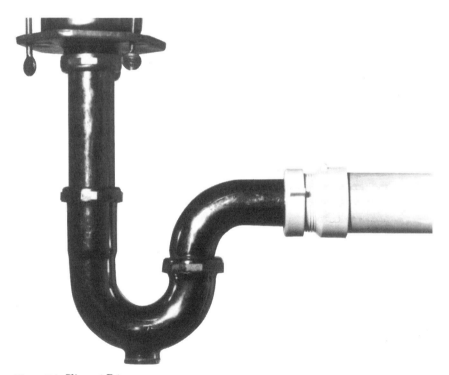

Figure 7.4 Slip-nut P-trap.

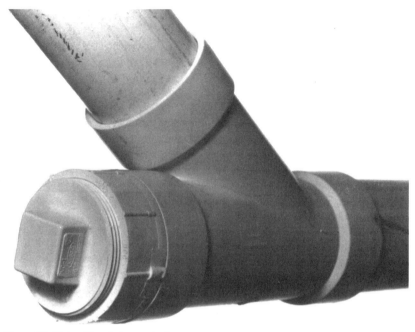

Figure 7.5 Raised-head cleanout plug.

the drain if the trap is allowed as a cleanout. Once the pipe size exceeds 4 in, the cleanouts used should have a minimum size of 4 in.

When cleanouts are installed, they must provide adequate clearance for drain cleaning (Fig. 7.5). The clearance required for pipes with diameters of 3 in, or more is 18 in. Smaller pipes require a minimum clearance of 12 in in front of their cleanouts. Many plumbers fail to remember this regulation. It is common to find cleanouts pointing toward floor joists or too close to walls. You will save yourself time and money by committing these clearance distances to memory.

Zone one takes the clearance rules a step further. In zone one, when a cleanout is installed in a floor, it must have a minimum height clearance of 18 in and a minimum horizontal clearance of 30 in. No underfloor cleanout is allowed to be placed more than 20 ft from an access opening.

Acceptable Types of Cleanouts

Cleanout plugs and plates must be easily removed. Access to the interior of the pipe should be available without undue effort or time. Cleanouts can take on many appearances (Fig. 7.6). The U bend of a

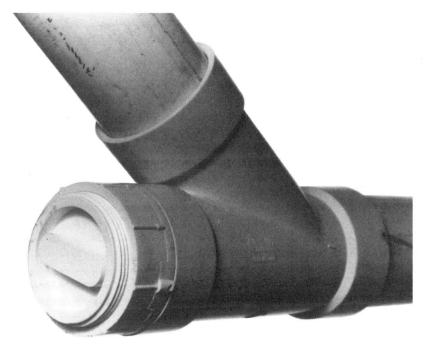

Figure 7.6 Recessed-head cleanout plug.

Figure 7.7 Rubber cap.

P-trap can be considered a cleanout, depending upon local interpretation. A rubber cap, held onto the pipe by a stainless steel clamp (Fig.
7.7), can serve as a cleanout. The standard female adapter and plug
are a fine cleanout. Test tees will work as cleanouts. Special

cleanouts, designed to allow rodding of a drain in either direction, are acceptable.

Very Big Cleanouts

The ultimate cleanout is a manhole. You can think of manholes as very big cleanouts. When a pipe's diameter exceeds 10 in in zone three, or 8 in in zone two, manholes replace cleanouts. Manholes are required every 400 ft in zone three and every 300 ft in zones one and two. In addition, they are required at all changes in direction, elevation, grade, and size. Manholes shall be protected against flooding and must be equipped with covers to prevent the escape of gases. Zone one requires connections with manholes to be made with flexible compression joints. These connections must not be made closer than 1 ft to the manhole and not farther than 3 ft away from it.

Traps

Traps are required on drainage-type plumbing fixtures. With some fixtures, like toilets, traps are not apparent because they are an integral part of the fixture. The following regulations do not apply to integral traps, which being a part of a fixture, are governed by regulations controlling the use of approved fixtures. We have already talked about trap seals, so now let's learn more about traps.

P-traps

P-traps are the traps most frequently used in modern plumbing systems. These traps are self-cleaning and frequently have removable U bends that may act as cleanouts, pending local approval. P-traps must be properly vented. Without adequate venting, the trap seal can be removed by back pressure.

S-traps

S-traps were very common when most plumbing drains came up through the floor instead of out of a wall. Many S-traps are still in operation, but they are no longer allowed in new installations. S-traps can lose their trap seal through self-siphoning.

Drum traps

Drum traps are not normally allowed in new installations without special permission from the code officer. The only occasion when drum traps are still used frequently is when they are installed with a combination waste and vent system.

Bell traps

Bell traps are not allowed for use in new installations.

House traps

House traps are no longer allowed; they represent a double trapping of all fixtures. House traps were once installed where the building drain joined with the sewer. Most house traps were installed inside the structure, but a fair number were installed outside, underground. Their purpose was to prevent sewer gas from coming out of the sewer and into the plumbing system, but house traps make drain cleaning very difficult and they create a double-trapping situation, which is not allowed. This regulation, like most regulations, is subject to amendment and variance by the local code official.

Crown-vented traps

Crown-vented traps are not allowed in new installations. These traps have a vent rising from the top of the trap. As you learned earlier, crown venting must be done at the trap arm, not the trap.

Other traps

Traps that depend on moving parts or interior partitions are not allowed in new installations.

Does Every Fixture
Require an Individual Trap?

Basically, every fixture requires an individual trap, but there are exceptions. One such exception is the use of a continuous waste to connect the drains from multiple sink bowls to a common trap. This is done frequently with kitchen sinks.

There are some restrictions involving the use of continuous wastes. Let's take a kitchen sink as an example. When you have a double-bowl sink, it is okay to use a continuous waste, as long as the drains from each bowl are no more than 30 in apart and neither bowl is more than 6 in deeper than the other bowl. Zone one requires that all sinks connected to a continuous waste must be of equal depth. Exceptions to this rule do exist.

What if your sink has three bowls? Three-compartment sinks may be connected with a continuous waste. You may use a single trap to collect the drainage from up to three separate sinks or lavatories as long as they are next to each other and in the same room. But, the trap must be in a location central to all sinks or lavatories.

TABLE 7.1   Recommended Trap Sizes for Zone One

Type of fixture	Trap size (in)
Bathtub	1½
Shower	2
Residential toilet	Integral
Lavatory	1¼
Bidet	1½
Laundry tub	1½
Washing machine standpipe	2
Floor drain	2
Kitchen sink	1½
Dishwasher	1½
Drinking fountain	1¼
Public toilet	Integral

TABLE 7.2   Recommended Trap Sizes for Zone Two

Type of fixture	Trap size (in)
Bathtub	1½
Shower	2
Residential toilet	Integral
Lavatory	1¼
Bidet	1½
Laundry tub	1½
Washing machine standpipe	2
Floor drain	2
Kitchen sink	1½
Dishwasher	1½
Drinking fountain	1
Public toilet	Integral

Trap Sizes

Trap sizes are determined by the local code. Tables 7.1, 7.2, and 7.3 give you examples of commonly accepted trap sizes. A trap may not be larger than the drain pipe it discharges into.

Tailpiece Length

The tailpiece between a fixture drain and the fixture's trap may not exceed 24 in.

Standpipe Height

A standpipe, when installed in zone three, must extend at least 18 in above its trap but may not extend more than 30 in above the trap.

TABLE 7.3   **Recommended Trap Sizes for Zone Three**

Type of fixture	Trap size (in)
Bathtub	1½
Shower	2
Residential toilet	Integral
Lavatory	1¼
Bidet	1¼
Laundry tub	1½
Washing machine standpipe	2
Floor drain	2
Kitchen sink	1½
Dishwasher	1½
Drinking fountain	1¼
Public toilet	Integral
Urinal	2

Zone two prohibits the standpipe from extending more than 4 ft from the trap. Zone one requires the standpipe not to exceed a height of more than 2 ft above the trap. Plumbers installing laundry stand-pipes often forget this regulation. When setting your fitting height in the drainage pipe, keep in mind the height limitations on your stand-pipe. Otherwise, your take-off fitting may be too low, or too high, to allow your standpipe receptor to be placed at the desired height. Traps for kitchen sinks may not receive the discharge from a laundry tub or clothes washer.

Proper Trap Installation

There is more to proper trap installation than location and trap selec-tion. Traps must be installed level in order for the trap seal to func-tion properly. An average trap seal will consist of 2 in of water. Some large traps may have a seal of 4 in, and where evaporation is a prob-lem, deep-sealing traps may have a deeper water seal. The position-ing of the trap is critical for the proper seal. If the trap is cocked, the water seal will not be uniform and may contribute to self-siphoning.

When a trap is installed below grade and must be connected from above grade, the trap must be housed in a box of a sorts (Fig. 7.8). An example of such a situation would be a trap for a tub waste. When installing a bathtub on a concrete floor, the trap is located below the floor. Since the trap cannot be reasonably installed until after the floor is poured, access must be made for the connection. This access, frequently called a tub box or trap box, must provide protection against water, insect, and rodent infiltration.

Figure 7.8 Trap box.

When Is a Trap Not a Trap?

One type of trap we have not yet discussed is a grease trap. The reason we haven't talked about grease traps is that grease traps are not really traps; they are interceptors. They are frequently called grease traps, but they are actually grease interceptors. There is a big difference between a trap and an interceptor.

Traps are meant to prevent sewer gas from entering a building. Traps do not restrict what goes down the drain, only what comes up the drain. Of course, traps do prevent objects larger than the trap from entering the drain, but this is not their primary objective.

Interceptors, on the other hand, are designed to control what goes down a drain. Interceptors are used to keep harmful substances from entering the sanitary drainage systems. Separators, because they separate the materials entering them and retain certain materials, while allowing others to continue into the drainage system, are also required in some circumstances. Interceptors are used to control grease, sand, oil, and other materials.

Interceptors and separators are required when conditions provide opportunity for harmful or unwanted materials to enter a sanitary drainage system. For example, a restaurant is required to be equipped with a grease interceptor because of the large amount of grease present in commercial food establishments. An oil separator would be required for a building where automotive repairs are made. Interceptors and separators must be designed for each individual situation. There is no rule-of-thumb method of choosing the proper interceptor or separator without expert design.

There are some guidelines provided in plumbing codes for interceptors and separators. The capacity of a grease interceptor is based on two factors: grease retention and flow rate. These determinations are typically made by a professional designer. The size of a receptor or separator is also normally determined by a design expert.

Interceptors for sand and other heavy solids must be readily accessible for cleaning. These units must contain a water seal of not less than 6 in, except in zone two, where the minimum water depth is 2 in. When an interceptor is used in a laundry, a water seal is not required. Laundry receptors, used to catch lint, string, and other objects, are usually made of wire, and they must be easily removed for cleaning. Their purpose is to prevent solids with a diameter of ½ in, or more, from entering the drainage system.

Other types of separators are used for various plants, factories, and processing sites. The purpose of all separators is to keep unwanted objects and substances from entering the drainage system. Vents are required if it is suspected that these devices will be subject to the loss of a trap seal. All interceptors and separators must be readily accessible for cleaning, maintenance, and repairs.

Back-Water Valves

Back-water valves are essentially check valves. They are installed in drains and sewers to prevent the backing-up of waste and water in the drain or sewer. Back-water valves are required to be readily accessible and installed whenever a drainage system is likely to encounter back-ups from the sewer.

The intent behind back-water valves is to prevent sewers from backing up into individual drainage systems. Buildings that have plumbing fixtures below the level of the street, where a main sewer is installed, are candidates for back-water valves.

This concludes our section on traps, cleanouts, interceptors, and other drainage-related regulations. While this is a short chapter, it is an important one. You may not have a need for installing manholes or back-water valves every day, but, as a plumber, you will frequently work with traps and cleanouts.

Reminder Notes

Zone one

1. Cleanouts are required in sewers when the pipe makes a change in direction of more than 135°.

2. Cleanouts in all horizontal drains may be spaced up to 100 ft apart, rather than 50 ft, as in zones two and three.

3. Cleanouts are not required in horizontal drains that are less than 5 ft in length, unless the pipe is draining urinals or sinks.

4. A change in direction from a vertical drain with a fifth bend does not require a cleanout.

5. Cleanouts are not required on pipes, other than building drains and their horizontal branches, that are above the first-floor level.

6. Cleanouts extending from a floor must have a vertical clearance of 18 in and a horizontal clearance of 30 in. No underfloor cleanout may be placed more than 20 ft from an access opening.

7. Manhole connections must be made with flexible connections. These connections must be made at least 1 ft from the manhole but not farther than 3 ft from it.

8. When multiple sinks are connected to a single trap, with a continuous waste, all sink bowls must be of equal depth.

9. Laundry standpipes may not extend more than 2 ft above the traps.

10. Tubular traps must have a minimum rating of 17 gauge; however, these thin-walled traps are not allowed for use with urinals.

Zone two

1. Cleanouts at the junction of a building drain and sewer should be located outside. If they are located inside, the cleanout should extend above the flood-level rim of the fixtures being served by the horizontal pipe. If getting the cleanout above the flood-level rim is not feasible, allowances may be made.

2. Cleanouts in a 4-in, or larger, sewer should be installed with intervals of not more than 75 ft between them. If the sewer pipe is smaller than 4 in, the interval should be reduced to 50 ft.

3. Manholes are required on sewers with diameters larger than 8 in.

4. Laundry standpipes may not extend more than 4 ft above the trap.

5. Tubular traps must have a minimum rating of 20 gauge.

6. Sand interceptors must have a minimum water seal of 2 in.

Zone three

1. Cleanouts at the junction of a building drain and building sewer may be eliminated when a cleanout with a minimum diameter of 3 in is within 10 ft of the junction.

2. Cleanouts are required in the sewer when the pipe takes a change in direction of more than 45°.

3. Cleanouts are required on all horizontal drains when a change in direction of more than 45° is needed.

4. A cleanout is required at the base of each drainage stack.

5. Manholes are required on sewers with diameters larger than 10 in.

6. Manholes must not be more than 400 ft apart.

7. Laundry standpipes must extend at least 18 in above the trap, but not more than 30 in.

8

Fixtures

There is much more to fixtures than meets the eye. Fixtures are a part of the final phase of plumbing. When you are planning a plumbing system, you must know which fixtures are required, what type of fixtures are needed, and how they must be installed. This chapter will guide you through the myriad of fixtures available and how they may be used.

What Fixtures Are Required?

The number and types of fixtures required will depend on local regulations and the use of the building in which they are being installed. Your code book will provide you with information on what is required and how many of each type of fixture is needed. These requirements are based on the use of the building housing the fixtures and the number of people that may be using the building. Here are some examples.

Single-family residence

When you are planning fixtures for a single-family residence, you must include certain fixtures. If you choose to install more than the minimum, that's fine, but you must install the minimum number of required fixtures, which are as follows:

- One toilet
- One lavatory
- One bathing unit
- One kitchen sink
- One hook-up for a clothes washer

Multifamily buildings

The minimum requirements for a multifamily building are the same as those for a single-family dwelling, but the requirements are that each dwelling in the multifamily building must be equipped with the minimum fixtures. There is one exception—the laundry hook-up. With a multifamily building, laundry hook-ups are not required in each dwelling unit. In zone three, it is required that a laundry hook-up be installed for common use when the number of dwelling units is 20. For each interval of 20 units, you must install a laundry hook-up. For example, in a building with 40 apartments, you would have to provide two laundry hook-ups. If the building had 60 units, you would need three hook-ups. In zone one, the dwelling-unit interval is 10 rental units. Zone two requires one hook-up for every 12 rental units but no less than two hook-ups for buildings with at least 15 units.

Nightclubs

When you get into businesses and places of public assembly, like nightclubs, the ratings are based on the number of people likely to use the facilities. In a nightclub, the minimum requirements for zone three are as follows:

- Toilets—one toilet for every 40 people
- Lavatories—one lavatory for every 75 people
- Service sinks—one service sink
- Drinking fountains—one drinking fountain for every 500 people
- Bathing units—none

Day-care facilities

The minimum number of fixtures for a day-care facility in zone three are as listed below:

- Toilets—one toilet for every 15 people
- Lavatories—one lavatory for every 15 people
- Bathing units—one bathing unit for every 15 people
- Service sinks—one service sink
- Drinking fountains—one drinking fountain for every 100 people

In contrast, zone two only requires the installation of toilets and lavatories in day-care facilities. The ratings for these two fixtures are the same as in zone three, but the other fixtures required by zone three are not required in zone two. This type of rating system will be found in your local code and will cover all the normal types of building uses.

In many cases, facilities will have to be provided in separate bathrooms, to accommodate each sex. When installing separate bathroom facilities, the number of required fixtures will be divided equally between the two sexes, unless there is cause and approval for a different appropriation.

Some types of buildings do not require separate facilities. For example, zone three does not require the following buildings to have separate facilities: residential properties and small businesses where less than 15 employees work or where less than 15 people are allowed in the building at the same time.

Zone two does not require separate facilities in the following buildings: offices with less than 1200 ft^2, retail stores with less than 1500 ft^2, restaurants with less than 500 ft^2, self-serve laundries with less than 1400 ft^2, and hair salons with less than 900 ft^2.

Employee and customer facilities

There are some special regulations pertaining to employee and customer facilities. For employees, toilet facilities must be available to employees within a reasonable distance and with relative ease of access. For example, zone three requires these facilities to be in the immediate work area; the distance an employee is required to walk to the facilities may not exceed 500 ft. The facilities must be located in a manner so that employees do not have to negotiate more than one set of stairs for access to the facilities. There are some exceptions to these regulations, but in general, these are the rules.

It is expected that customers of restaurants, stores, and places of public assembly shall have toilet facilities. In zone three, this is based on buildings capable of holding 150 or more people. Buildings in zone three with an occupancy rating of less than 150 people are not required to provide toilet facilities, unless the building serves food or beverages. When facilities are required, they may be placed in individual buildings or in a shopping mall situation, in a common area, not more than 500 ft from any store or tenant space. These central toilets must be placed so that customers will not have to use more than one set of stairs to reach them.

Zone two uses a square-footage method to determine minimum requirements in public places. For example, retail stores are rated as having an occupancy load of one person for every 200 ft^2 of floor space. This type of facility is required to have separate facilities when the store's square footage exceeds 1500 ft^2. A minimum of one toilet is required for each facility when the occupancy load is up to 35 people.

One lavatory is required in each facility, for up to 15 people. A drinking fountain is required for occupancy loads up to 100 people.

Handicap Fixtures

Handicap fixtures are not cheap; you cannot afford to overlook them when bidding a job. The plumbing code will normally require specific minimum requirements for handicap-accessible fixtures in certain circumstances. It is your responsibility to know when handicap facilities are required. There are also special regulations pertaining to how handicap fixtures shall be installed. We are about to embark on a journey into handicap fixtures and their requirements.

When you are dealing with handicap plumbing, you must mix the local plumbing code with the local building code. These two codes work together in establishing the minimum requirements for handicap plumbing facilities. When you step into the field of handicap plumbing, you must play by a different set of rules. Handicap plumbing is like a different code of its own.

Where are handicap fixtures required?

Most buildings frequented by the public are required to have handicap-accessible plumbing fixtures. The following handicap examples are based on zone three requirements. Zones one and two do not go into as much detail on handicap requirements in their plumbing codes.

Single-family homes and most residential multifamily dwellings are exempt from handicap requirements. A rule-of-thumb standard for most public buildings is the inclusion of one toilet and one lavatory for handicap use.

Hotels, motels, inns, and the like are required to provide a toilet (Fig. 8.1), lavatory (Fig. 8.2), bathing units (Fig. 8.3), and kitchen sink, where applicable, for handicap use. Drinking fountains may also be required. This provision will depend on the local plumbing and building codes. If plumbing a gang shower arrangement, such as in a school gym, at least one of the shower units must be handicap-accessible. Door sizes and other building code requirements must be observed when dealing with handicap facilities. There are local exceptions to these rules; check with your local code officers for current, local regulations.

Installation considerations

When it comes to installing handicap plumbing facilities, you must pay attention to the plumbing and building codes. In most cases, approved blueprints will indicate the requirements of your job, but in

4094 Atlas Elongated Rim
- 18″ rim height handicapped.
- 12″ rough-in.
- Anti-Siphon ballock.
- 3.5 G.P.F.

Figure 8.1 Handicap toilet. (*Courtesy Universal-Rundle Corporation*)

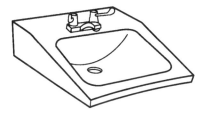

4682 8″cc/4683 4″cc — 27″ x 20″ Wheelchair

Figure 8.2 Handicap lavatory. (*Courtesy Universal-Rundle Corporation*)

rural areas, you may not enjoy the benefit of highly detailed plans and specifications. When it comes time for a final inspection, the plumbing must pass muster along with the open space around the fixtures. If the inspection is failed, your pay is held up and you are likely to incur unexpected costs. This section will apprise you of what you may need to know. It is not all plumbing, but it is all needed information when working with handicap facilities.

Handicap toilet facilities

When you think of installing a handicap toilet, you probably think of a toilet that sits high off the floor. But, do you think of the grab bars

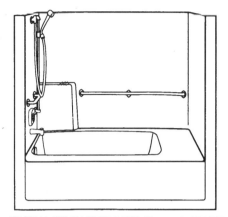

6266-H RHO/6267 LHO Summit 75 TS
- Molded-in seat.
- One-piece seamless construction.
- 1½" diameter safety grab bars.
- Slip resistant bottom.

Figure 8.3 Handicap bathtub. (*Courtesy Universal-Rundle Corporation*)

and partition dimensions required around the toilet? Some plumbers don't, but they should. The door to a privacy stall for a handicap toilet must provide a minimum of 32 in of clear space.

The distance between the front of the toilet and the closed door must be at least 48 in. It is mandatory that the door open outward, away from the toilet. Think about it; how could a person in a wheelchair close the door if the door opened into the toilet? These facts may not seem like your problem, but if your inspection doesn't pass, you don't get paid.

The width of a water closet compartment for handicap toilets must be a minimum of 5 ft. The length of the privacy stall shall be at least 56 in for wall-mounted toilets and 59 in for floor-mounted models. Unlike regular toilets that require a rough-in of 15 in to the center of the drain from a side wall, handicap toilets require the rough-in to be at least 18 in from the side wall.

Then, there are the required grab bars. Sure, you may know that grab bars are required, but do you know the mounting requirements for the bars? Two bars are required for each handicap toilet. One bar should be mounted on the back wall and the other should be installed on the side wall. The bar mounted on the back wall must be at least 3 ft long. The first mounting bracket of the bar must be mounted no more than 6 in from the side wall. Then, the bar must extend at least 24 in past the center of the toilet's drain.

The bar mounted on the side wall must be at least 42 in long. The bar should be mounted level and with the first mounting bracket located no more than 1 ft from the back wall. The bar must be mounted on the side wall that is closest to the toilet. This bar must extend to a point at least 54 in from the back wall. If you do your math, you will see that a 42-in bar is pushing the limits on both ends. A longer bar will allow more assurance of meeting the minimum requirements.

When a lavatory will be installed in the same toilet compartment, the lavatory must be installed on the back wall in a way that its closest point to the toilet is no less than 18 in from the center of the toilet's drain. When a privacy stall of this size and design is not suitable, there is an option. Another way to size the compartment to house a handicap toilet and lavatory is available. There may be times when space restraints will not allow a stall with a width of 5 ft. In these cases, you may position the fixture differently and use a stall with a width of only 3 ft. In these situations, the width of the privacy stall may not exceed 4 ft.

The depth of the compartment must be at least 66 in when wall-mounted toilets are used. The depth extends to a minimum of 69 in with the use of a floor-mounted water closet. The toilet requires a minimum distance from side walls of 18 in to the center of the toilet drain. If the compartment is more than 3 ft wide, grab bars are required, with the same installation methods as described before.

It the stall is made at the minimum width of 3 ft, grab bars, with a minimum length of 42 in, are required on each side of the toilet. These bars must be mounted no more than 1 ft from the back wall, and they must extend a minimum of 54 in from the back wall. If a privacy stall is not used, the side wall clearances and the grab bar requirements are the same as listed in these two examples. To determine which set of rules to use, you must assess the shape of the room when no stall is present.

If the room is laid out in a fashion like the first example, use the guidelines for grab bars as listed there. If, on the other hand, the room tends to meet the description of the last example, use the specifications in that example. In both cases, the door to the room may not swing into toilet area.

Handicap fixture design

Handicap fixtures are specially designed for people with less physical ability than the general public. The differences in handicap fixtures may appear subtle, but they are important. Let's look at the requirements a fixture must meet to be considered a handicap fixture.

Toilets. Toilets will have a normal appearance, but they will sit higher above the floor than a standard toilet. A handicap toilet will rise to a height of between 16 and 20 in off the finished floor; 18 in is a common height for most handicap toilets. There are many choices in toilet style; they include the following:

- Siphon jet
- Siphon wash
- Siphon vortex
- Reverse trap
- Blowout

Sinks and lavatories.   Visually, handicap sinks, lavatories, and faucets may appear to be standard fixtures, but their method of installation is regulated and the faucets are often unlike a standard faucet. Handicap sinks and lavatories must be positioned to allow a person in a wheelchair to use them easily.

The clearance requirements for a lavatory are numerous. There must be at least 30 in of clearance in front of the lavatory. This clearance must extend 30 in from the front edge of the lavatory or countertop, whichever protrudes the farthest, and to the sides. If you can sit a square box, with a 30- by 30-in dimension, in front of the lavatory or countertop, you have adequate clearance for the first requirement. This applies to kitchen sinks and lavatories.

The next requirement calls for the top of the lavatory to be no more than 35 in from the finished floor. For a kitchen sink, the maximum height is 34 in. Then, there is knee clearance to consider. The minimum allowable knee clearance requires 29 in in height and 8 in in depth. This is measured from the face of the fixture, lavatory, or kitchen sink. Toe clearance is another issue. A space 9 in high and 9 in deep is required, as a minimum, for toe space. The last requirement deals with hot-water pipes. Any exposed hot-water pipes must be insulated or shielded to prevent users of the fixture from being burned.

Sink and lavatory faucets.   Handicap faucets frequently have blade handles. The faucets must be located no more than 25 in from the front edge of the lavatory or counter, whichever is closest to the user. The faucets could use wing handles, single-handles, or push buttons to be operated, but the operational force required by the user shall not be more than 5 lb.

Bathing units.   Handicap bathtubs and showers must meet the requirements of approved fixtures, like any other fixture, but they are also required to have special features and installation methods. The special features are required under the code for approved handicap fixtures. The clear space in front of a bathing unit is required to be a minimum of 1440 in^2. This is achieved by leaving an open space of 30 in in front of the unit and 48 in to the sides. If the bathing unit is not accessible from the side, the minimum clearance is increased to an area with a dimension of 48 by 48 in.

Handicap bathtubs are required to be installed with seats and grab bars. A grab bar, for handicap use, must have a diameter of at least 1¼ in. The diameter may not exceed 1½ in. All handicap grab bars are meant to be installed 1½ in from walls. The design and strength of these bars are set forth in the building codes.

The seat may be an integral part of the bathtub, or it may be a removable, after-market seat. The grab bars must be at least 2 ft long. Two of these grab bars are to be mounted on the back wall, one above the other. The bars are to run horizontally. The lowest grab bar must be mounted 9 in above the flood-level rim of the tub. The top grab bar must be mounted a minimum of 33 in, but no more than 36 in, above the finished floor. The grab bars should be mounted near the seat of the bathing unit.

Additional grab bars are required at each end of the tub. These bars should be mounted horizontally and at the same height as the highest grab bar on the back wall. The bar over the faucet must be at least 2 ft long. The bar on the other end of the tub may be as short as 1 ft.

The faucets in these bathing units must be located below the grab bars. The faucets used with a handicap bathtub must be able to operate with a maximum force of 5 lb. A personal, hand-held shower is required in all handicap bathtubs. The hose for the hand-held shower must be at least 5 ft long.

Two types of showers are normally used for handicap purposes. The first type allows the user to leave a wheelchair and shower while sitting on a seat (Fig. 8.4). The other style of shower stall is meant for the user to roll a wheelchair into the stall and shower while seated in the wheelchair (Fig. 8.5).

If the shower is intended to be used with a shower seat, its dimensions should form a square, with 3 ft of clearance. The seat should be no more than 16 in wide and mounted along the side wall. This seat should run the full length of the shower. The height of the seat should be between 17 and 19 in above the finished floor. There should be two grab bars installed in the shower. These bars should be located between 33 and 36 in above the finished floor. The bars are intended to be mounted in an L shape. One bar should be 36 in long and run the length of the seat, mounted horizontally. The other bar should be installed on the side wall of the shower. This bar should be at least 18 in long.

The faucet for this type of shower must be mounted on the wall across from the seat. The faucet must be at least 38 in but not more than 48 in above the finished floor. There must be a hand-held shower installed in the shower. The hand-held shower can be in addition to a

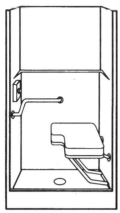

6066-H RHO/6067 LHO Summit 36S

- One-piece seamless construction.
- Fold-down bench.
- 1½″ diameter safety grab bars
- Meets ANSI standard A117.1-80.
- Slip resistant floor.

Figure 8.4 Handicap shower with seat. (*Courtesy Universal-Rundle Corporation*)

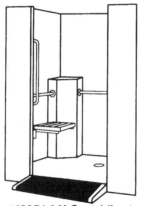

6950 RH Seat/6951 LH Seat Liberte

- Has fold-down seat. Placed at 18″ height for easy transfer from wheelchair to seat.
- Two built-in soap shelves.
- One vertical and three horizontal grab bars.
- Inside diameter of 5′ for easy wheelchair turn inside stall.
- Entry ramp 36″ wide with gentle 8.3% grade.
- Lipped door ledge to prevent rolling out of stall.
- Anti-skid floor mat included.
- White.
- Optional dome (6951) available.

Figure 8.5 Handicap shower with seat and ramp. (*Courtesy Universal-Rundle Corporation*)

fixed shower head, but there must be a hand-held shower on a hose at least 5 ft long installed. The faucet must be able to operate with a maximum force of 5 lb.

Drinking units. The distribution of water from a water cooler or drinking fountain must occur at a maximum height of 36 in above the finished floor. The outlet for drinking water must be located at the front of the unit and the water must flow upward for a minimum distance of 4 in. Levers or buttons to control the operation of the drinking unit may be mounted on the front of the unit or on the side, near the front.

Clearance requirements call for an open space of 30 in in front of the unit and 48 in to the sides. Knee and toe clearances are the same as required for sinks and lavatories. If the unit is made so that the drinking spout extends beyond the basic body of the unit, the width clearance may be reduced from 48 in to 30 in so long as knee and toe requirements are met.

Standard Fixture Installation Regulations

Standard fixtures must also be installed according to local code regulations. There are space limitations, clearance requirements, and predetermined, approved methods for installing standard plumbing fixtures. First, let's look at the space and clearance requirements for common fixtures (Fig. 8.6).

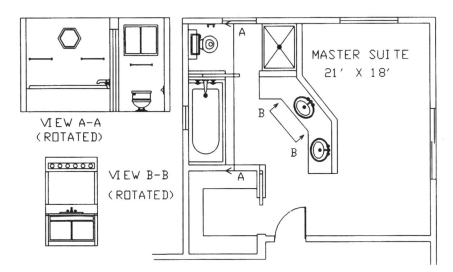

Figure 8.6 A typical bathroom layout.

4944 Hygiene II Bidet

Figure 8.7 Bidet. (*Courtesy Universal-Rundle Corporation*)

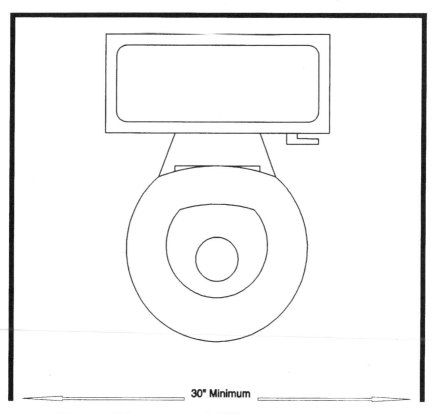

30" Minimum

Figure 8.8 Minimum width requirements for WC.

Standard fixture placement

Toilets and bidets require a minimum distance of 15 in from the center of the fixture's drain to the nearest side wall. These fixtures must have at least 15 in of clear space between the center of their drains and any obstruction, such as a wall, cabinet, or other fixture. With this rule in mind, a toilet or bidet (Fig. 8.7) must be centered in a space of at least 30 in. Figure 8.8 illustrates this placement. Zone one further requires that there be a minimum of 18 in of clear space in front of these fixtures (Fig. 8.9) and that when toilets are placed in privacy stalls, the stalls must be at least 30 in wide and 60 in deep.

Zones one and two require urinals (Fig. 8.10) to be installed with a minimum clear distance of 12 in from the center of their drains to the nearest obstacle on either side. When urinals are installed side by side in zones one and two, the distance between the centers of their drains must be at least 24 in. Zone three requires urinals to have minimum side-wall clearances of at least 15 in. In zone three, the center-to-center distance is a minimum of 30 in. Urinals in zone three must also have a minimum clearance of 18 in in front of them.

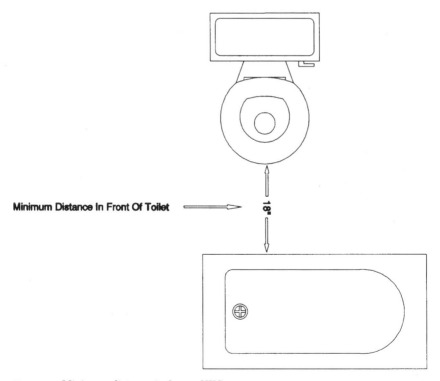

Figure 8.9 Minimum distance in front of WC.

4981 Siphon Jet Extended Lip
• Wall hung urinal.
• Flush valve not included.

Figure 8.10 Urinal. (*Courtesy Universal-Rundle Corporation*)

Standard fixtures, as with all fixtures, must be installed level and with good workmanship. The fixture should normally be set with an equal distance from walls to avoid a crooked or cocked installation. See Figs. 8.11 and 8.12 for examples of the right and wrong ways to position a toilet. All fixtures should be designed and installed with proper cleaning in mind.

Bathtubs, showers, vanities, and lavatories should be placed in a manner to avoid violating the clearance requirements for toilets, urinals, and bidets. See Fig. 8.13 for an example of a legal bathroom layout in zone three. Figure 8.14 shows an illegal bathroom grouping.

Securing and sealing fixtures

Some fixtures hang on walls (Fig. 8.15), and others sit on floors (Fig. 8.16). When securing fixtures to walls and floors, there are some rules you must obey. Floor-mounted fixtures, like most residential toilets, should be secured to the floor with the use of a closet flange. The flange is first screwed or bolted to the floor. A wax seal is then placed on the flange, and closet bolts are placed in slots on both sides of the flange. Then, the toilet is set into place.

The closet bolts should be made of brass or some other material that will resist corrosive action. The closet bolts are tightened until the toilet will not move from side to side or front to back. In some cases, a flange is not used, in which case the toilet should be secured with corrosion-resistant lag bolts.

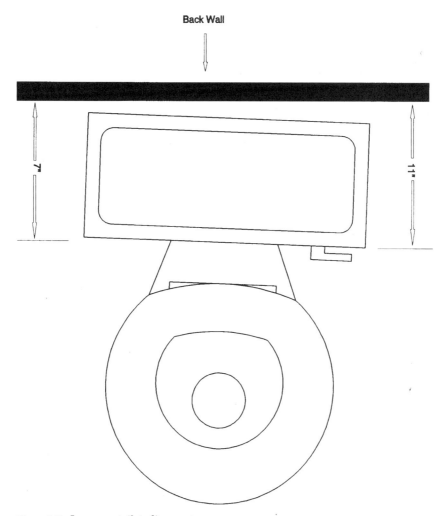

Back Wall

Figure 8.11 Improper toilet alignment.

When toilets or other fixtures are being mounted on a wall, the procedure is a little different. The fixture must be installed on, and supported by, an approved hanger. These hangers are normally packed with the fixture. The hanger must assume the weight placed in and on the fixture itself to avoid stress on the fixture.

In the case of a wall-hung toilet, the hanger usually has a pattern of bolts extending from the hanger to a point outside of the wall. The hanger is concealed in the wall cavity. A watertight joint is made at the point of connection, usually with a gasket ring, and the wall-hung toilet is bolted to the hanger.

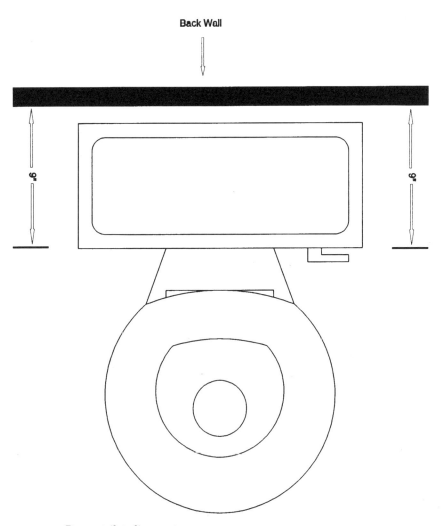

Figure 8.12 Proper toilet alignment.

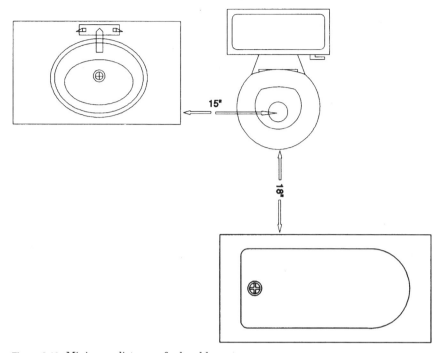

Figure 8.13 Minimum distances for legal layout.

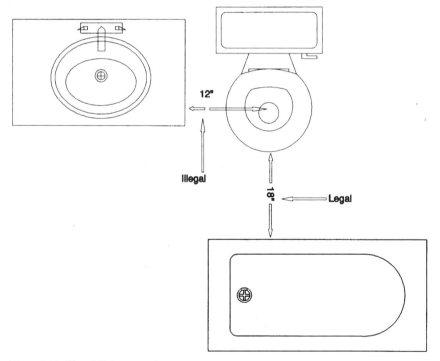

Figure 8.14 Illegal fixture spacing.

Figure 8.15 Wall-hung toilet. (*Courtesy Crane Plumbing*)

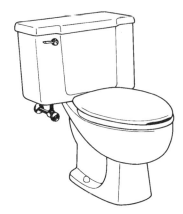

Figure 8.16 Floor-mount toilet. (*Courtesy Crane Plumbing*)

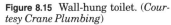

With lavatories, the hanger is usually mounted on the outside surface of the finished wall. A piece of wood blocking is typically installed in the wall cavity to allow a solid surface for mounting the bracket. The bracket is normally secured to the blocking with lag bolts. The hanger is put in place and lag bolts are screwed through the bracket and finished wall into the wood blocking. Then, the lavatory is hung on the bracket.

The space where the lavatory meets the finished wall must be sealed. This is true of all fixtures coming into contact with walls, floor, or cabinets. The crevice caused by the fixture meeting the finished surface must be sealed to protect against water damage. A caulking compound, such as silicone, is normally used for this purpose. This seal does more than prevent water damage. It eliminates hard-to-clean areas and makes the plumbing easier to keep free of dirt and germs.

When bathtubs are installed, they must be installed level, and they must be properly supported. The support for most one-piece units is the floor. These units are made to be set into place, leveled, and secured. Other types of tubs, like cast-iron tubs, require more support than the floor will give. They need a ledger or support blocks placed under the rim, where the edge of the tub meets the back wall.

The ledger can be a piece of wood, like a wall stud. The ledger should be about the same length as the tub. This ledger is installed horizontally and level. It should be at a height that will support the tub in a level fashion or with a slight incline, so excess water on the rim of the tub will run back into the tub. The ledger is nailed to wall studs.

If blocks are used, they are cut to a height that will put the bathtub into the proper position. Then, the blocks are placed at the two ends, and often in the middle, of where the tub will sit. The blocks should be installed vertically and nailed to the stud wall.

When the tub is set into place, the rim, at the back wall, rests on the blocks or ledger for additional support. This type of tub has feet on the bottom so that the floor supports most of the weight. The edges where the tub meets the walls must be caulked. If shower doors are installed on a bathtub or shower, they must meet safety requirements set forth in the building codes.

Showers today are usually one-piece units. These units are meant to sit in their place, be leveled, and be secured to the wall. The securing process for one-piece showers and bathtubs is normally accomplished by placing nails or screws through a nailing flange, which is molded as part of the unit, into the stud walls. If only a shower base is being installed, it must also be level and secure. Now, let's look at some of the many other regulations involved in installing plumbing fixtures.

The Facts about Fixture Installations

When it is time to install fixtures, there are many rules and regulations to adhere to. Water supply is one issue. Access is another. Air gaps and overflows are factors. There are a host of requirements governing the installation of plumbing fixtures. We will start with the fixtures most likely to be found in residential homes. Then, we will look at the fixtures normally associated with commercial applications.

Typical Residential Fixture Installation

Typical residential fixture installations could include everything from hose bibbs to bidets. This section is going to take each fixture that could be considered a typical residential fixture and tell you more about how they must be installed.

With most plumbing fixtures you have water coming into the fixture and water going out of the fixture. The incoming water lines must be protected against freezing and back-siphonage. Freeze protection is usually accomplished with the placement of the piping. In cold climates it is advisable to avoid putting pipes in outside walls. Insulation is often applied to water lines to reduce the risk of freezing. Back-siphonage is typically avoided with the use of air gaps and back-flow preventers.

Some fixtures, like lavatories and bathtubs, are equipped with overflow routes. These overflow paths must be designed and installed to prevent water from remaining in the overflow after the fixture is drained. They must also be installed in a manner that back-siphonage cannot occur. This normally means nothing more than having the faucet installed so that it is not submerged in water if the fixture floods. By keeping the faucet spout above the high-water mark, you have created an air gap. The path of a fixture's overflow must carry the overflowing water into the trap of the fixture. This should be done by integrating the overflow path with the same pipe that drains the fixture.

Bathtubs must be equipped with wastes and overflows. Zone one and zone three require these wastes and overflows to have a minimum diameter of 1½ in. The method for blocking the waste opening must be approved. Common methods for holding water in a tub include the following:

- Plunger-style stoppers
- Lift and turn stoppers
- Rubber stoppers
- Push and pull stoppers

Some fixtures, like hand-held showers pose special problems. Since the shower is on a long hose, it could be dropped into a bathtub full of water. If a vacuum was formed in the water pipe while the shower head was submerged, the unsanitary water from the bathtub could be pulled back into the potable water supply. This is avoided with the use of an approved back-flow preventer.

When a drainage connection is made with removable connections, like slip nuts and washers, the connection must be accessible. This normally isn't a problems for sinks and lavatories, but it can create some problems with bathtubs. Many builders and home buyers despise having an ugly access panel in the wall where their tub waste is located. To eliminate the need for this type of access, the tub waste can be connected with permanent joints. This could mean soldering a brass tub waste or gluing a plastic one. But, if the tub waste is connected with slip nuts, an access panel is required.

Washing machines generally receive their incoming water from boiler drains or laundry faucets. There is a high risk of a cross-connection when these devices are used with an automatic clothes washer. This type of connection must be protected against back-siphonage. The drainage from a washing machine must be handled by an indirect waste. An air break is required and is usually accomplished by placing the washer's discharge hose into a 2-in pipe as an indirect-waste receptor. The water supply to a bidet must also be protected against back-siphonage.

Dishwashers are another likely source of back-siphonage. These appliances must be equipped with either a back-flow protector or an air gap that is installed on the water-supply piping. The drainage from dishwashers is handled differently in each zone.

Zone one requires the use of an air gap on the drainage of a dishwasher. These air gaps are normally mounted on the countertop or in the rim of the kitchen sink. The air gap forces the waste discharge of the dishwasher through open air and down a separate discharge hose. This eliminates the possibility of back-siphonage or a back-up from the drainage system into the dishwasher.

Zone two requires dishwasher drainage to be separately trapped and vented or to be discharged indirectly into a properly trapped and vented fixture.

Zone three allows the discharge hose from a dishwasher to enter the drainage system in several ways. It may be individually trapped. It may discharge into a trapped fixture. The discharge hose could be connected to a wye tailpiece (Figs. 8.17 and 8.18) in the kitchen sink drainage. Further, it may be connected to the waste connection provided on many garbage disposers.

While we are on the subject of garbage disposers, be advised that garbage disposers require a drain of at least 1½ in and must be

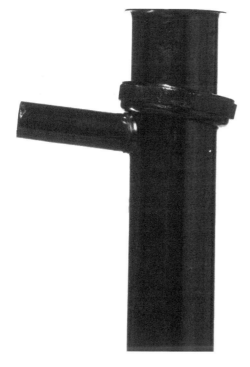

Figure 8.17 Wye tailpiece.

Figure 8.18 Dishwasher drain adapter.

trapped. It may seem to go without saying, but garbage disposers must have a water source. This doesn't mean you have to pipe a water supply to the disposer; a kitchen faucet provides adequate water supply to satisfy the code.

Floor drains must have a minimum diameter of 2 in. Remember, piping run under a floor may never be smaller than 2 in in diameter. Floor drains must be trapped, usually must be vented, and must be equipped with removable strainers. It is necessary to install floor drains so that the removable strainer is readily accessible.

Laundry trays are required to have 1½-in drains. These drains should be equipped with cross-bars (Fig. 8.19) or a strainer. Laundry trays may act as indirect-waste receptors for clothes washers. In the case of a multiple-bowl laundry tray, the use of a continuous waste is acceptable.

Lavatories are required to have drains of at least 1¼ in in diameter. The drain must be equipped with some device to prevent foreign objects from entering the drain. These devices could include pop-up assemblies, cross-bars, or strainers.

When installing a shower, it is necessary to secure the pipe serving the shower head with water. This riser is normally secured with a drop-ear ell and screws. Figure 8.20 shows this type of installation. It is, however, acceptable to secure the pipe with a pipe clamp.

When we talk of showers here, we are speaking only of showers, not tub-shower combinations. The use of tub-shower combinations confuses many people. A shower has different requirements than those of a tub-shower combination. A shower drain must have a diameter of at least 2 in. The reason for this is simple. In a tub-shower combination, a 1½-in drain is sufficient, because the walls of the bathtub will retain

Figure 8.19 Cross-bar drain.

water until the smaller drain can remove it. A shower doesn't have high retaining walls; therefore, a larger drain is needed to clear the shower base of water more quickly. Shower drains must have removable strainers. The strainers should have a diameter of at least 3 in.

In zone three, all showers must contain a minimum of 900 in^2 of shower base. This area must not be less than 30 in in any direction. These measurements must be taken at the top of the threshold, and they must be interior measurements. A shower advertised as a 30-in shower may not meet code requirements. If the measurements are taken from the outside dimensions, the stall will not pass muster. There is one exception to the above ruling. Square showers with a rough-in of 32 in may be allowed. But the exterior of the base may not measure less than 31½ in.

Zone one requires the minimum interior area of a shower base to be at least 1024 in^2. When determining the size of the shower base, the measurements should be taken from a height equal to the top of the

Figure 8.20 Drop-ear ell.

threshold. The minimum size requirements must be maintained for a vertical height equal to 70 in above the drain. The only objects allowed to protrude into this space are grab bars, faucets, and shower heads.

The waterproof wall enclosure of a shower or a tub-shower combination must extend from the finished floor to a height of no less than 6 ft. Another criteria for these enclosures is that they must extend at least 70 in above the height of the drain opening. The enclosure walls must be at the higher of the two determining factors. An example of when this might come into play is a deck-mounted bathing unit. With a tub mounted in an elevated platform, an enclosure that extends 6 ft above the finished floor might not meet the criteria of being 70 in above the drain opening.

Although not as common as they once were, built-up shower stalls are still popular in high-end housing. These stalls typically use a concrete base, covered with tile. You may never install one of these classic shower bases, but you need to know how, just in case the need arises. These bases are often referred to as shower pans. Cement is poured into the pan to create a base for ceramic tile.

Before these pans can be formed, attention must be paid to the surface that will be under the pan. The subfloor, or other supporting surface, must be smooth and able to accommodate the weight of the shower. When the substructure is satisfactory, you are ready to make your shower pan.

Shower pans must be made from a waterproof material. In the old days, these pans were made of lead or copper. Today, they are generally made with coated papers or vinyl materials. These flexible materials make the job much easier. When forming a shower pan, the edges of the pan material must extend at least 2 in above the height of the threshold. Zone one requires the material to extend at least 3 in above the threshold. The pan material must also be securely attached to the stud walls.

Zone one goes deeper with its shower regulations. In zone one, the shower threshold must be 1 in lower than the other sides of the shower base, but the threshold must never be lower than 2 in. The threshold must also never be higher than 9 in. When installed for handicap facilities, the threshold may be eliminated.

Zone one goes on to require the shower base to slope toward the drain with a minimum pitch of ¼ in/ft, but not more than ½ in/ft. The opening into the shower must be large enough to accept a shower door with minimum dimensions of 22 in.

The drains for this type of shower base are new to many young plumbers; they may attempt to use standard shower drains for these types of bases, which you cannot do if you don't want the pan to leak. This type of shower base requires a drain that is similar to some floor drains.

The drain must be installed in a way that will not allow water that might collect in the pan to seep around the drain and down the exterior of the pipe. Any water entering the pan must go down the drain. The proper drain will have a flange that sits beneath the pan material, which will be cut to allow water into the drain. Then, another part of the drain is placed over the pan material and bolted to the bottom flange. The compression of the top piece and the bottom flange, with the pan material wedged between them, will create a watertight seal. Then, the strainer portion of the drain will screw into the bottom flange housing. Since the strainer is on a threaded extension, it can be screwed up or down to accommodate the level of the finished shower pan.

Sinks are required to have drains with a minimum diameter of 1½ in. Strainers or cross-bars are required in the sink drain. If you look, you will see that basket strainers have the basket part, as a strainer, and cross-bars below the basket. This provides protection from foreign objects even when the basket is removed. If a sink is equipped with a garbage disposer, the drain opening in the sink should have a diameter of at least 3½ in.

Toilets installed in zone three are required to be water-saver models. The older models that use 5 gallons (gal) per flush are no longer allowed in zone three for new installations.

The seat on a residential water closet must be smooth and sized for the type of water closet it is serving. This usually means that the seat will have a round front.

The fill valve or ballcock for toilets must be of the antisiphon variety. There are still older ballcocks being sold that are not of the antisiphon style. Just because these units are available doesn't make them acceptable. Don't use them; you will be putting your license and yourself on the line.

Toilets of the flush-tank type are required to be equipped with overflow tubes, which do double duty as refill conduits. The overflow tube must be large enough to accommodate the maximum water intake entering the water closet at any given time.

Whirlpool tubs must be installed as recommended by the manufacturer. All whirlpool tubs should be installed to allow access to the unit's pump. The pump's drain should be pitched to allow the pump to empty its volume of water when the whirlpool is drained. The whirlpool pump should be positioned above the fixture's trap.

All plumbing faucets and valves using both hot and cold water must be piped in a uniform manner. This manner calls for the hot water to be piped to the left side of the faucet or valve. Cold water should be piped to the right side of the faucet or valve. This uniformi ty reduces the risk of unwarranted burns from hot water.

In zone three, valves or faucets used for showers must be designed to provide protection from scalding. This means that any valve or faucet used in a shower must be pressure-balanced or contain a thermostatic-mixing valve. The temperature control must not allow the water temperature to exceed 110°F. This provides safety, especially to the elderly and the very young, against scalding injuries from the shower. Zones one and two do not require these temperature-controlled valves in residential dwellings. When zone one requires temperature-controlled shower valves, the maximum allowable temperature is 120°F.

Commercial Fixture Applications

Drinking fountains are a common fixture in commercial applications. Restaurants use garbages disposers that are so big it can take two plumbers to move them. Gang showers are not uncommon in school gyms and health clubs. Urinals are another common commercial fixture. Then, there are water closets. Water closets are in homes, but the ones installed for commercial applications often differ from residential toilets. Special fixtures and applications exist for some unusual plumbing fixtures, like baptismal pools in churches. This section is going to take you into the commercial field and show you how plumbing needs vary from residential uses to commercial applications.

Let's start with drinking fountains and water coolers. The main fact to remember about water coolers and fountains is this: they are not allowed in toilet facilities. You may not install a water fountain in a room that contains a water closet. If the building for which a plumbing diagram is being designed will serve water, such as a restaurant, or if the building will provide access to bottled water, drinking fountains and water coolers may not be required.

Commercial garbage disposers can be big. These monster grinding machines require a drain with a diameter of no less than 2 in. Commercial disposers must have their own drainage piping and trap. As with residential disposers, commercial disposers must have a cold-water source. In zone two, the water source must be of an automatic type. These large disposers may not be connected to a grease interceptor.

Garbage-can washers are not something you will find in the average home, but they are not uncommon in commercial applications. Due to the nature of this fixture, the water supply to the fixture must be protected against back-siphonage. This can be done with either a back-flow preventer or an air gap. The waste pipe from these fixtures must have individual traps. The receptor that collects the residue from the garbage-can washer must be equipped with a removable strainer, capable of preventing the entrance of large particles into the sanitary drainage system.

Special fixtures are just that, special. Fixtures that might fall into this category include church baptismal pools, swimming pools, fish ponds, and other such arrangements. The water pipes to any of these special fixtures must be protected against back-siphonage.

Showers for commercial or public use can be very different from those found in a residence. It is not unusual for showers in commercial-grade plumbing to be gang showers. This amounts to one large shower enclosure with many shower heads and shower valves. In gang showers, the shower floor must be properly graded toward the shower drain or drains. The floor must be graded in a way to prevent water generated at one shower station from passing through the floor area of another shower station.

The methods employed to divert water from each shower station to a drain are up to the designer, but it is imperative that water used by one occupant may not pass into another bather's space. Zone one requires the gutters of gang showers to have rounded corners. These gutters must have a minimum slope toward the drains of 2 percent. The drains in the gutter must not be more than 8 ft from side walls and not more than 16 ft apart.

Urinals are not a common household item, but they are typical fixtures in public toilet facilities. The amount of water used by a urinal, in a single flush, should be limited to a maximum of 1½ gal. Water supplies to urinals must be protected from back-flow. Only one urinal may be flushed by a single flush valve. When urinals are used, they must not take the place of more than one-half of the water closets normally required. Public-use urinals are required to have a water trap seal that is visible and unobstructed by strainers.

Floor and wall conditions around urinals are another factor to be considered. These areas are required to be waterproof and smooth. They must be easy to clean, and they may not be made from an absorbent material. In zone three, these materials are required around a urinal in several directions. They must extend to at least 1 ft on each side of the urinal. This measurement is taken from the outside edge of the fixture. The material is required to extend from the finished floor to a point 4 ft off the finished floor. The floor under a urinal must be made of this same type of material, and the material must extend to a point at least 1 ft in front of the farthest portion of the urinal.

Commercial-grade water closets can present some of their own variations on residential requirements. The toilets used in public facilities must have elongated bowls. These bowls must be equipped with elongated seats. Further, the seats must be hinged and they must have open or split fronts.

Flush valves are used almost exclusively with commercial-grade fixtures (Fig. 8.21). They are used on water closets, urinals, and some special sinks. If a fixture depends on trap siphonage to empty itself, it must be equipped with a flush valve or a properly rated flush tank. These valves or tanks are required for each fixture in use.

4347 Mercury
- 18″ rim height.
- Floor mount bowl.
- 12″ rough-in.
- 3.5 G.P.F.
- Flush valve not included.

Figure 8.21 Flush-valve toilet.

Flush valves must be equipped with vacuum breakers that are accessible. Flush valves, in zone three, must be rated as water-conserving valves. These valves must be able to be regulated for water pressure, and they must open and close fully. If water pressure is not sufficient to operate a flush valve, other measures, such as a flush tank, must be incorporated into the design. All manually operated flush tanks should be controlled by an automatic filler, designed to refill the flush tank after each use. The automatic filler will be equipped to cut itself off when the trap seal is replenished and the flush tank is full. If a flush tank is designed to flush automatically, the filler device will be controlled by a timer.

Special Fixtures

There is an entire group of special fixtures that are normally found only in facilities providing health care. The requirements for these fixtures are extensive. While you may never have a need to work with these specialized fixtures, you should know the code requirements for them. This section is going to provide you with the information you may need.

Many special fixtures are required to be made of materials providing a higher standard than normal fixture materials. They may be required to endure excessive heat or cold. Many of these special fixtures are also required to be protected against back-flow. The fear of back-flow extends to the drainage system, as well as to the potable water supply. All special fixtures must be of an approved type.

Sterilizers

Any concealed piping that serves special fixtures and that may require maintenance or inspection must be accessible. All piping for sterilizers must be accessible. Steam piping to a sterilizer should be installed with a gravity system to control condensation and to prevent moisture from entering the sterilizer. Sterilizers must be equipped with a means to control the steam vapors. The drains from sterilizers are to be piped as indirect wastes. Sterilizers are required to have leak detectors. These leak detectors are designed to expose leaks and to carry unsterile water away from the sterilizer. The interior of sterilizers may not be cleaned with acid or other chemical solutions while the sterilizers are connected to the plumbing system.

Clinical sinks

Clinical sinks are sometimes called bedpan washers. Clinical sinks are required to have an integral trap. The trap seal must be visible and the contents of the sink must be removed by siphonic or blow-out action. The trap seal must be automatically replenished, and the sides of the fixture must be cleaned by a flush rim at every flushing of the sink. These special fixtures are required to connect to the drain waste and vent (DWV) system in the same manner as a water closet. When clinical sinks are installed in utility rooms, they are not meant to be a substitute for a service sink. On the other hand, service sinks may never be used to replace a clinical sink. Devices for making or storing ice shall not be placed in a soiled utility room.

Vacuum fluid-suction systems

Vacuum system receptacles are to be built into cabinets or cavities, but they must be visible and readily accessible. Bottle suction systems used for collecting blood and other human fluids must be equipped with overflow prevention devices at each vacuum receptacle. Secondary safety receptacles are recommended as an additional safeguard. Central fluid-suction systems must provide continuous service. If a central suction system requires periodic cleaning or maintenance, it must be installed so that it can continue to operate, even while cleaning or maintenance is being performed. When central systems are installed in hospitals, they must be connected to emergency power facilities. The vent discharge from these systems must be piped separately to the outside air, above the roof of the building.

Waste originating in a fluid suction system that is to be drained into the normal drainage piping must be piped into the drainage system with a direct-connect, trapped arrangement; an indirect-waste connection of this type of unit is not allowed.

Piping for these fluid suction systems must be noncorrosive and have a smooth interior surface. The main pipe shall have a diameter of no less than 1 in. Branch pipes must not be smaller than ½ in. All piping is required to have accessible cleanouts and must be sized according to manufacturer's recommendations. The air flow in a central fluid-suction system should not be allowed to exceed 5000 ft/min.

Special vents

Institutional plumbing uses different styles of vents for some equipment than what is encountered with normal plumbing. One such vent is called a local vent. One example of use for a local vent pertains to bedpan washers. A bedpan washer must be connected to at least one vent, with a minimum diameter of 2 in and that vent must extend to the outside air, above the roof of the building.

These local vents are used to vent odors and vapors. Local vents may not tie in with vents from the sanitary plumbing or sterilizer vents. In multistory buildings, a local vent stack may be used to collect the discharge from individual local vents for multiple bedpan washers, located above each other. A 2-in stack can accept up to three bedpan washers. A 3-in stack can handle six units, and a 4-in stack will accommodate up to twelve bedpan washers. These local vent stacks are meant to tie into the sanitary drainage system, and they must be vented and trapped if they serve more than one fixture.

Each local vent must receive water to maintain its trap seal. The water source shall come from the water supply for the bedpan washer being served by the local vent. A minimum of ¼-in tubing shall be run to the local vent, and it shall discharge water into the vent each time the bedpan washer is flushed.

Vents serving multiple sterilizers must be connected with inverted wye fittings, and all connections must be accessible. Sterilizer vents are intended to drain to an indirect waste. The minimum diameter of a vent for a bedpan sterilizer shall be 1½ in. When serving a utensil sterilizer, the minimum vent size shall be 2 in. Vents for pressure-type sterilizers must be at least 2½ in in diameter. When serving a pressure instrument sterilizer, a vent stack must be at least 2 in in diameter. Up to two sterilizers of this type may be on a 2-in vent. A 3-in stack can handle four units.

Water supply

Hospitals are required to have at least two water services. These two water services may, however, connect to a single water main. Hot water must be made available to all fixtures, as required by the fixture manufacturer. All water heaters and storage tanks must be of a type approved for the intended use.

Zone two requires the hot-water system to be capable of delivering 6½ gal of 125°F water per hour for each bed in a hospital. Zone two further requires hospital kitchens to have a hot water supply of 180°F water equal to 4 gal/h for each bed. Laundry rooms are required to have a supply of 180°F water at a rate of 4½ gal/h, for each bed. Zone two continues its hot-water regulations by requiring hot-water storage tanks to have capacities equal to no less than 8 percent of the water heating capacity.

Zone two continues with its hot-water requirements by dictating the use of copper in submerged steam heating coils. If a building is higher than three levels, the hot-water system must be equipped to circulate. Valves are required on the water distribution piping to fixture groups.

Back-flow prevention

When back-flow prevention devices are installed, they must be installed at least 6 in above the flood-level rim of the fixture. In the case of hand-held showers, the height of installation shall be 6 in above the highest point at which the hose can be used.

In most cases, hospital fixtures will be protected against back-flow by the use of vacuum breakers. However, a boiling-type sterilizer should be protected with an air gap. Vacuum suction systems may be protected with either an air gap or a vacuum breaker.

This has been a long chapter, but it was necessary to give you all the pertinent details on fixtures. As you now know, fixtures are not as simple as they may first appear. There are numerous regulations to learn and apply when installing plumbing fixtures. Your local jurisdiction may require additional or different code compliance. As always, check with your local authority before installing plumbing.

Reminder Notes

Zone one

1. Minimum fixture requirements vary from other zones.

2. Tub wastes and overflows must have a minimum diameter of 1½ in.

3. Dishwashers must drain through an air gap.

4. Requirements for shower stalls vary between the zones.

Zone two

1. Minimum fixture requirements vary from other zones.

2. Dishwashers must drain into a trap that is vented and used only for the dishwasher waste, or the waste hose must discharge indirectly into the piping of a properly trapped and vented fixture.

3. Commercial garbage disposers must be equipped with an automatic water source.

4. See the text for requirements on hot water in health care facilities.

Zone three

1. Minimum fixture requirements vary from other zones.

2. Urinals must have minimum clearances as follows: side-wall clearance of at least 15 in, center-to-center clearance of at least 30 in, and front clearance of at least 18 in.

3. Privacy stalls for water closets must have minimum dimensions of 30 in in width and 60 in in depth.

4. Tub wastes and overflows must have a minimum diameter of 1½ in.

5. Requirements for shower stalls vary between the zones.

6. Toilets are required to be water-saver models.

7. All shower valves must be of the type to prevent scalding.

8. Urinals may not use more than 1½ gal of water for a single flush.

9. Flush-valves must be of a water-conserving type.

9

After the Installation

After installing various phases of plumbing, the work must be inspected and approved. The inspections are generally performed by local plumbing inspectors. The permit holder is responsible for all costs and efforts required to test the plumbing. It is not uncommon to have as many as three or four inspections. These inspections might include inspection of water services, sewers, underground plumbing, rough-in plumbing, and final plumbing.

Before a plumber's job is finished, the work must be inspected and approved. Time and money spent on reinspections is lost time and lost profits. When you are installing plumbing for profit, it is especially important to get it right the first time. This chapter is going to take you through the general requirements for testing each plumbing phase.

It is often acceptable to test plumbing phases in sections. But, common procedure calls for testing entire phases simultaneously. Generally, air or water may be used for testing plumbing. Occasionally, special tests, like smoke or peppermint tests, will be required. Let's take a look at permissible ways to test your plumbing.

Testing Sewers

There are two common methods for testing building sewers. The first method uses water; the second uses air. In either case, the building sewer should be capped or plugged at the point where it will connect with the main sewer. Test-tee fittings are commonly installed in this portion of the sewer to allow for the test. Sewers must be tested, inspected, and approved before they are covered. Sewers should be covered by a minimum of 12 in of earth.

When testing with water, the sewer must be filled with water to a point equal to a 10-ft head. In simple terms, this means extending a pipe, like a cleanout riser, to a point 10 ft higher than the sewer. The pipe rising to allow for the 10-ft head should have water resting at its upper limit. The water must be visible.

When testing with water, water pressure must be maintained for at least 15 min before an official inspection is made. If the water level goes down, you've got a problem. All joints must be watertight.

If leaks are present, plan on cutting them out and replacing them. You should not patch the leaks with wax or glue. This is done from time to time, but it is not right. Also, don't try to pull a fast one on the inspector. Some plumbers put a plastic test cap on the pipe rising above the sewer, to give a false impression. With the plastic cap in place, the water level in the head riser would not fall, but the sewer had no water in it. The plastic cap, at the fitting entering the sewer is not visible, but a smart inspector will require you to release the test water so the flow of water can be seen and identified as having filled the sewer and test riser.

When testing with air, you must rig the sewer to accept a pressure gauge. The sewer must be pumped with air until the contents reach a pressure of at least 5 psi. If a mercury gauge is used, the pressure must balance 10 in of mercury. The time requirements for an air test are the same as those for a water test.

Testing the Water Service

The test of a water service is sometimes waived. If the water service comprises a single pipe, with no joints, a pressure test may not be required. If a test is required, the pipe can be tested with potable water or air. The water service must be tested, inspected, and approved before being buried. Water services must be located deep enough to prevent freezing.

Zones one and three require water services to be tested at a pressure equal to their maximum working pressure. Zone two requires the test pressure to be set at a pressure of at least 25 psi higher than the maximum working pressure.

Testing Groundworks

Underground plumbing is tested in essentially the same way that a building sewer is tested. An air pressure of 5 psi or a 10-ft head of water is required. When a mercury gauge is used, the test must balance a 10-in column of mercury. The test must be maintained for at least 15 min prior to inspection.

Testing the Drain Waste
and Vent (DWV) Rough-In

All DWV rough-ins must be inspected before being concealed. When testing with air, vent terminals, fixture outlets, and the building drain must be capped or plugged (Fig. 9.1). The DWV system must be subjected to a 15-min test with either air or water. If air is used, the system must be tested with a minimum pressure of 5 psi or a 10-in column of mercury.

When testing a DWV system with water, the test-water level is usually required to extend to the top of the roof vents. Some areas will allow the test to terminate at the flood-level rim of the highest bathing unit in the premises, but normally, the water must be to the top of the vents.

During the DWV test, inspectors will look for pipe protection. When a pipe is installed in a way that it may be penetrated by nails or screws, the pipe must be protected with nail plates. If structural members have been substantially weakened by your plumbing installation, your job will not pass inspection. Pipe hangers will also be inspected.

Figure 9.1 Plastic test cap.

Testing the Water-Distribution Rough-In

All water-distribution pipes must be tested, inspected, and approved before being concealed (Fig. 9.2). The test pressure required for a potable water system is usually the same as the maximum working pressure for the system. However, zone two requires the test pressure to be 25 psi higher than the working pressure.

Another consideration in a water-pipe inspection includes pipe protection from punctures and freezing. Pipe hangers are also inspected. Back-flow preventers, air gaps, and all other code requirements are examined in these inspections, as they are with other inspections.

Figure 9.2 Test rig on water pipes.

Testing the Final Plumbing

The job is not done until the final approval is issued from the code enforcement office. What is involved in a final plumbing inspection? Well, typically, the inspection is a matter of a visual tour of the plumbing. This tour normally includes the inspector's use and observation of all plumbing fixtures. For example, an inspector will test to see that the hot water is piped to the left side of a faucet. The inspector will check traps and other connections for leaks.

In the final inspection, inspectors put all the plumbing fixtures through their paces. Cut-offs are inspected, aerators are checked, back-flow preventers are checked, fixtures are filled and drained, and water heaters may be tested. In general, all plumbing is checked to assure proper installation procedures and working conditions.

If an inspector has reason to suspect a plumbing system is not up to snuff, the inspector may require a smoke or peppermint test. These tests are designed to expose leaks in the DWV system. In these tests, all traps are filled with water. The DWV system is filled with a colored smoke or an oil of peppermint. When the smoke is visible at a vent or the peppermint is noticeable, the vents are capped. Then, the inspector will check each trap for evidence of a leak. The colored smoke or aromatic peppermint makes it easy to find traps that are not doing their jobs.

Interior Rain Leaders and Downspouts

Interior rain leaders and downspouts should be tested, inspected, and approved in the same manner used for DWV systems. These pipes should not be concealed before testing, inspecting, and approval.

The Approval

When the proper installation and testing methods are used, approvals come easily.

10

Working with Gas

Depending upon the area you work in, your plumbing duties may extend over into working with gas. While gas fitting is not a plumber's job, many plumbers are also gas fitters. Some jurisdictions don't require special licensing for working with gas; others do. Anyone working with gas should be licensed and required to pass a strict examination for the privilege.

Working with gas can be very dangerous. Unlike most plumbing, where a mistake will get you wet, a mistake while working with gas could get you killed. This is not to say that plumbing doesn't present its own set of potentially dangerous circumstances, but the risks of serious injury seem more apparent when working with gas.

Gas work is not usually regulated by the plumbing code, but it is often referred to within the code book. In most jurisdictions the gas code is governed by the mechanical code. Depending upon where you are, plumbing and mechanical codes may overlap. The states in zone one are regulated for gas installations under the plumbing code. This chapter is going to give you a basic understanding of the requirements for working with gas. It is based on a combination of acceptable gas practices in zone one and in other parts of the country. Since zone one is the only major area where gas piping is covered under the plumbing code, there will not be reminder notes for the various regions. Instead special requirements will be identified as they pertain to the states in zone one.

No one without the proper training and experience should work with gas. Even if your jurisdiction does not require a special gas-fitting license, do your homework. Try to learn from your mistakes since your work with gas could be deadly. The risk of personal injury extends beyond the installer. If gas pipe or gas equipment is not installed properly, the people injured or killed as a result of the faulty work could be staggering. When you work with gas, you hold the safety of many people in your hands. Don't take this part of your work lightly.

The two types of gas most often worked with are natural gas and propane gas. There are some differences between them. Equipment that is meant for use with natural gas is not necessarily compatible with propane. Before you make a gas connection to any appliance or equipment, verify the type of gas the unit is intended to work with.

Approved Materials

Several types of piping materials are approved for gas work. All piping used must meet minimum requirements, as established by local codes. The two materials most often used for gas piping in buildings are steel pipe and copper pipe. When copper is used, it should be either type L or type K, and it must be approved for use with gas. Polyvinyl chloride (PVC) and polyethylene (PE) pipe are usually allowed for gas pipe in buried installations outside a building.

Metallic pipe can be used in buildings and above ground so long as the gas being conveyed will not corrode the pipe. Steel pipe, approved copper pipe, and yellow brass pipe are the three types of pipes required for use in zone one. Aluminum pipe, where it is approved, may not be used below ground. It must not be used outside, and when used inside, it may not come into contact with masonry, plaster, or insulation. Further, it must be protected from contact with moisture.

Ductile iron pipe, when approved, is only allowed for underground use, outside of buildings. If any pipe is subject to corrosive action from surrounding conditions, the pipe must be protected to avoid it.

The fittings used with gas pipe must be compatible with the pipe. They must also be approved fittings. When working with gas, bushings are not generally allowed. Increasers and reducers are normally fine. Zone one allows the use of bushings if they are not concealed.

Flexible connectors are often used to connect an appliance to a gas source. These connectors must be approved and marked to prove it. Flex connectors may not be longer than 6 ft. Zone one requires appliance connectors for all appliances, except ranges and dryers, to be no more than 3 ft long. Flex connectors may not be concealed in walls, floors, or partitions. Further, flex connectors may not penetrate walls, floors, or partitions. Flex connectors must be properly sized. They may not be smaller than the inlet of the device they are serving.

Gas hose is not a flex connector. Gas hose is generally prohibited, except for special circumstances. Such circumstances could include a biology lab, where gas burners need to be moved around. If gas hose is approved for use, it must be as short as reasonably possible, and it may not exceed 6 ft in length. This length restriction does not apply to items like hand-held torches.

Gas hose may not be concealed, and it may not penetrate walls, floors, or partitions. If the hose will be exposed to high temperatures, temperatures above 125°F, it may not be used. If allowed for use, gas hose must be connected to a cut-off valve, at the gas pipe supply. This type of hose may be used on outdoor appliances that are designed to be portable. In these uses, the length of the hose may not exceed 15 ft. The hose still must connect to a cut-off valve at the gas pipe supply.

When flex connectors are not used, soft copper tubing often is. In zone one, quick-disconnect connectors are approved. These devices allow the connection to be broken by hand, and the gas is shut off automatically. When copper is used at an appliance connection, it should be type L or type K, and it must not be bent in a manner to damage the structural qualities of the tubing. All pipe bending must be done with approved equipment.

Installing Gas Pipe

Installing gas pipe is not the same as installing plumbing pipes. There are similarities, but the procedure is not the same. One difference is in the way pipe and fittings are put together. All joints must be made gastight. The joints should be tested with a mercury gauge, at the required pressure, to ensure good joints. Zone one requires the test to maintain 6 in of mercury. If tested with a pressure gauge, the test must maintain 10 psi. Air is commonly used to provide the pressure test on gas pipe. The pipe must maintain its test pressure for at least 15 min.

In pipes carrying gas at high pressure, the test pressure is required to be 60 psi. These high-pressure tests are often required to be maintained for 30 min.

All pipe ends are to be cut squarely and with a full diameter. Any burrs on the pipe must be removed. The surfaces of a gas joint must be clean. If flux is used to make a joint, the flux must be approved for the purpose. When installing threaded pipe, only the male threads are allowed to be sealed with pipe dope or tape. Mechanical joints, when used, must be used according to the manufacturer's specification.

The only two types of pipes allowed to have heat-fusion joints are PE and polybutylene (PB), where approved. However, these two types of pipe may not be used with a cement or glue joint. When PVC pipe is used, it must be primed and glued with approved materials.

When more than one type of piping is used, the joint between the opposing pipe types must be made with an approved adapter fitting.

In the case of matching metallic pipes together, a dielectric fitting is generally required.

In general, all underground gas piping must be installed at a depth of at least 18 in. Zone one allows buried metallic pipe to be covered by a minimum of 12 in of dirt. The pipe must not be installed in a way to hinder maintenance or to place the pipe in jeopardy of damage. There is, of course, an exception to this rule. Many states allow gas lines serving an individual outside appliance to be buried 8 in deep, but zone one does not recognize this exception. This exception, as usual, is subject to local inspection and approval.

Any underground gas pipe penetrating a foundation must be protected by a pipe sleeve. A rule-of-thumb sizing for the sleeve is two pipe sizes larger than the gas pipe. The additional space in the sleeve must be sealed to prevent water, insect, or vermin invasion. Just as with plumbing, gas pipe located in flood areas must be protected against flooding and the complications associated with flooding.

Piping for gas, other than dry gas, must be graded with a pitch of ¼-in fall for every 15 ft the pipe runs. At any point where the pipe is low or condensation may occur, a drip leg is required. Drip legs must be accessible and must be protected from freezing temperatures. See Fig. 10.1 for a typical drip-leg installation.

The connection of branch piping to a main distribution pipe must be made either on top of the main or on the side but not on the bottom. Bottom connections are not the only prohibited practices. Gas pipe may not be installed in or through heat ducts, air ducts, chimneys, laundry drops, vents, dumbwaiters, or elevator shafts. The reasoning behind this regulation is self-evident. Would you really want your gas pipe running up your chimney?

Concealed piping may not have union connections. Tubing fittings and running threads are also prohibited in concealed locations. There is yet another rule pertaining to concealed gas piping. Unless the pipe is made of steel, it must be protected from punctures. This is most easily accomplished with the use of nail plates. Nail plates are required when a pipe, other than steel, is positioned within 1¼ in from the surface of a wood member, such as a stud or floor joist. The nail plate must have a minimum thickness of ⅟₁₆ in. The plate must be large enough to protect the pipe from punctures. Usually, this means installing a plate that extends at least 4 in beyond a normal nailing surface.

When installing gas pipe in concrete, there are still more regulations to observe. Gas pipe buried in concrete must be covered by no less than 1½ in of concrete. The pipe may not make contact with metal objects, and the concrete must not contain materials that will have an adverse effect on the piping.

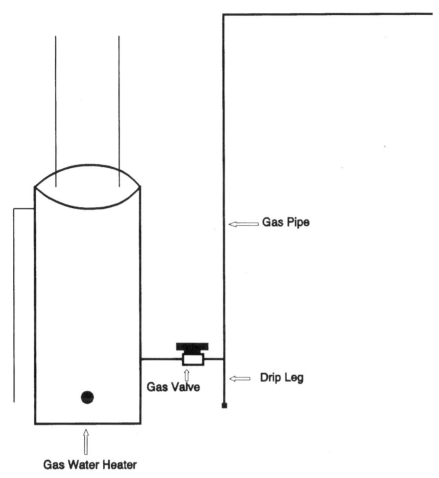

Figure 10.1 Gas drip leg.

Gas-Pipe Supports

Gas pipe can get heavy, and the means of support for the piping must be capable of supporting this weight. Hangers and supports should be made of approved materials that are intended for use with the type of pipe being supported. The required supporting distances are different from plumbing requirements.

All gas pipe installed above ground must be supported in an approved manner and protected from damage. Zone one's requirements for support are as follows: ½-in pipe must be supported at intervals not to exceed 6 ft, ¾- and 1-in pipe at 8-ft intervals, and larger pipe at 10-ft intervals. Pipe with a diameter of at least 1¼ in is required to be supported at each floor level, when installed vertically.

The intervals for pipe support in other states also depends upon the type of pipe used and the size of the pipe. For example, when supporting tubing with a diameter of 1½ in or more, support must be present every 10 ft. This is the same distance allowed for supports holding rigid pipe with a diameter of ¾ in or less. Rigid pipe with a diameter of 1 in or larger only needs to be supported every 12 ft. Smaller tubing, tubing 1¼ in in diameter or less, requires support at minimum intervals of 6 ft.

The Rest of the Common Rules

Every building housing gas piping must have a cut-off valve located on the outside of the building. This is a big help in the event of a fire. All gas meters are required to be equipped with shut-offs on the incoming side of the meter. Cut-off valves are required at all locations where appliances connect to gas supply pipes. These valves must not only be accessible, they must be adjacent to the appliance. Of course, all cut-offs must be of an approved type. The connection between an appliance and a supply pipe must be equipped with an approved union fitting. If an appliance is removed or a gas pipe is not in use, the pipe must be capped to prevent any gas from escaping.

When gas is provided by a bulk dispenser, the dispenser must have an emergency cut-off switch. A back-flow preventer must be installed on the supply side of the dispenser. All gas-dispensing systems that are located inside a building must be vented in an approved fashion. Back-flow preventers are also required on systems using a back-up or a supplemental gas source.

Additional Zone One Requirements for Liquid Propane (LP) Gas

The relief valves for LP gas must discharge into the open air. These valves must not be located closer than 5 ft, measured horizontally, from any opening into a building. LP gas may not be piped to water heaters in locations where gas might collect and provide opportunity for fire or explosion.

Testing Your Work

All concealed gas piping must be tested and approved before it is concealed. A standard test pressure is a pressure equal to 1½ times the normal working pressure of the system. However, the test pressure must never be less than 3 psi. With LP gas, the test pressure must equate to an 18-in water column. The test must be conducted for a minimum of 10 min. To be approved, the system must not lose pressure during the test. A mercury gauge is the most common way of testing gas pipe.

If the piping loses pressure, leaks should be located with soapy water, not fire or acid. When leaks are located, defective pipe or fittings should be removed and replaced, not repaired.

Once done with the test, the only job left is to purge the system and to get it on line. It is not permissible to purge the gas system through an appliance. The purging must be done in a safe location, where combustion is not a potential threat.

Sizing Gas Pipe

If you thought sizing a potable water system strained your brain, wait until you try sizing gas pipe. No, it's not all that bad, at least not when you use sizing tables. There are formulas available for sizing gas piping, but unless you are a math wizard, they will only serve to frustrate and confuse you.

It is permissible to use a less than exact method when sizing gas pipe. The local gas code will contain tables for your use in selecting the proper pipe sizing. These tables will be based on a few factors that include maximum capacity of the pipe, gas pressure, pressure drop, gravity, and pipe length. The sizing tables provided with gas codes are illustrated and described to make sizing relatively easy.

Most buildings are restricted to a maximum operating pressure of 5 psig. There are exceptions to this rating, but 5 psig is an average rating. This rating is based on the gas in the pipe being natural gas. If the gas is propane, the numbers change; LP gas is meant for a maximum operating pressure of 20 psig. As usual, there will be exceptions to this rule, but 20 psig is normal.

Regulators

Regulators are often needed to regulate gas pressure. If a regulator is used outside, it must be approved for exterior use. Some regulators require an individual vent. When such a vent is required, it must be piped independently to the outside of the building. The vent must be protected against damage and the influx of foreign objects.

Gas regulators must be installed in accessible locations. All regulators must be installed in a manner to prevent them from being damaged. A regulator is required when a gas appliance is designed to work at a lower gas pressure than the pressure present in the piping. If a second-stage regulator is required for LP gas, it must be an approved regulator.

Some More Regulations for Zone One

Zone one has more gas regulations. Used pipe, unless it was used for gas, may not be used in gas installations. If gas pipe is welded, it must be welded by a certified pipeline welder. Exposed gas pipe must be installed in a way to keep it at least 6 in above the ground or other obstructions. It is not permissible to install gas piping below grade within the confines of a building. When special cavities are provided, gas pipe may be concealed and unprotected.

Underground ferrous gas piping must be protected from electricity with isolation fittings that are installed at least 6 in above ground. If unions are installed in gas pipe, they must be installed with right and left nipples and couplings. Unions may be used when they are exposed. When gas pipe serves multiple buildings or tenants, there must be an individual cut-off valve installed for each user. These valves must be installed outside, and they must be readily accessible at all times.

When more than one type of gas has access to a gas pipe, the pipe must be protected against back-flow. Gas-fired barbecues and fireplaces must be controlled with approved valves. The valves must be in the same room as the gas-fired unit. However, the valve may not be in the unit or on a hearth that serves the unit. When installing these valves, they must be installed within 4 ft of the gas outlet for the gas-fired unit. The pipe going from the valve to the unit may be installed in concrete or masonry if the pipe is a standard-weight brass or galvanized steel and there will be at least 2 in of concrete or masonry around the pipe.

Cut-offs for appliances are required to be within 3 ft of the appliance. The cut-offs must be of an approved type. They must be installed on the gas supply pipe. These cut-offs must be installed in front of unions installed between the gas supply pipe and an appliance. These cut-offs can be placed adjacent to, in, or under appliances, so long as the appliance can be moved without affecting the cut-off. When cut-offs are installed in or under gas-fired units, they must be accessible. Appliances may not be piped in a way to allow a gas supply from more than one gas piping system.

When installing underground gas pipe that is not metallic, a number 18 copper wire must also be installed. The wire shall run with and be attached to the gas pipe. The wire must be exposed above grade at both ends of the pipe run.

There you have it; you know the basic requirements for gas piping. However, reading this chapter does not qualify you to work with gas. But if you retain what you have read, you are well on your way to having a working knowledge of gas piping.

Installations

11

Understanding
the Vent System

Most people don't think much about vents when they consider the plumbing in their homes or offices, but vents play a vital role in the scheme of sanitary plumbing. Many plumbers underestimate the importance of vents. The sizing and installation of vents often cause more confusion than the same tasks applied to drains. This chapter will teach you the role and importance of vents. It will also instruct you in the proper methods of sizing and installing them.

Whether you are working with simple individual vents or complex island vents, this chapter will improve your understanding and installation of them. Why do we need vents? They perform three easily identified functions. The most obvious function of a vent is its capacity to carry sewer gas out of a building and into the open air. A less obvious, but equally important, aspect of the vent is its ability to protect the seal in the trap it serves. The third characteristic of the vent is its ability to enable drains to drain faster and better. Let's look more closely at each of these factors.

Transportation of Sewer Gas

Vents transport sewer gas through a building, without exposing occupants of the building to the gas, to an open air space. Why is this important? Sewer gas can cause health problems. The effect of sewer gas on individuals will vary, but it should be avoided by all individuals. In addition to health problems caused by sewer gas, explosions are also possible when sewer gas is concentrated in a poorly ventilated area. Yes, sewer gas can create an explosion when it is concentrated, confined, and ignited. As you can see, just from looking at this single purpose of vents, vents are an important element of a plumbing system.

Protecting Trap Seals

Another job plumbing vents perform is the protection of trap seals. The water sitting in a fixture's trap blocks the path of sewer gas trying to enter the plumbing fixture. Without a trap seal, sewer gas could rise through the drainage pipe and enter a building through a plumbing fixture. As mentioned above, this could result in health problems and the risk of explosion. Good trap seals are essential to sanitary plumbing systems.

Vents protect trap seals by regulating the atmospheric pressure applied to them. It is possible for pressures to rise in unvented traps to a point where the trap contents actually expel into the fixture it serves. This is not a common problem, but if it occurs, the plumbing fixture could become contaminated.

A more likely problem is when the pressure on a trap seal is reduced and becomes something of a vacuum. When this happens, the water creating the trap seal is sucked out of the trap and down the drain. Once the water is taken from the trap, there is no trap seal. The trap will remain unsealed until water is replaced in the trap. Without water in it, a trap is all but useless. Vents prevent these extreme atmospheric pressure changes, therefore, protecting the trap seal.

Tiny Tornados

Have you ever drained your sink or bathtub and watched the tiny water tornados? When you see the fast swirling action of water being pulled down a drain, it usually indicates that the drain is well vented. If water is sluggish and moves out of the fixture like a lazy river, the vent for the fixture, if there is one, is not performing at its best.

Vents help fixtures to drain faster. The air allowed from the vent keeps the water moving at a more rapid pace. This not only entertains us with tiny tornados, but it aids in the prevention of clogged pipes. It is possible for drains to drain too quickly, removing the liquids and leaving hair, grease, and other potential pipe blockers present. However, if a pipe is properly graded and does not contain extreme vertical drops into improper fittings, such problems should not occur.

Do All Plumbing Fixtures Have Vents?

Most local plumbing codes require all fixture traps to be vented, but there are exceptions. In some jurisdictions, combination waste and vent systems are used. In a combination waste and vent system, vertical vents are rare. Instead of vertical vents being used, larger drainage pipes are used. The larger diameter of the drain allows air to circulate in the pipe, eliminating the need for a vent, as far as satisfactory drainage is concerned. Experience with both types of systems has shown that vented systems perform much better than combination waste and vent systems.

Combination waste and vent systems do not have vents on each fixture, so how is the trap seal protected? Trap seals in a combination waste and vent system are protected through the use of antisyphon traps or drum traps. Vented systems normally use P-traps. By using an antisiphon or drum trap, the trap is not susceptible to back-siphonage. Since these traps are larger, deeper, and made so that the water in the trap is not replaced by fresh water with each use of the fixture, they are not required to be vented, subject to local code requirements.

Most jurisdictions prohibit the use of drum traps and require traps to be vented. Before you install your plumbing, check with the local code officer for the facts pertinent to your location. The following tables show piping requirements for each zone. Table 11.1 shows approved above-ground vent materials for zone one. Table 11.2 does the same for zone two, and Table 11.3 covers approved vent materials for zone three. Table 11.4 gives a listing of approved underground venting materials for zone one. Table 11.5 covers underground materials approved for use in zone two. Table 11.6 shows approved underground materials for zone three. Fittings for vent piping must be compatible with the piping used.

TABLE 11.1   Materials Approved for Above-Ground Vents in Zone One

Cast iron
ABS*
PVS*
Copper
Galvanized
Lead
Brass

* These materials may not be used with buildings having more than three floors above grade.

TABLE 11.2   Materials Approved for Above-Ground Vents in Zone Two

Cast iron
ABS
PVC
Copper
Galvanized
Lead
Aluminum
Borosilicate Glass
Brass

TABLE 11.3   Materials Approved for Above-Ground Vents in Zone Three

Cast iron
ABS
PVC
Copper
Galvanized
Lead
Aluminum
Brass

TABLE 11.4   Materials Approved for Underground Vents in Zone One

Cast iron
ABS*
PVC*
Copper
Brass
Lead

* These materials may not be used with buildings having more than three floors above grade.

TABLE 11.5   Materials Approved for Underground Vents in Zone Two

Cast iron
ABS
PVC
Copper
Aluminum
Borosilicate glass

TABLE 11.6   Materials Approved for Underground Vents in Zone Three

Cast iron
ABS
PVC
Copper

Individual Vents

Individual vents are, as the name implies, vents that serve individual fixtures. These vents only vent one fixture, but they may connect into another vent that will extend to the open air. Individual vents do not have to extend from the fixture being served to the outside air, without joining another part of the venting system, but they must vent to open air space. See Fig. 11.1 for an example of an individual vent.

Sizing an individual vent is easy. The vent must be at least one-half the size of the drain it serves, but it may not have a diameter of less

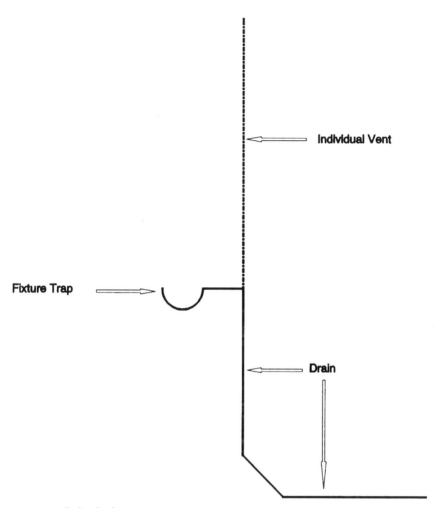

Figure 11.1 Individual vent.

than 1¼ in. For example, a vent for a 3-in drain could, in most cases, have a diameter of 1½ in. A vent for a 1½-in drain may not have a diameter of less than 1¼ in.

Relief Vents

Relief vents are used in conjunction with other vents. Their purpose is to provide additional air to the drainage system when the primary vent is too far from the fixture. See Fig. 11.2 for an example of a relief vent. Relief vents must be at least one-half the size of the pipe it is

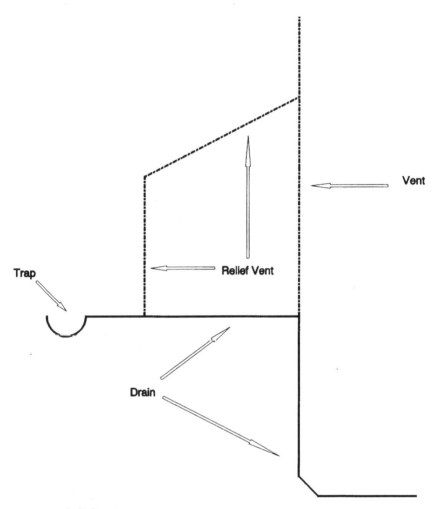

Figure 11.2 Relief vent.

venting. For example, if a relief vent is venting a 3-in pipe, the relief vent must have a 1½-in or larger diameter.

Circuit Vents

Circuit vents are used with a battery of plumbing fixtures. They are normally installed just before the last fixture of the battery. Then, the circuit vent is extended upward to the open air or tied into another vent that extends to the outside. Circuit vents may tie into stack vents or vent stacks. See Fig. 11.3 for an example of a circuit vent. When sizing a circuit vent, you must account for its developed length.

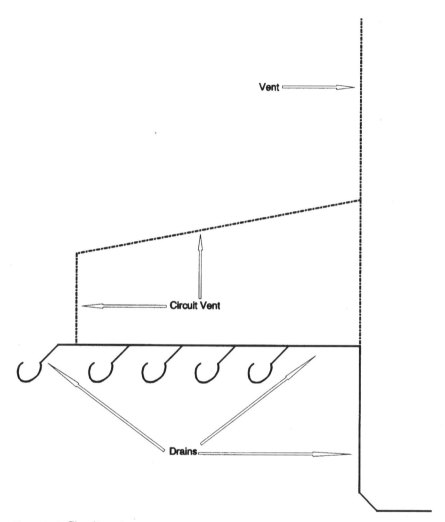

Figure 11.3 Circuit vent.

But in any event, the diameter of a circuit vent must be at least one-half the size of the drain it is serving.

Vent Sizing Using Developed Length

What effect does the length of the vent have on the vent's size? The developed length, the total linear footage of pipe making up the vent, is used in conjunction with factors provided in code books to determine vent sizes. To size circuit vents, branch vents, and individual vents for horizontal drains, you must use this method of sizing.

TABLE 11.7   Vent Sizing Table for Zone Three (For Use with Individual, Branch, and Circuit Vents for Horizontal Drain Pipes)

Drain pipe size (in)	Drain pipe grade (in/ft)	Vent pipe size (in)	Maximum developed length of vent pipe (ft)
1½	¼	1¼	Unlimited
1½	¼	1½	Unlimited
2	¼	1¼	290
2	¼	1½	Unlimited
3	¼	1½	97
3	¼	2	420
3	¼	3	Unlimited
4	¼	2	98
4	¼	3	Unlimited
4	¼	4	Unlimited

The criteria needed for sizing a vent, based on developed length, are the grade of the drainage pipe, the size of the drainage pipe, the developed length of the vent, and the factors allowed by local code requirements. Let's look at a few examples of how to size a vent using this method.

For our first example, assume the drain you are venting is a 3-in pipe with a ¼-in/ft grade. This sizing exercise is done using zone three requirements. Look at Table 11.7 and find the proper pipe size and grade. Now, looking at the table, notice the number listed under the 1½-in vent column. You will see the number is 97. This means that a 3-in drain, running horizontally, with a ¼-in/ft grade, can be vented with 1½-in vent that has a developed length of 97 ft. It would be rare to extend a vent anywhere near 97 ft, but if your vent needed to exceed this distance, you could go to a larger vent. A 2-in vent would allow you to extend the vent for a total length of 420 ft. A vent larger than 2 in would allow you to extend the vent indefinitely.

For the second example, still using zone three's rules, assume the drain is a 4-in pipe, with a ¼-in/ft grade. In this case you could not use a 1½-in vent. Remember, the vent must be at least one-half the size of the drain it is venting. A 2-in vent would allow a developed vent length of 98 ft, and a 3-in vent would allow the vent to extend to an unlimited length. As you can see, this type of sizing is not difficult.

Now, let's size a vent with the use of zone one's rules. In zone one, vent sizing is based on the vent's length and the number of fixture units on the vent. If you were sizing a vent for a lavatory, you would need to know how many fixture units the lavatory represents. Lavatories are rated as 1 fixture unit. By using a table in the code book, you would find that a vent serving 1 fixture unit can have a diameter of 1¼ in and extend for 45 ft. A bathtub, rated at 2 fixture units, would require a 1½-in vent. The bathtub vent could run for 60 ft.

Branch Vents

Branch vents are vents extending horizontally that connect multiple vents together. Figure 11.4 shows an example of a branch vent. Branch vents are sized with the developed-length method, just as you were shown in the examples above. A branch vent or individual vent that is the same size as the drain it serves is unlimited in the developed length it may obtain. Be advised, zone two and zone three use different tables and ratings for sizing various types of vents; zone one uses the same rating and table for all normal venting situations.

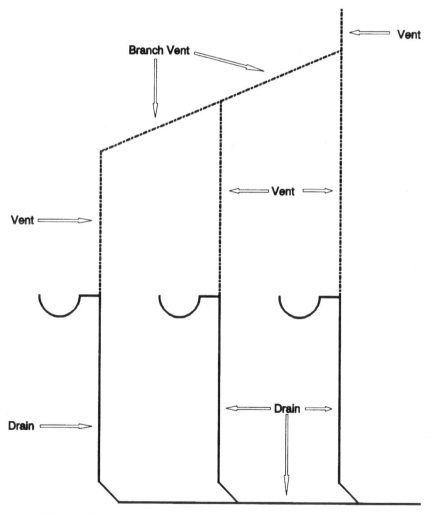

Figure 11.4 Branch vent.

Vent Stacks

A vent stack is a pipe used only for the purpose of venting. Vent stacks extend upward from the drainage piping to the open air, outside of a building. Vent stacks are used as connection points for other vents, such as branch vents. A vent stack is a primary vent that accepts the connection of other vents and vents an entire system. Refer to Fig. 11.5 for an example of a vent stack. Vent stacks run vertically and are sized a little differently.

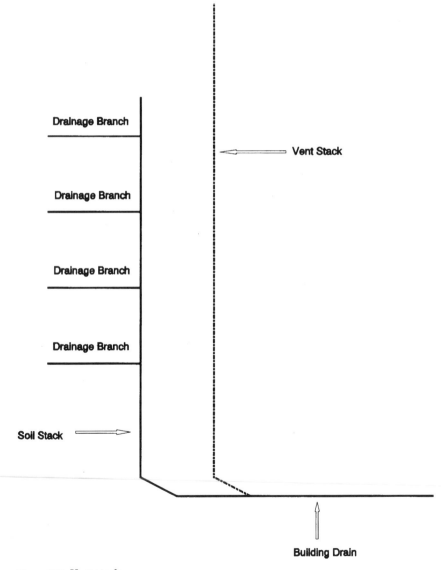

Figure 11.5 Vent stack.

TABLE 11.8   Vent Sizing Table for Zone Three (For Use with Vent Stacks
and Stack Vents)

Drain pipe size (in)	Fixture-unit load on drain pipe	Vent pipe size (in)	Maximum developed length of vent pipe (ft)
1½	8	1¼	50
1½	8	1½	150
1½	10	1¼	30
1½	10	1½	100
2	12	1½	75
2	12	2	200
2	20	1½	50
2	20	2	150
3	10	1½	42
3	10	2	150
3	10	3	1040
3	21	1½	32
3	21	2	110
3	21	3	810
3	102	1½	25
3	102	2	86
3	102	3	620
4	43	2	35
4	43	3	250
4	43	4	980
4	540	2	21
4	540	3	150
4	540	4	580

The basic procedure for sizing a vent stack is similar to that used
with branch vents, but there are some differences. Refer to Table
11.8 for an example of the criteria needed to size a vent stack in zone
three. Zone two uses a very similar table, but the numbers vary in
some instances. You must know the size of the soil stack, the number
of fixture units carried by the soil stack, and the developed length of
your vent stack. With this information and the regulations of your
local plumbing code, you can size your vent stack. Let's work on an
example.

Assume your system has a soil stack with a diameter of 4 in. This
stack is loaded with 43 fixture units. Your vent stack will have a
developed length of 50 ft. What size pipe will you have to use for your
vent stack? When you look at the table, you will see that a 2-in pipe,
used as a vent for the described soil stack, would allow a developed
length of 35 ft. Your vent will have a developed length of 50 ft, so you
can rule out 2-in pipe. In the column for 2½-in pipe, you see a rating
for up to 85 ft. Since your vent is only going 50 ft, you could use a 2½-
in pipe. However, since 2½-in pipe is not common, you would probably
use a 3-in pipe. This same sizing method is used when computing the
size of stack vents.

Stack Vents

Stack vents are really two pipes in one. The lower portion of the pipe is a soil pipe, and the upper portion is a vent. This is the type of primary vent most often found in residential plumbing. Figure 11.6 shows you what a stack vent looks like. Stack vents are sized with the same methods used on vent stacks.

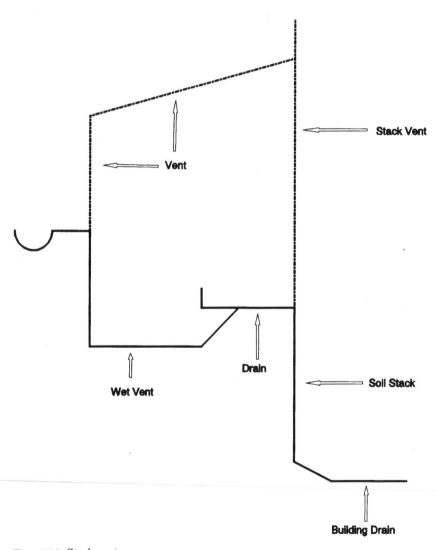

Figure 11.6 Stack vent.

Common Vents

Common vents are single vents that vent multiple traps. Figure 11.7 shows a diagram of a typical common vent. Common vents are only allowed when the fixtures being served by the single vent are on the same floor level. Zone one requires the drainage of fixtures being vented with a common vent to enter the drainage system at the same level. Normally, not more than two traps can share a common vent, but there

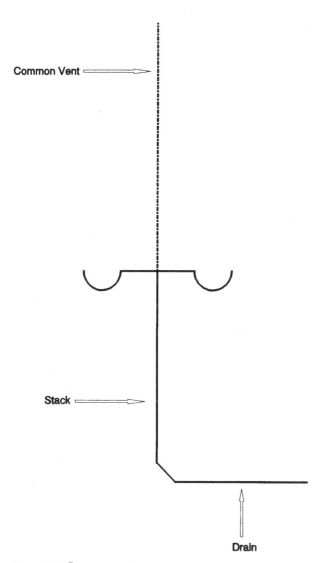

Figure 11.7 Common vent.

is an exception in zone three. Zone three allows you to vent the traps of up to three lavatories with a single common vent. Common vents are sized with the same technique applied to individual vents.

Island Vents

Island vents are unusual looking vents. They are allowed for use with sinks and lavatories. The primary use for these vents is with the trap of a kitchen sink, when the sink is placed in an island cabinet. Sometimes pictures speak louder than words; refer to Fig. 11.8 for a detail of how island venting works.

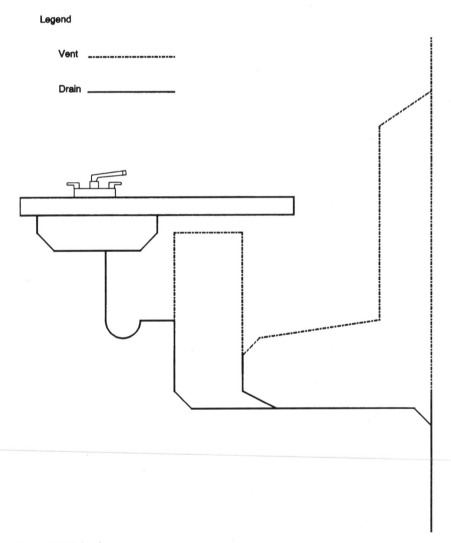

Figure 11.8 Island vent.

As you can see from the figure, island venting may take a little getting used to. Notice that the vent must rise as high as possible under the cabinet before it takes a U-turn and heads back downward. Since this piping does not rise above the flood-level rim of the fixture, it must be considered a drain. Fittings approved for drainage must be used in making an island vent.

The vent portion of an island vent must be equipped with a cleanout. The vent may not tie into a regular vent until it rises at least 6 in above the flood-level rim of the fixture.

Wet Vents

Wet vents are pipes that serve as a vent for one fixture and a drain for another. Wet vents, once you know how to use them, can save you a lot of money and time. By effectively using wet vents you can reduce the amount of pipe, fittings, and labor required to vent a bathroom group or two.

The sizing of wet vents is based on fixture units. The size of the pipe is determined by how may fixture units it may be required to carry. A 3-in wet vent can handle 12 fixture units. A 2-in wet vent is rated for 4 fixture units, and a 1½-in wet vent is allowed only 1 fixture unit. It is acceptable to wet vent two bathroom groups, six fixtures, with a single vent, but the bathroom groups must be on the same floor level. Figures 11.9 and 11.10 show some examples of wet venting. Zone two makes provisions for wet venting bathrooms on different floor levels. Zone one takes a different approach to wet venting.

Zone two has some additional regulations that pertain to wet venting; here they are. The horizontal branch connecting to the drainage stack must enter at a level equal to, or below, the water-closet drain. However, the branch may connect to the drainage at the closet bend. When wet venting two bathroom groups, the wet vent must have a minimum diameter of 2 in.

Table 11.9 shows the ratings used to size a wet-vented stack. Kitchen sinks and washing machines may not be drained into a 2-in combination waste and vent. Water closets and urinals are restricted on vertical combination waste and vent systems.

As for zone two's allowance in wet venting on different levels, here are the facts. Wet vents must have at least a 2-in diameter. Water closets that are not located on the highest floor must be back-vented. If, however, the wet vent is connected directly to the closet bend, with a 45° bend, the toilet being connected is not required to be back-vented, even if it is on a lower floor. Table 11.10 shows the ratings used to size a vent stack for a wet-vented application.

Zone one limits wet venting to vertical piping. These vertical pipes are restricted to receiving only the waste from fixtures with fixture-

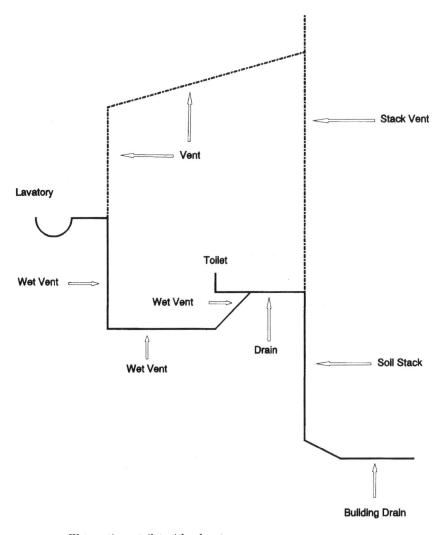

Figure 11.9 Wet venting a toilet with a lavatory.

unit ratings of 2, or less, and that serve to vent no more than four fixtures. Wet vents must be one pipe size larger than normally required, but they must never be smaller than 2 in in diameter.

Crown Vents

A crown vent is a vent that extends upward from a trap or trap arm. Crown-vented traps are not allowed. When crown vents are used, they are normally used on trap arms, but even then, they are not common. Figure 11.11 shows how an approved crown vent would look.

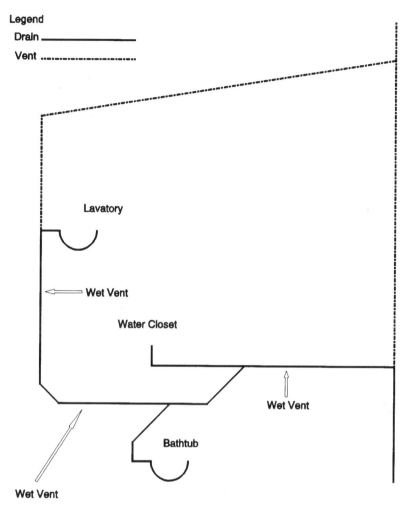

Figure 11.10 Wet venting a bathroom group.

TABLE 11.9   Table for Sizing a Wet Stack Vent in Zone Two

Stack pipe size	Fixture-unit load on stack	Maximum length of stack (ft)
2	4	30
3	24	50
4	50	100
6	100	300

The vent must be on the trap arm, and it must be behind the trap by a distance equal to twice the pipe size. For example, on a 1½-in trap, the crown vent would have to be 3 in behind the trap, on the trap arm.

TABLE 11.10   Table for Sizing a Vent Stack for Wet Venting in Zone Two

No. of fixtures	Vent-stack size requirements (in)
1–2 bathtubs or showers	2
3–5 bathtubs or showers	2½
6–9 bathtubs or showers	3
10–16 bathtubs or showers	4

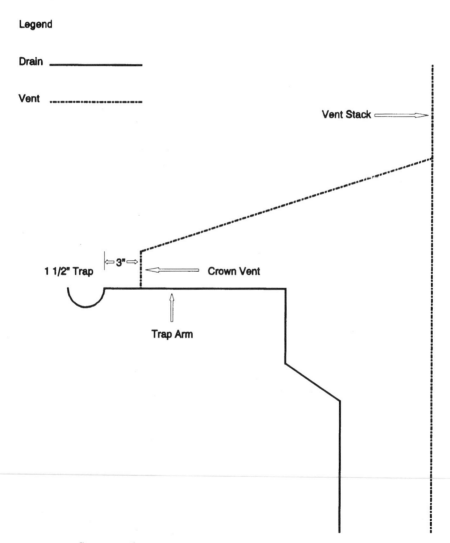

Figure 11.11 Crown venting.

Vents for Sumps and Sewer Pumps

When sumps and sewer pumps are used to store and remove sanitary waste, the sump must be vented. Zones one and two treat these vents about the same as vents installed on gravity systems.

If you will be installing a pneumatic sewer ejector, you will need to run the sump vent to outside air, without tying it into the venting system for the standard sanitary plumbing system. This ruling on pneumatic pumps applies to all three zones. If your sump will be equipped with a regular sewer pump, you may tie the vent from the sump back into the main venting system for the other sanitary plumbing.

Zone three has some additional rules. The following is an outline of the requirements in zone three. Sump vents may not be smaller than a 1¼-in pipe. The size requirements for sump vents are determined by the discharge of the pump. For example, a sewer pump capable of producing 20 gpm could have its sump vented for an unlimited distance with a 1½-in pipe. If the pump was capable of producing 60 gpm, a 1½-in pipe could not have a developed length of more than 75 ft.

In most cases, a 2-in vent is used on sumps, and the distance allowed for developed length is not a problem. However, if your pump will pump more than 100 gpm, you had better take the time to do some math. Your code book will provide you with the factors you need to size your vent, and the sizing is easy. You simply look for the maximum discharge capacity of your pump and match it with a vent that allows the developed length you need.

This concludes the general description and sizing techniques for various vents. Next, we are going to look at regulations dealing with the methods of installation for vents.

Vent-Installation Requirements

Since there are so many types of vents and their role in the plumbing system is so important, there are many regulations affecting the installation of vents. What follows are specifics for installing various vents.

In zone two, any building equipped with plumbing must also be equipped with a main vent. Zone three requires any plumbing system that receives the discharge from a water closet to have either a main vent stack or stack vent. This vent must originate at a 3-in drainage pipe and extend upward until it penetrates the roof of the building and meets outside air. The vent size requirements for both zones two and three call for a minimum diameter of 3 in. However, zone two does allow the main stack in detached buildings, where the only plumbing is a washing machine or laundry tub, to have a diameter of

1½ in. Zone one requires all plumbing fixtures, except for exceptions, to be vented.

When a vent penetrates a roof, it must be flashed or sealed to prevent water from leaking past the pipe and through the roof. Metal flashings with rubber collars are normally used for flashing vents, but more modern flashings are made from plastic rather than metal.

The vent must extend above the roof to a certain height. The height may fluctuate between geographical locations. Average vent extensions are between 12 and 24 in; check with your local regulations to determine the minimum height in your area. Zones one and two generally have height requirements for vent terminations set at 6 in above the roof. Zone three requires the vent to extend at least 12 in above the roof.

When vents terminate in the open air, the proximity of their location to windows, doors, or other ventilating openings must be considered. If a vent were placed too close to a window, sewer gas might be drawn into the building when the window was open. Vents should be kept 10 ft from any window, door, opening, or ventilation device. If the vent cannot be kept at least 10 ft from the opening, the vent should extend at least 2 ft above the opening. Zone one requires these vents to extend at least 3 ft above the opening.

If the roof being penetrated by a vent is used for activities other than just weather protection, such as a patio, the vent must extend 7 ft above the roof in zone three. Zone two requires these vents to rise at least 5 ft above the roof. In cold climates, vents must be protected from freezing. Condensation can collect on the inside of vent pipes. In cold climates this condensation may turn to ice. As the ice mass grows, the vent becomes blocked and useless.

This type of protection is usually accomplished by increasing the size of the vent pipe. This ruling normally applies only in areas where temperatures are expected to be below 0°F. Zone three requires vents in this category to have a minimum diameter of 3 in. If this requires an increase in pipe size, the increase must be made at least 1 ft below the roof. In the case of side-wall vents, the change must be made at least 1 ft inside the wall.

Zone one's rules for protecting vents from frost and snow are a little different. All vents must have diameters of at least 2 in but never less than the normally required vent size. Any change in pipe size must take place at least 12 in before the vent penetrates into open air, and the vent must extend to a height of 10 in.

There may be occasions when it is better to terminate a plumbing vent out the side of a wall rather than through a roof. Zone one prohibits side-wall venting. Zone two prohibits side-wall vents from terminating under any building's overhang. When side-wall vents are installed, they must be protected against birds and rodents with a wire mesh or similar cover. Side-wall vents must not extend closer than 10 ft to the property boundary of the building lot. If the building is equipped with soffit vents, side-wall vents may not be used if they terminate under the soffit vents. This rule is in effect to prevent sewer gas from being sucked into the attic of the home.

Zone three requires buildings having soil stacks with more than five branch intervals to be equipped with a vent stack. Zone one requires a vent stack with buildings having at least 10 stories above the building drain. The vent stack will normally run up near the soil stack. The vent stack must connect into the building drain at or below the lowest branch interval. Figure 11.12 shows you an example of an approved vent stack installation. The vent stack must be sized according to the instructions given earlier. In zone three, the vent stack must be connected within 10 times its pipe size on the downward side of the soil stack. This means that a 3-in vent stack must be within 30 in of the soil stack on the downward side of the building drain.

Zone one further requires these stack vents to be connected to the drainage stack at intervals of every five stories. The connection must be made with a relief yoke vent. The yoke vent (Fig. 11.13) must be at least as large as either the vent stack or soil stack, whichever is smaller. This connection must be made with a wye fitting that is at least 42 in off the floor.

In large plumbing jobs, where there are numerous branch intervals, it may be necessary to vent offsets in the soil stack. Normally, the offset must be more than 45° to warrant an offset vent. Zones two and three require offset vents when the soil stack offsets and has five or more branch intervals above it. See Fig. 11.14 for an example of this procedure in zone three.

Just as drains are installed with a downward pitch, vents must also be installed with a consistent grade (Fig. 11.15). Vents should be graded to allow any water entering the vent pipe to drain into the drainage system. A typical grade for vent piping is ¼-in/ft. Zone one allows vent pipes to be installed level, without pitch (Fig. 11.16).

Dry vents must be installed in a manner to prevent clogging and blockages. You may not lay a fitting on its side and use a quarter bend to turn the vent up vertically. Dry vents should leave the drainage pipe in a vertical position. An easy way to remember this is that if you need an elbow to get the vent up from the drainage, you are doing it the wrong way.

Most vents can be tied into other vents, such as a vent stack or stack vent. But the connection for the tie-in must be at least 6 in above the flood-level rim of the highest fixture served by the vent.

Zone two allows the use of circuit vents to vent fixtures in a battery.

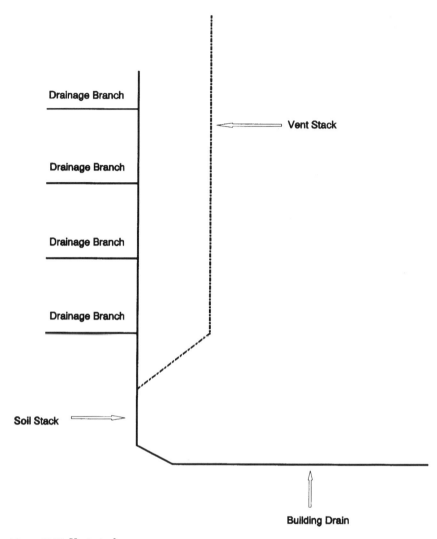

Figure 11.12 Vent stack.

The drain serving the battery must be operating at one-half of its fixture-unit rating. If the application is on a lower-floor battery with a minimum of three fixtures, relief vents are required. You must also pay attention to the fixtures draining above these lower-floor batteries.

When a fixture with a fixture rating of 4 or less and a maximum drain size of 2 in is above the battery, every vertical branch must have a continuous vent. If a fixture with a fixture-unit rating exceeding 4 is present, all fixtures in the battery must be individually vent-

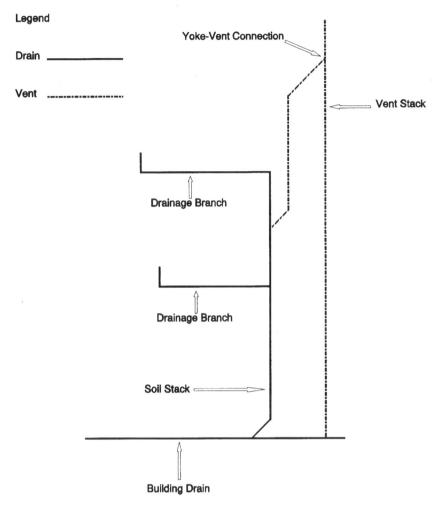

Figure 11.13 Yoke vent.

ed. Circuit-vented batteries may not receive the drainage from fixtures on a higher level.

Circuit vents should rise vertically from the drainage. However, the
vent can be taken off the drainage horizontally if the vent is washed
by a fixture with a rating of no more than 4 fixture units. The washing cannot come from a water closet. The pipe being washed must be
at least as large as the horizontal drainage pipe it is venting.

In zone three, circuit vents may be used to vent up to eight fixtures
using a common horizontal drain. Circuit vents must be dry vents,
and they should connect to the horizontal drain in front of the last fix-

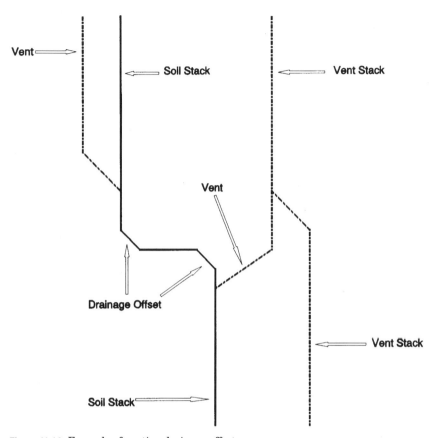

Figure 11.14 Example of venting drainage offsets.

ture on the branch. The horizontal drain being circuit-vented must not have a grade of more than 1 in/ft. Zone three interprets the horizontal section of drainage being circuit-vented as a vent. If a circuit vent is venting a drain with more than four water closets attached to it, a relief vent must be installed in conjunction with the circuit vent. Figure 11.17 shows you how this would be done.

Vent placement in relation to the trap it serves is important and regulated. The maximum allowable distance between a trap and its vent will depend on the size of the fixture drain and trap. Table 11.11 shows you allowable distances for zone one. Table 11.12 depicts the allowable distances for zone two, and Table 11.13 covers the requirements of zone three.

All vents, except those for fixtures with integral traps, should connect above the trap seal. A sanitary-tee fitting should be used when going from a vertical stack vent to a trap. Other fittings, with a

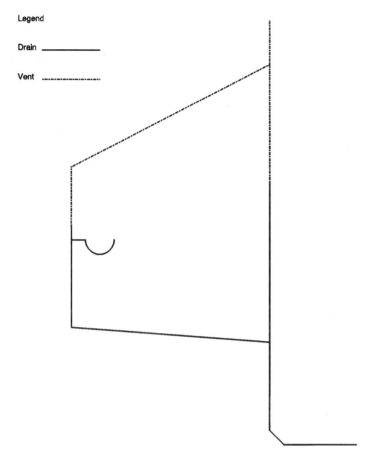

Figure 11.15 Graded-vent connection.

longer turn, like a combination wye and eighth bend, will place the trap in more danger of back-siphonage. This goes against the common sense of a smoother flow of water, but the sanitary tee reduces the risk of a vacuum.

Supporting Your Pipe

Vent pipes must be supported. Vents may not be used to support antennas, flag poles, and similar items. Depending upon the type of material you are using, and whether the pipe is installed horizontally or vertically, the spacing between hangers will vary. Both horizontal and vertical pipes require support. The regulations in the plumbing code apply to the maximum distance between hangers. Tables 11.14 and 11.15 give you the minimums for zone one. Tables 11.16 and

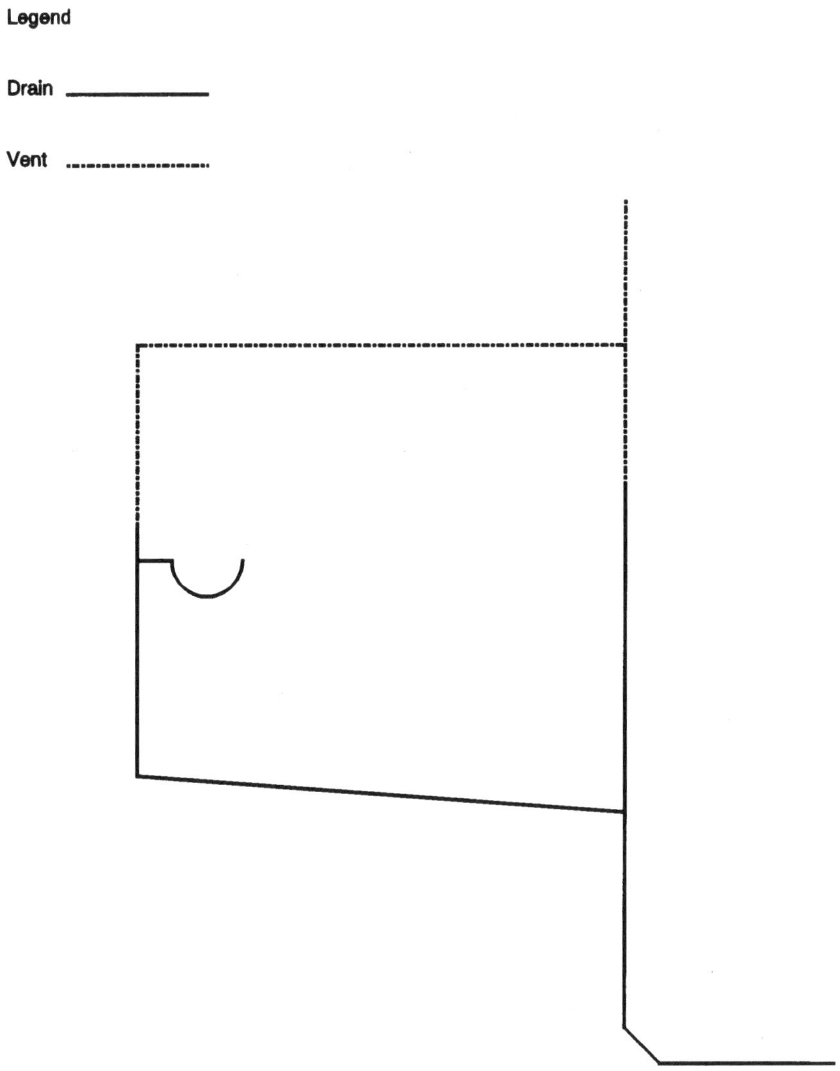

Legend

Drain _____

Vent ....

Figure 11.16 Zone one's level-vent rule.

11.17 provide similar information for zone two. Tables 11.18 and 11.19 give you maximum hanger intervals for zone three.

Some More Venting Regulations for Zone One

Some interceptors, like those used as a settling tank that discharges through a horizontal indirect waste, are not required to be vented.

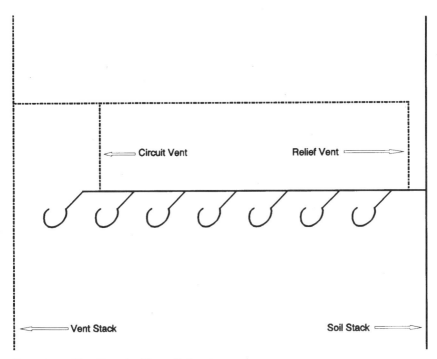

Figure 11.17 Circuit vent with a relief vent.

TABLE 11.11   Trap-to-Vent Distances in Zone One

Grade on drain pipe (in)	Size of trap arm (in)	Maximum distance between trap and vent (ft)
¼	1¼	2½
¼	1½	3½
¼	2	5
¼	3	6
¼	4 and larger	10

TABLE 11.12   Trap-to-Vent Distances in Zone Two

Grade on drain pipe (in)	Fixture's drain size (in)	Trap size (in)	Maximum distance between trap and vent (ft)
¼	1¼	1¼	3½
¼	1½	1¼	5
¼	1½	1½	5
¼	2	1½	8
¼	2	2	6
⅛	3	3	10
⅛	4	4	12

TABLE 11.13   Trap-to-Vent Distances in Zone Three

Grade on drain pipe (in)	Fixture's drain size (in)	Trap size (in)	Maximum distance between trap and vent (ft)
¼	1¼	1¼	3½
¼	1½	1¼	5
¼	1½	1½	5
¼	2	1½	8
¼	2	2	6
⅛	3	3	10
⅛	4	4	12

TABLE 11.14   Horizontal Pipe-Support Intervals in Zone One

Support material	Maximum distance between supports (ft)
ABS	4
Cast iron	At each pipe joint*
Galvanized	12
Copper (1½ in and smaller)	6
PVC	4
Copper (2 in and larger)	10

* Cast-iron pipe must be supported at each joint, but supports may not be more than 10 ft apart.

TABLE 11.15   Vertical Pipe-Support Intervals in Zone One*

Type of vent pipe	Maximum distance between supports
Lead pipe	4 ft
Cast iron	At each story
Galvanized	At least every other story
Copper	At each story†
PVC	Not mentioned
ABS	Not mentioned

* All stacks must be supported at their bases.
† Support intervals may not exceed 10 ft.

TABLE 11.16   Horizontal Pipe-Support Intervals in Zone Two

Type of vent pipe	Maximum distance between supports (ft)
ABS	4
Cast iron	At each pipe joint
Galvanized	12
PVC	4
Copper (2 in and larger)	10
Copper (1½ in and smaller)	6

TABLE 11.17   Vertical Pipe-Support Intervals in Zone Two*

Type of vent pipe	Maximum distance between supports (ft)
Lead pipe	4
Cast iron	At each story[†]
Galvanized	At each story[‡]
Copper (1¼ in)	4
Copper (1½ in and larger)	At each story
PVC (1½ in and smaller)	4
PVC (2 in and larger)	At each story
ABS (1½ in and smaller)	4
ABS (2 in and larger)	At each story

* All stacks must be supported at their bases.
[†] Support intervals may not exceed 15 ft.
[‡] Support intervals may not exceed 30 ft.

TABLE 11.18   Horizontal Pipe-Support Intervals in Zone Three

Type of vent pipe	Maximum distance between supports (ft)
Lead pipe	Continuous
Cast iron	5*
Galvanized	12
Copper tube (1¼ in)	6
Copper tube (1½ in and larger)	10
ABS	4
PVC	4
Brass	10
Aluminum	10

* Or at every joint.

TABLE 11.19   Vertical Pipe-Support Intervals in Zone Three

Type of vent pipe	Maximum distance between supports (ft)
Lead pipe	4
Cast iron	15
Galvanized	15
Copper tubing	10
ABS	4
PVC	4
Brass	10
Aluminum	15

However, the interceptor receiving the discharge from the unvented interceptor must be properly vented and trapped.

Traps for sinks that are a part of a piece of equipment, like a soda fountain, are not required to be vented when venting is impossible.

But these drains must drain through an indirect waste to an approved receptor.

Other Venting Requirements for Zone Two

All soil stacks that receive the waste of at least two vented branches must be equipped with a stack vent or a main stack vent. Except when approved, fixture drainage may not be allowed to enter a stack at a point above a vent connection. Side-inlet closet bends are allowed to accept the connection of fixtures that are vented. However, these connections may not be used to vent a bathroom unless the connection is washed by a fixture. All fixtures dumping into a stack below a higher fixture must be vented except when special approval is granted for a variance. Stack vents and vent stacks must connect to a common vent header prior to vent termination.

Traps for sinks that are a part of a piece of equipment, like a soda fountain, are not required to be vented when venting is impossible. But, these drains must be piped in accordance with the combination waste and vent regulations for zone two.

Up to two fixtures, set back to back or side by side, within the allowable distance between the traps and their vents may be connected to a common horizontal branch that is vented by a common vertical vent. However, the horizontal branch must be one pipe size larger than normal. When applying this rule, the following ratings apply: shower drains, 3-in floor drains, 4-in floor drains, pedestal urinals, and water closets with fixture-unit ratings of 4 shall be considered to have 3-in drains.

Some fixture groups are allowed to be stack-vented without individual back vents. These fixture groups must be located in one-story buildings or must be located on the top floor of the building, with some special provisions. Fixtures located on the top floor must connect independently to the soil stack, and the bathing units and water closets must enter the stack at the same level. Table 11.20 shows the fixtures allowed to be vented in this fashion.

TABLE 11.20   Stack-Venting Without Individual Vents in Zone Two

Fixtures allowed to be stack-vented without individual vents*
Water closets
Basins
Bathtubs
Showers
Kitchen sinks, with or without dishwasher and garbage disposer

* Restrictions apply to this type of installation.

This same stack-venting procedure can be adapted to work with fixtures on lower floors. The stack being stack-vented must enter the main soil stack though a vertical eighth bend and wye combination. The drainage must enter above the eighth bend. A 2-in vent must be installed on the fixture group. This vent must be 6 in above the flood-level rim of the highest fixture in the group.

Some fixtures are allowed to be served by a horizontal waste that is within a certain distance of a vent. When piped in this manner, bathtubs and showers are both required to have 2-in P-traps. These drains must run with a minimum grade of ¼-in/ft. A single drinking fountain can be rated as a lavatory for this type of piping. On this type of system, fixture drains for lavatories may not exceed 1¼ in, and sink drains cannot be larger than 1½ in, in diameter.

In multistory situations, it is possible to drain up to three fixtures into a soil stack above the highest water closet or bathtub connection, without reventing. To do this, certain requirements must be met. These requirements are as follows:

- Minimum stack size of 3 in is required.

- Approved fixture-unit load on stack is met.

- All lower fixtures must be properly vented.

- All individually unvented fixtures are within allowable distances of the main vent.

- Fixture openings shall not exceed the size of their traps.

- All code requirements must be met and approved.

Working with a Combination Waste and Vent System

Most jurisdictions limit the extent to which fixtures can be served by a combination waste and vent system, but not all. In many locations it is a code violation to include a toilet on a combination system, but Maine, for example, will allow toilets on a combination waste and vent system. Since combination waste and vent systems can get you into a sticky situation, you should consult your local code officer before using such a system. However, this is how the system works, in general.

The types of fixtures you are allowed to connect to with a combination waste and vent system may be limited. In some areas the only fixtures allowed on the combination system are floor drains, standpipes, sinks, and lavatories. Other areas allow showers, bathtubs, and even toilets to be installed with the combo system. You will have to check your local regulations to see how they affect your choice in types of plumbing systems.

It is intended that the combination waste and vent system will be mainly made up of horizontal piping. Generally, the only vertical piping is the vertical risers to lavatories, sinks, and standpipes. These vertical pipes may not normally exceed 8 ft in length. This type of system relies on an oversized drain pipe to provide air circulation for drainage. The pipe is often required to be twice the size required for a drain vented normally. The combination system typically must have at least one vent. The vent should connect to a horizontal drain pipe.

Any vertical vent must rise to a point at least 6 in above the highest fixture being served before it may take a horizontal turn. In a combination system the pipes are rated for fewer fixture units. A 3-in pipe connecting to a branch or stack may only be allowed to carry 12 fixture units. A 4-in pipe, under the same conditions, could be restricted to conveying 20 fixture units. Similarly, a 2-in pipe might only handle 3 fixture units, and a 1½-inch pipe may not be allowed. The ratings for these pipes can increase when the pipes are connecting to a building drain.

Stack vents are allowed but not always in the normal way. All fixtures on a combo system may be required to enter the stack vent individually, as opposed to on a branch, as would normally be the case. A stack vent used in a combo system generally must be a straight vertical vent, without offsets. The stack vent usually cannot even be offset vertically; it simply cannot be offset. This rule is different in some locations, so check with your local plumbing inspector to see if you are affected by the no-offset rule.

Since stack vents are common, and often required, in a combination system, you must know how to size these pipes. The sizing is generally done based on the number of fixture units entering the stack. Here is an example of how a stack vent for a combo system might be sized in zone three.

Since not all pipes run in conjunction with a combination waste and vent system have to apply to the combo rules, it is possible that you would have a 1½-in pipe entering a stack. The 1½-in pipe could only be used if it had an individual vent. It is also possible that the stack vent would be a 1½-in pipe.

First, let's look at the maximum number of fixture units (fu) allowed on a stack; they are as follows:

- 1½-in stack = 2 fu
- 2-in stack = 4 fu
- 3-in stack = 24 fu

- 4-in stack = 50 fu
- 5-in stack = 75 fu
- 6-in stack = 100 fu

When you are concerned with the size of a drain dumping into the stack, there are only two pipe sizes to contend with. All pipe sizes larger than 2-in may dump an unlimited number of fixture units into the stack. A 1½-in pipe may run one fixture unit into the stack, and a 2-in pipe may deliver two fixture units to the stack. Sizing your stack is as simple as finding your fixture-unit load on the chart in your local code book. Compare your fixture-unit load to the chart, and select a pipe size rated for that load.

Again, remember that combination waste and vent systems vary a great deal in what is, and is not, allowed. To show that contrast, here are how the regulations in Maine differ from the ones already described. In Maine, at the time of this writing, there are regulations that allow a much different style of combination waste and vent plumbing. The Maine regulations allow water closets, showers, and bathtubs to be included on the combo system. These regulations are under consideration for change, but at the moment, they are still in effect.

This alternate form of plumbing is only applicable to buildings with two stories or less. The regulations require that the only fixture located in a basement may be a clothes washer. Not only can a water closet be used on this system, but up to three water closets are allowed to enter a single 3-in stack from a branch. You are also allowed to pipe up to two of these branches, each carrying the discharge from up to three toilets, to the stack at the same location. If a toilet is installed in the building, there must be at least one 3-in vent.

Vertical rises from the building drain are allowed to extend 10 feet. A branch may contain up to eight water closets if vented by a circuit vent. Up to three fixtures, not counting floor drains, that are on the same floor level, or a level not exceeding 30 in apart, may be tied into the building drain at a distance greater than 10 ft. Basically, this means that an entire bathroom group can be piped to the building drain without limitation on the distance.

When it comes to sizing on the Maine system, there are some big differences. With a grade of a ¼-in/ft, a 1½-in pipe can carry up to 5 fixture units; a 2-in pipe is allowed up to 12 fixture units. These numbers are much higher than the ones given earlier. As you can see, there are some major differences to be found in the rulings for combination waste and vent systems. Maine currently operates on a plumbing code that is different from the ones used in zones one, two, and three.

Reminder Notes

Zone one

1. All normal venting is sized using the same ratings and table.

2. Fixtures being vented with a common vent must enter the drainage at the same level.

3. Vent venting is restricted to vertical piping.

4. All plumbing fixtures, except for exceptions, must be vented.

5. Vents terminating closer than 10 ft from a building opening must rise at least 3 ft above the opening.

6. Vents subject to freezing must be no less than 2 in in diameter and must extend at least 10 in above the roof. Any change in pipe size must be made at least 12 in before the pipe penetrates into open air.

7. Side-wall vents are prohibited.

8. Buildings with at least 10 stories above the building drain must be equipped with vent stacks. These vent stacks are required to have yoke vents at every five-story interval.

9. Vents may be installed level, without pitch.

Zone two

1. Wet venting is allowed for multiple bathrooms on different floor levels.

2. Any building equipped with plumbing must be equipped with a main vent.

3. Vents terminating above a roof used for purposes other than weather protection must rise at least 5 ft above the roof.

4. Side-wall vents may not terminate under overhangs.

5. Offsets of more than 45° in soil stacks, when there are at least five stories in the building, must be vented.

6. There are many rules on circuit venting; refer to the text for examples.

Zone three

1. Common vents may serve up to three lavatory traps.

2. Zone three has specific requirements for venting sewer sumps.

3. Any plumbing system receiving the discharge from toilets must be equipped with a main stack for venting.

4. Vents must extend at least 12 in above the roof they penetrate.

5. Vents subject to freezing must have minimum diameters of 3 in. Any change in pipe sizing must occur at least 12 in prior to penetration into open air.

6. Buildings with at least five branch intervals must have vent stacks.

7. In buildings with at least five branch intervals, horizontal offsets in drainage piping must be vented.

8. Circuit vents can serve up to eight fixtures.

9. Circuit vents venting more than four water closets must be equipped with relief vents.

10. Refer to the text for more rules on circuit venting.

12

Design Criteria

What are the design criteria for plumbing systems? The plumbing code is certainly a big part of the criteria. Any system being designed must conform to applicable code requirements. We've already talked about pipe sizing, which is part of designing a system. In this chapter we focus on basic design rules.

We've talked about why commercial plumbers are not normally required to design their own layouts. Residential plumbers, however, are often required to create their own layouts. The designer must keep many factors in mind. The cost of material is one consideration. Efficient designs use minimal amounts of material to create the highest profit possible. Another consideration in layout is the amount of effort required to install it. If you can reduce the labor required to make an installation, you can make more money.

What else should you keep in mind when working out a design? Code requirements must be met. Customer satisfaction is also a factor to consider. Just making your work easier is reason enough to put some thought into making a good design. With this in mind, let's talk about some design issues.

Grading Pipe

When you install horizontal drainage piping, you must install it so that it falls toward the waste disposal site. A typical grade for drainage pipe is a fall of $\frac{1}{4}$ in/ft. This means the lower end of a 20-ft pipe is 5 in lower than the upper end, when properly installed. While the $\frac{1}{4}$ in/ft grade is typical, it is not the only acceptable grade for all pipes.

If you are working with pipe that has a diameter of $2\frac{1}{2}$ in or less, the minimum grade for the pipe is $\frac{1}{4}$ in/ft. Pipes with diameters between 3 and 6 in are allowed a minimum grade of $\frac{1}{8}$ in/ft. In zone 1, special permission must be granted prior to installation of pipe with a $\frac{1}{8}$ in/ft grade. In zone 3, for pipes with diameters of 8 in or more, an acceptable grade is $\frac{1}{16}$ in/ft.

Supporting Pipe

Pipe support is also regulated by the plumbing code (Tables. 12.1, 12.2, and 12.3). There are requirements about the types of materials you may use and how. One concern with the type of hangers used is their compatibility with the pipe they are supporting. You must use a hanger that will not harm the piping. For example, you may not use a galvanized-steel strap hanger to support copper pipe. As a rule of thumb, the hangers used to support a pipe should be made of the same material as the pipe. For example, copper pipe should be hung with copper hangers. This eliminates the risk of corrosive action between two different types of materials. A plastic or plastic-coated hanger can be used with all types of pipe. An exception to this rule might be when the piping is carrying a liquid with a temperature that might affect or melt the plastic hanger.

TABLE 12.1   Support Intervals for Water Pipe in Zone 1

Type of pipe, in	Vertical support interval	Horizontal support interval, ft
Threaded pipe ($\frac{3}{4}$-in and smaller)	Every other story	10
Threaded pipe (1-in and larger)	Every other story	12
Copper tube ($1\frac{1}{2}$-in and smaller)	Every story, not to exceed 10 ft	6
Copper tube (2-in and larger)	Every story, not to exceed 10 ft	10
Plastic pipe	Not mentioned	4

TABLE 12.2   Support Intervals for Water Pipe in Zone 2

Type of pipe	Vertical support interval	Horizontal support interval, ft
Threaded pipe	90 ft	12
Copper tube ($1\frac{1}{4}$ in and smaller)	4 ft	6
Copper tube ($1\frac{1}{2}$ in)	Every story	6
Copper tube (larger than $1\frac{1}{2}$ in)	Every story	10
Plastic pipe (2 in and larger)	Every story	4
Plastic pipe ($1\frac{1}{2}$ in and smaller)	4 ft	4

TABLE 12.3   Pipe Support Intervals in Zone 3

Type of vent pipe	Maximum distance between supports
Horizontal	
PB	32 in
Lead	Continuous
Cast iron	5 ft or at each joint
Galvanized steel	12 ft
Copper tube ($1\frac{1}{4}$ in)	6 ft
Copper tube ($1\frac{1}{2}$ in and larger)	10 ft
ABS	4 ft
PVC	4 ft
Brass	10 ft
Aluminum	10 ft
Vertical	
Lead	4 ft
Cast iron	15 ft
Galvanized steel	15 ft
Copper tubing	10 ft
ABS	4 ft
PVC	4 ft
Brass	10 ft
PB	4 ft
Aluminum	15 ft

The hangers used to support pipe must be capable of supporting the pipe at all times. Hangers must be attached to the pipe and to the member holding the hanger in a satisfactory manner. For example, it is not acceptable to wrap a piece of wire around a pipe and then wrap the wire around the bridging between two floor joists. Hangers should be securely attached to the member supporting it. For example, a hanger should be attached to the pipe and then nailed to a floor joist. The nails used to hold a hanger in place should be made of the same material as the hanger, if corrosive action is a possibility.

Both horizontal and vertical pipes require support. The intervals between supports will vary, depending upon the type of pipe being used and whether it is installed vertically or horizontally. The following examples will show you how often to support various types of pipes when they are hung horizontally. These are examples of the maximum distances allowed between supports for zone 3:

ABS—every 4 ft   Cast iron—every 5 ft

Galvanized steel—every 12 ft   PVC—every 4 ft

DWV copper—every 10 ft

When these same types of pipes are installed vertically, in zone 3, they must be supported at no less than the following intervals:

ABS—every 4 ft   Cast iron—every 15 ft

Galvanized steel—every 15 ft   PVC—every 4 ft

DWV copper—every 10 ft

When cast-iron stacks are installed, the base of each stack must be supported. Pipes with flexible couplings, bands, or unions must be installed and supported so as to prevent the flexible connections from moving. In larger pipes—pipes larger than 4 in—all flexible couplings must be supported to prevent the force of the pipe's flow from loosening the connection at changes in direction.

Pipe Size Reduction

You may not reduce the size of a drainage pipe as it heads for the waste disposal site. The pipe size may be enlarged, but it may not be reduced. There is one exception to this rule. Reducing closet bends, such as a 4-by-3 closet bend, is allowed.

Underground Pipes

A drainage pipe installed underground must have a minimum diameter of 2 in. When you are installing a horizontal branch fitting near the base of a stack, keep the branch fitting away from the point where the vertical stack turns to a horizontal run. The branch fitting should be installed at least 30 in back on a 3-in pipe and 40 in back on a 4-in pipe. Multiply the pipe size by 10 to determine how far back the branch fitting should be installed.

All drainage piping must be protected from the effects of flooding. When a stub of pipe is left to connect with fixtures planned for the future, the stub must not be more than 2 ft long, and it must be capped. Some exceptions are possible on the prescribed length of a pipe stub. If you need a longer stub, consult the local code officer. Cleanout extensions are not affected by the 2-ft rule.

Fittings

Fittings are also a part of the drainage system. Knowing when, where (Fig. 12.1), and how to use the proper fittings is essential to the installation of a drainage system. Fittings are used to make branches and to change direction. The use of fittings to change direction is where we will start. When you wish to change direction with pipe, you may change it from a horizontal run to a vertical rise. You may be going from a vertical position to a horizontal one, or you might only want to offset the pipe in a horizontal run. Each of these three cate-

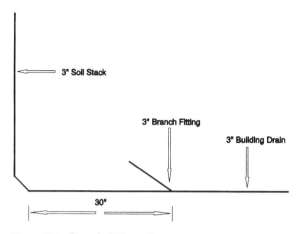

Figure 12.1 Branch-fitting rule.

gories requires the use of different fittings. Let's take each circumstance and examine the fittings allowed.

Offsets in horizontal piping

When you want to change the direction of a horizontal pipe, you must use fittings approved for that purpose. You have six choices in zone 3:

Sixteenth bend	Eighth bend
Sixth bend	Long-sweep fittings
Combination wye and eighth bend	Wye

Any of these fittings is generally approved for changing direction with horizontal piping; but as always, it is best to check with your local code officer for current regulations.

Horizontal to vertical

There is a wider choice in fittings for going from a horizontal position to a vertical position. There are nine possible candidates in zone 3:

Sixteenth bend	Eighth bend
Sixth bend	Long-sweep fittings
Combination wye and eighth bend	Wye
Quarter bend	Short-sweep fittings
Sanitary tee	

You may not use a double sanitary tee in a back-to-back situation if the fixtures being served are of a blowout or pump type. For example, you could not use a double sanitary tee to receive the discharge of two washing machines, if the machines were positioned back to back. The sanitary tee's throat is not deep enough to keep drainage from feeding back and forth between the fittings. In a case like this, use a double combination wye and eighth bend. The combination fitting has a much longer throat and will prohibit waste water from transferring across the fitting to the other fixture.

Vertical to horizontal

Seven fittings can be used to change direction from vertical to horizontal:

Sixteenth bend	Eighth bend
Sixth bend	Long-sweep fittings
Combination wye and eighth bend	Wye
Short-sweep fittings 3-in or larger	

Indirect Wastes

Indirect-waste requirements can pertain to a number of types of plumbing fixtures and equipment. These might include a clothes washer drain, a condensate line, a sink drain, or the blow-off pipe from a relief valve, just to name a few. These indirect wastes are piped in this manner to prevent the possibility of contaminated matter backing up the drain into a potable water or food source, among other things.

Most indirect-waste receptors are trapped. If the drain from the fixture is more than 2 ft long, the indirect-waste receptor must have a trap. However, this rule applies to fixtures like sinks, not to an item such as a blow-off pipe from a relief valve. The rule is different in zone 1. In zone 1, if the drain is more than 5 ft long, it must have a trap.

The safest disposal method for an indirect waste is accomplished by using an air gap. When an air gap is used, the drain from the fixture terminates above the indirect-waste receptor, with open airspace between the waste receptor and the drain. This prevents any backup or back-siphonage.

Some fixtures, depending on local code requirements, may be piped with an air break, rather than an air gap. With an air break, the drain may extend below the flood-level rim and terminate just above the trap's seal. The risk with an air break is the possibility of a backup. Since the drain is run below the flood-level rim of the waste receptor, the waste receptor could overflow and back up into the drain. This could create contamination, but in cases where contamination is

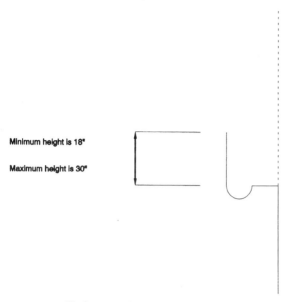

Minimum height is 18"

Maximum height is 30"

Figure 12.2 Washing machine receptor rule.

likely, an air gap will be required. Check with the local code officer before you use an air break.

Standpipes, like those used for washing machines, are a form of indirect-waste receptor. A standpipe (Fig. 12.2) used for this purpose in zones 1 and 3 must extend at least 18 in above the trap's seal, but it may not extend more than 30 in above the trap's seal. If a clear-water waste receptor is located in a floor, in zone 3 the lip of the receptor must extend at least 2 in above the floor. This eliminates the waste receptor from being used as a floor drain.

Buildings used for food preparation, storage, and similar activities are required to have their fixtures and equipment discharge drainage through an air gap. In zone 3 there is an exception to this rule. In zone 3, dishwashers and open culinary sinks are excepted. In zone 2 the discharge pipe must terminate at least 2 in above the receptor. In zone 1 the distance must be a minimum of 1 in. In zones 2 and 3 the air-gap distance must be a minimum of twice the size of the pipe discharging the waste. For example, a $\frac{1}{2}$-in discharge pipe requires a 1-in air gap.

In zones 2 and 3 the installation of an indirect-waste receptor is prohibited in any room containing toilet facilities. The same holds in zone 1 but there is one exception. The exception is the installation of a receptor for a clothes washer, when the clothes washer is installed in the same room. Indirect-waste receptors may not be allowed to be installed in closets and other unvented areas. Indirect-waste receptors must be accessible. In zone 2 all receptors must be equipped with a means of preventing solids with diameters of $\frac{1}{2}$ in or larger from entering the drainage system. These straining devices must be removable, to allow for cleaning.

When you are dealing with extreme water temperatures in waste water, such as with a commercial dishwasher, the dishwasher drain must be piped to an indirect waste. The indirect waste will be connected to the sanitary plumbing system, but the dishwasher drain may not connect to the sanitary system directly, if the waste water temperature exceeds 140°F. Steam pipes may not be connected directly to a sanitary drainage system. Local regulations may require the use of special piping, sumps, or condensers to accept high-temperature water. In zone 1 direct connection of any dishwasher to the sanitary drainage system is prohibited.

Clear-water waste, from a potable source, must be piped to an indirect-waste receptor with the use of an air gap. Sterilizers and swimming pools are two examples of when this rule applies. Clear water from nonpotable sources, such as a drip from a piece of equipment, must be piped to an indirect waste. In zone 3, an air break is allowed in place of an air gap. In zone 2 any waste entering the sanitary drainage system from an air conditioner must do so through an indirect waste.

Special Wastes

Special wastes are those wastes that may be harmful to a plumbing system or the waste disposal system. Possible locations for special-waste piping might include photographic labs, hospitals, or buildings where chemicals or other potentially dangerous wastes are dispersed. Small, personal-type photograhic darkrooms do not generally fall under the scrutiny of these regulations. Buildings that are considered to have a need for special-wastes plumbing are often required to have two plumbing systems, one system for normal sanitary discharge and a separate system for the special wastes. Before many special wastes are allowed to enter a sanitary drainage system, the wastes must be neutralized, diluted, or otherwise treated.

Depending upon the nature of the special wastes, special materials may be required. When you venture into the plumbing of special wastes, it is always best to consult the local code officer before proceeding with your work.

Vent Installation Requirements

Since there are so many types of vents (Fig. 12.3) and their role in the plumbing system is so important, there are many regulations affecting their installation. What follows are the specifics for installing various vents.

In zone 2, any building equipped with plumbing must also be equipped with a main vent (Fig. 12.4). In zone 3 any plumbing system that receives the discharge from a water closet must have either a main vent stack or a stack vent. This vent must originate at a 3-in drain pipe and extend upward until it penetrates the roof of the build-

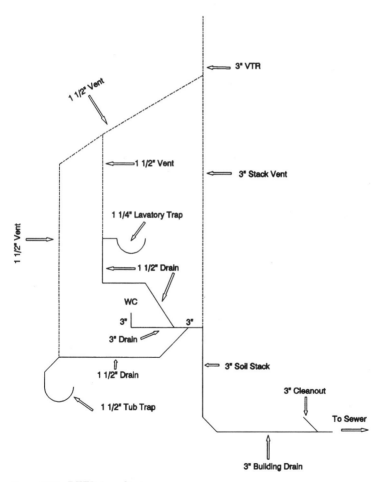

Figure 12.3 DWV riser diagram.

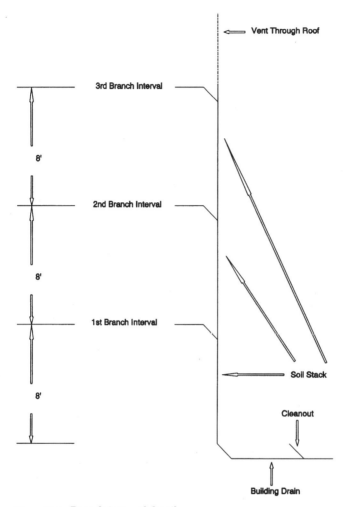

Figure 12.4 Branch-interval detail.

ing and meets outside air. The vent size requirements for both zones 2 and 3 call for a minimum diameter of 3 in. However, in zone 2, the main stack in detached buildings, where the only plumbing is a washing machine or laundry tub, may have a diameter of $1\frac{1}{2}$ in. In zone 1, all plumbing fixtures must be vented.

When a vent penetrates a roof, it must be flashed or sealed to prevent water from leaking past the pipe and through the roof. Metal flashings with rubber collars are normally used for flashing vents, but more modern flashings are made of plastic, rather than metal.

The vent must extend above the roof to a certain height. The height may vary from one geographical location to another. Average vent extensions are between 12 and 24 in, but check the local regulations to determine the minimum height in your area. In zones 1 and 2, generally height requirements for vent terminations are set at 6 in above the roof. In zone 3 the vent must extend at least 12 in above the roof.

When vents terminate in open air, the proximity of their location to windows, doors, or other ventilating openings must be considered. If a vent were placed too close to a window, sewer gas could be drawn into the building when the window was open. Vents should be kept 10 ft from any window, door, opening, or ventilation device. If the vent cannot be kept at least 10 ft from the opening, the vent should extend at least 3 ft above the opening. In zone 1, these vents must extend at least 3 ft above the opening.

If the roof being penetrated by a vent is used for activities other than just weather protection, such as a patio, the vent must extend 7 ft above the roof in zone 3. In zone 2 these vents must rise at least 5 ft above the roof. In cold climates, vents must be protected from freezing. Condensation can collect on the inside of vent pipes. In cold climates this condensation may turn to ice. As the ice mass grows, the vent becomes blocked and useless.

This type of protection is usually afforded by increasing the size of the vent pipe. This rule normally applies only in areas where temperatures are expected to be below 0°F. In zone 3, vents in this category must have a minimum diameter of 3 in. If this requires an increase in pipe size, the increase must be made at least 1 ft below the roof. In the case of sidewall vents, the change must be made at least 1 ft inside the wall.

In zone 1, the rules for protecting vents from frost and snow are a little different. All vents must have diameters of at least 2 in, but never less than the normally required vent size. Any change in pipe size must occur at least 12 in before the vent penetrates open air, and the vent must extend to a height of 10 in.

On some occasions it is better to terminate a plumbing vent out the side of a wall than through a roof. In zone 1 sidewall venting is prohibited. In zone 2, sidewall vents may not terminate under any building's overhang. When sidewall vents are installed, they must be protected against birds and rodents by a wire mesh or similar cover. Sidewall vents must not extend closer than 10 ft to the property boundary of the building lot. If the building is equipped with soffit vents, sidewall vents may not be used in such a way that they terminate under the soffit vents. This rule is, in effect, to prevent sewer gas from being sucked into the attic of the home.

In zone 3, buildings having soil stacks with more than five branch intervals (Fig. 12.5) must be equipped with a vent stack. In zone 1, a vent stack is required with buildings having at least 10 stories above

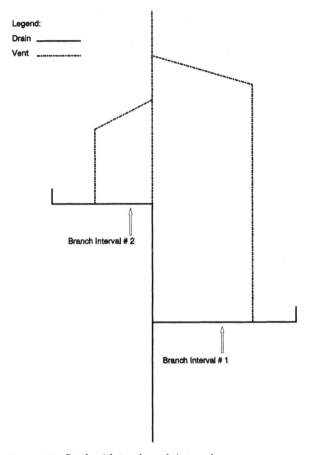

Figure 12.5 Stack with two branch intervals.

the building drain. The vent stack normally runs up near the soil stack. The vent stack must connect to the building drain at or below the lowest branch interval. The vent stack must be sized according to the instructions given earlier. In zone 3, the vent stack must be connected within 10 times its pipe size on the downward side of the soil stack. This means that a 3-in vent stack must be within 30 in of the soil stack, on the downward side of the building drain.

In zone 1, these stack vents also must be connected to the drainage stack at intervals of every five stories. The connection must be made with a relief yoke vent. The yoke vent (Fig. 12.6) must be at least as large as either the vent stack or the soil stack, whichever is smaller. This connection must be made with a wye fitting, at least 42 in off the floor.

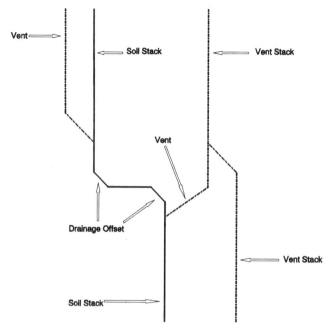

Figure 12.6   Example of venting drainage offsets.

In large plumbing jobs where there are numerous branch intervals, it may be necessary to vent offsets in the soil stack. Normally, the offset must be more than 45° to warrant an offset vent. In zones 2 and 3, offset vents are required when the soil stack offsets and has five, or more, branch intervals above it.

Just as drains are installed with a downward pitch, vents must be installed with a consistent grade. Vents should be graded to allow any water entering the vent pipe to drain into the drainage system. A typical grade for vent piping is a $\frac{1}{4}$ in/ft. In zone 1, vent pipes may be installed level, without pitch.

Dry vents must be installed in a manner to prevent clogging and blockages. You may not lay a fitting on its side and use a quarter bend to turn the vent up vertically. Dry vents should leave the drain pipe in a vertical position. An easy way to remember this is that if you need an elbow to get the vent up from the drainage, you are doing it wrong.

Most vents can be tied into other vents, such as a vent stack or stack vent (Table 12.4). But the connection for the tie-in must be at least 6 in above the flood-level rim of the highest fixture served by the vent.

TABLE 12.4   Stack-Venting without Individual Vents in Zone 2

Fixtures allowed to be stack-vented without individual vents*
Water closets
Basins
Bathtubs
Showers
Kitchen sinks, with or without dishwasher and garbage disposal

Note: Restrictions apply to this type of installation.

In zone 2, circuit vents can be used to vent fixtures in a battery. The drain serving the battery must be operating at one-half its fixture-unit rating. If the application is on a lower-floor battery with a minimum of three fixtures, relief vents are required. You must also pay attention to the fixtures draining above these lower-floor batteries.

When a fixture with a fixture-unit rating (Table 12.5) of 4 or less and a maximum drain size of 2 in is above the battery, every vertical branch must have a continuous vent. If a fixture with a fixture-unit rating exceeding 4 is present, all fixtures in the battery must be individually vented. Circuit-vented batteries may not receive the drainage from fixtures on a higher level.

Circuit vents should rise vertically from the drainage. However, the vent can be taken off the drainage horizontally, if the vent is washed by a fixture with a rating of no more than 4 fixture units. The washing cannot come from a water closet. The pipe being washed must be at least as large as the horizontal drainage pipe it is venting (Table 12.6).

In zone 3, circuit vents may be used to vent up to eight fixtures utilizing a common horizontal drain. Circuit vents must be dry vents, and they should connect to the horizontal drain in front of the last fixture on the branch. The horizontal drain being circuit-vented must not have a grade of more than 1 in/ft. In zone 3, the horizontal section of drainage being circuit-vented is interpreted as a vent. If a circuit vent is venting a drain with more than four water closets attached to it, a relief vent must be installed in conjunction with the circuit vent.

TABLE 12.5   Fixture-Unit Ratings in Zone 3

Bathtub	2
Shower	2
Residential toilet	4
Lavatory	1
Kitchen sink	2
Dishwasher	2
Clothes washer	3
Laundry tub	2

TABLE 12.6   Vent Sizing for Zone 3

(For Use with Individual, Branch, and Circuit Vents for Horizontal Drainpipes)

Drainpipe size, in	Drainpipe grade, in/ft	Vent pipe size, in	Maximum developed length of vent pipe
$1\frac{1}{2}$	$\frac{1}{4}$	$1\frac{1}{4}$	Unlimited
$1\frac{1}{2}$	$\frac{1}{4}$	$1\frac{1}{2}$	Unlimited
2	$\frac{1}{4}$	$1\frac{1}{4}$	290 ft
2	$\frac{1}{4}$	$1\frac{1}{2}$	Unlimited
3	$\frac{1}{4}$	$1\frac{1}{2}$	97 ft
3	$\frac{1}{4}$	2	420 ft
3	$\frac{1}{4}$	3	Unlimited
4	$\frac{1}{4}$	2	98 ft
4	$\frac{1}{4}$	3	Unlimited
4	$\frac{1}{4}$	4	Unlimited

All vents, except those for fixtures with integral traps, should connect above the trap's seal. A sanitary-tee fitting should be used going from a vertical stack vent to a trap. Other fittings, with a longer turn, such as a combination wye and eighth bend, will place the trap in greater danger of back-siphonage. I know this goes against the common sense of a smoother flow of water, but the sanitary tee reduces the risk of a vacuum.

The Main Water Pipe

The main water pipe delivering potable water to a building is called a *water service*. A water service pipe must have a diameter of at least $\frac{3}{4}$ in. The pipe must be sized, according to code requirements, to provide adequate water volume and pressure to the fixtures.

Ideally, a water service pipe should be run from the primary water source to the building in a private trench. By private trench, I mean a trench not used for any purpose, except the water service. However, it is normally allowed to place the water service in the same trench used by a sewer or building drain, when specific installation requirements are followed. The water service pipe must be separated from the drain pipe. The bottom of the water service pipe may not be closer than 12 in to the drain pipe at any point.

A shelf must be made in the trench to support the water service (Fig. 12.7). The shelf must be made solid and stable, at least 12 in above the drain pipe. It is not acceptable to have a water service located in an area where pollution is probable. A water service should never run through, above, or under a waste disposal system, such as a septic field.

If a water service is installed in an area subject to flooding, the pipe

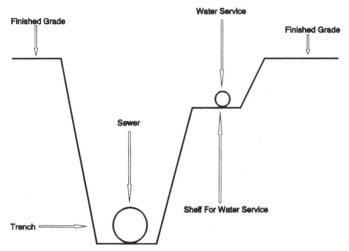

Figure 12.7   Water service and sewer in common trench.

must be protected against flooding. Water services must also be protected against freezing. The depth of the water service will depend on the climate of the location. Check with the local code officer to see how deep a water service pipe must be buried. Care must be taken when backfilling a water service trench. The backfill material must be free of objects, such as sharp rocks, that may damage the pipe.

When a water service enters a building through or under the foundation, the pipe must be protected by a sleeve. This sleeve is usually a pipe with a diameter at least twice that of the water service. Once through the foundation, the water service may need to be converted to an acceptable water distribution pipe. As you learned while reading about approved materials, some materials approved for water service piping are not approved for interior water distribution.

If the water service pipe is not an acceptable water distribution material, it must be converted to an approved material, generally within the first 5 ft of entry into the building. Once inside a building, the maze of hot and cold water pipes is referred to as water distribution pipes (Table 12.7).

Pressure-Reducing Valves

Pressure-reducing valves must be installed on water systems when the water pressure coming to the water distribution pipes is in excess of 80 pounds per square inch (psi). The only time this regulation is generally waived is when the water service is bringing water to a device requiring high pressure.

TABLE 12.7   Common Minimum Fixture Supply Sizes

Fixture	Minimum supply size, in
Bathtub	$\frac{1}{2}$
Bidet	$\frac{3}{8}$
Shower	$\frac{1}{2}$
Toilet	$\frac{3}{8}$
Lavatory	$\frac{3}{8}$
Kitchen sink	$\frac{1}{2}$
Dishwasher	$\frac{1}{2}$
Laundry tub	$\frac{1}{2}$
Hose bib	$\frac{1}{2}$

Booster Pumps

Not all water sources are capable of providing optimum water pressure. When this is the case, a booster pump may be needed to increase water pressure. If water pressure fluctuates heavily, the water distribution system must be designed to operate on the minimum water pressure anticipated.

When calculating the water pressure needs of a system, you can use information provided by the code book. There are ratings for all common fixtures that show the minimum pressure requirements for each type of fixture. A water distribution system must be sized to operate satisfactorily under peak demands.

Booster pumps must be equipped with low-water cutoffs. These safety devices are required to prevent the possibility of a vacuum, which may cause back-siphonage.

Water Tanks

When booster pumps are not a desirable solution, water storage tanks are a possible alternative. Water storage tanks must be protected from contamination. They may not be located under soil or waste pipes. If the tank is a gravity type, it must be equipped with overflow provisions.

The water supply to a gravity-style water tank must be automatically controlled. This may be accomplished with a ball cock or other suitable and approved device. The incoming water should enter the tank by way of an air gap. The air gap should be at least 4 in above the overflow.

Water tanks must have provisions that allow them to be drained. The drainpipe must have a valve, to prevent draining, except when desired.

Pressurized water tanks are the type most commonly encountered in modern plumbing. These tanks are the type used with well systems. All pressurized water tanks should be equipped with a vacuum breaker. The vacuum breaker is installed on top of the tank, and it should have at least a ½-in diameter. These vacuum breakers should be rated for proper operation up to maximum temperatures of 200°F.

It is also necessary to equip these tanks with pressure relief valves. These safety valves must be installed on the supply pipe that feeds the tank or on the tank itself. The relief valve discharges when pressure builds to a point of endangering the integrity of the pressure tank. The valve's discharge must be carried, by gravity, to a safe and approved disposal location. The piping carrying the discharge of a relief valve may not be connected directly to the sanitary drainage system.

Water Conservation

Water conservation continues to grow as a major concern. When the flow rates for various fixtures are set, water conservation is a factor. The flow rates of many fixtures must be limited to no more than 3 gallons per minute (gpm). In zone 3, these fixtures include the following:

| Showers | Lavatories |
| Kitchen sinks | Other sinks |

The rating of 3 gpm is based on a water pressure of 80 psi.

When installed in public facilities, lavatories must be equipped with faucets producing no more than ½ gpm. If the lavatory is equipped with a self-closing faucet, it may produce up to ¼ gpm per use. Water closets are restricted to a use of no more than 4 gal of water, and urinals must not exceed a usage of 1½ gal of water, with each use.

Antiscald Precautions

It is easy for the very young or the elderly to suffer serious burns from plumbing fixtures. In an attempt to reduce accidental burns, mixed water to gang showers must be controlled by thermostatic means or by pressure-balanced valves. All showers, except for showers in residential dwellings in zones 1 and 2, must be equipped with pressure-balanced valves or thermostatic controls. These temperature control valves may not allow water with a temperature of more than 120°F to enter the bathing unit. In zone 3, the maximum water temperature is 110°F. In zone 3 these safety valves are required on all showers.

Valve Regulations

Gate valves and ball valves are examples of full-open valves, as required under valve regulations. These valves do not depend on rubber washers, and when they are opened to their maximum capacity, there is full flow through the pipe. Many locations along the water distribution installation require the installation of full-open valves. In zone 1, these types of valves are required in the following locations:

- On the water service, before and after the water meter
- On each water service for each building served
- On discharge pipes of water supply tanks, near the tank
- On the supply pipe to every water heater, near the heater
- On the main supply pipe to each dwelling

In zone 3, the mandatory locations of full-open valves are as follows:

- On the water service pipe, near the source connections
- On the main water distribution pipe, near the water service
- On water supplies to water heaters
- On water supplies to pressurized tanks, such as like well-system tanks
- On the building side of every water meter

In zone 2, full-open valves must be used in all water distribution locations, except as cutoffs for individual fixtures, in the immediate area of the fixtures. Other local regulations may apply to specific building uses; check with the local code officer to confirm where full-open valves are required in your system. All valves must be installed so that they are accessible.

Cutoffs

Cutoff valves do not have to be full-open valves. Stop-and-waste valves are an example of cutoff valves that are not full-open valves. Every sill cock must be equipped with an individual cutoff valve. Appliances and mechanical equipment that have water supplies are required to have cutoff valves installed in the service piping. Generally, with only a few exceptions, cutoffs are required on all plumbing fixtures. Check with the local code officer for fixtures not requiring cutoff valves. All valves installed must be accessible.

Backflow Prevention

Backflow and back-siphonage are genuine health concerns. When a backflow occurs, it can pollute entire water systems. Without backflow and back-siphonage protection, municipal water services could become contaminated. There are many ways to degrade the quality of potable water.

All potable water systems must be protected against back-siphonage and backflow with approved devices. Numerous types are available to provide this type of protection. The selection of devices is governed by the local plumbing inspector. It is necessary to choose the proper device for the use intended.

An air gap is the most positive form of protection from backflow. However, air gaps are not always feasible. Since air gaps cannot always be used, there are a number of devices available for the protection of potable water systems.

Some backflow preventers are equipped with vents. When these devices are used, the vents must not be installed so that they may become submerged. Also these units must be capable of performing their function under continuous pressure.

Some backflow preventers are designed to operate in a manner similar to an air gap. With these devices, when conditions arise that may cause a backflow, the devices open and create an open airspace between the two pipes connected to it. Reduced-pressure backflow preventers perform this action very well. Another type of backflow preventer is an atmospheric vent backflow preventer.

Vacuum breakers are frequently installed on water heaters, hose bibs, and sill cocks. They are also generally installed on the faucet spout of laundry tubs. These devices either mount on a pipe or screw onto a hose connection. Some sill cocks are equipped with factory-installed vacuum breakers. These devices open, when necessary, and break any siphonic action with the introduction of air.

In some specialized cases, a barometric loop is used to prevent back-siphonage. In zone 3, these loops must extend at least 35 ft high and can only be used as a vacuum breaker. These loops are effective because they rise higher than the point at which a vacuum suction can occur. Barometric loops work on the principle that since they are 35 ft high, suction is not achieved.

Double-check valves are used in some instances to control backflow. When used in this capacity, doublecheck valves must be equipped with approved vents. This type of protection would be used on a carbonated beverage dispenser, for example.

Backflow preventers must be inspected from time to time and so must be installed in accessible locations.

Some fixtures require an air gap as protection from backflow, such as lavatories, sinks, laundry tubs, bathtubs, and drinking fountains. This air gap is achieved through the design and installation of the faucet or spout serving these fixtures.

When vacuum breakers are installed, they must be installed at least 6 in above the flood-level rim of the fixture. Vacuum breakers, because of the way they are designed to introduce air into the potable water piping, may not be installed where they may suck in toxic vapors or fumes. For example, it is not acceptable to install a vacuum breaker under the exhaust hood of a kitchen range.

When potable water is connected to a boiler, for heating purposes, the potable water inlet should be equipped with a vented backflow preventer. If the boiler water contains chemicals, the potable water connection should be made with an air gap or a reduced-pressure backflow preventer.

Connections between a potable water supply and an automatic fire sprinkling system should be made with a check valve. If the potable water supply is being connected to a nonpotable water source, the connection should be made through a reduced-pressure backflow preventer.

Lawn sprinklers and irrigation systems must be installed with backflow prevention in mind. Vacuum breakers are a preferred method for backflow prevention, but other types of backflow preventers are allowed.

Hot Water Installations

When hot water pipe is installed, it is often expected to maintain the temperature of the hot water for a distance of up to 100 ft from the fixture it serves. If the distance between the hot water source and the fixture being served is more than 100 ft, a recirculating system is frequently required. When a recirculating system is not appropriate, other means may be used to maintain water temperature. These means could include insulation or heating tapes. Check with the local code officer for approved alternatives to a recirculating system, if necessary.

If a circulator pump is used on a recirculating line, the pump must be equipped with a cutoff switch. The switch may operate manually or automatically.

Water Heaters

The standard working pressure for a water heater is 125 psi. The maximum working pressure of a water heater is required to be permanently marked in an accessible location. Every water heater must have a drain, located at the lowest possible point on the water heater. Some exceptions may be allowed for very small, under-the-counter water heaters.

All water heaters must be insulated. The insulation factors are determined by the heat loss of the tank in 1 hour's time. These regulations are required of a water heater before it is approved for installation.

Relief valves are mandatory equipment on water heaters. These safety valves are designed to protect against excessive temperature and pressure. The most common type of safety valve used will protect against both temperature and pressure, from a single valve. The blow-off rating for these valves must not exceed 210°F and 150 psi. The pressure relief valve must not have a blow-off rating of more than the maximum working pressure of the water heater it serves; this is usually 125 psi.

When temperature and pressure relief valves are installed, their sensors should monitor the top 6 in of water in the water heater. No valves may be located between the water heater and the temperature and pressure relief valves.

The blow-off from relief valves must be piped down, to protect bystanders, in the event of a blow-off. The pipe used for this purpose must be rigid and capable of sustaining temperatures of up to 210°F. The discharge pipe must be the same size as the relief valve's discharge opening, and it must run, undiminished in size, to within 6 in of the floor. If a relief valve discharge pipe is piped into the sanitary drainage system, the connection must be through an indirect waste. The end of a discharge pipe may not be threaded, and no valves may be installed in the discharge pipe.

When the discharge from a relief valve could damage property or people, safety pans should be installed. These pans typically have a minimum depth of $1\frac{1}{2}$ in. Plastic pans are commonly used for electric water heaters, and metal pans are used for fuel-burning heaters. These pans must be large enough to accommodate the discharge flow from the relief valve.

The pan's drain may be piped to the outside of the building or to an indirect waste, where people and property will not be affected. The discharge location should be chosen so that it will be obvious to building occupants when a relief valve discharges. Traps should not be installed on the discharge piping from safety pans.

Water heaters must be equipped with an adjustable temperature control. This control must be automatically adjustable from the lowest to the highest temperatures allowed. Some locations restrict the maximum water temperature in residences to 120°F. There must be a switch to shut off the power to electric water heaters. When the water heater uses a fuel, such as gas, there must be a valve to cut off the fuel source. Both the electric and the fuel shutoffs must be able to be used without affecting the remainder of the building's power or fuel. All water heaters requiring venting must be vented in compliance with local code requirements.

Common Sense

Common sense plays a big role in the successful design of a plumbing system. If you know the local plumbing code, you should be able to design residential systems with ease. To create designs that are cost-effective, use common sense.

Lay out your jobs on paper (Fig. 12.8), and then look at what you've done. Could you combine some vents that are shown as individual vents? Would it be better to run all of a bathroom's water pipe in the partition walls, or would it be easier and cheaper to bring the pipes up under each fixture? Just study your initial design and look for ways to improve it. I'll bet you'll find some.

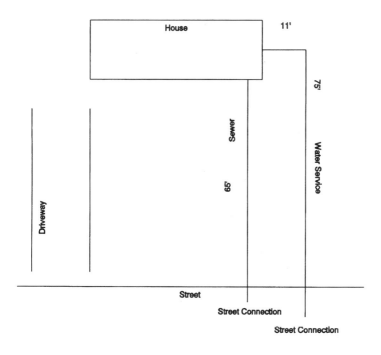

Figure 12.8 Site plan with city water and sewer.

13

Sensible Pipe Sizing

Sensible pipe sizing turns off many plumbers. Few plumbers enjoy sizing pipe systems. Part of the explanation may be the complicated methods used by some people to size systems. Does sizing have to be complicated? Is it really necessary to use friction-loss charts to size a water distribution system? What do you think?

I don't see any reason for pipe sizing to be difficult. Charts and tables in code books can be troublesome to understand. But sizing doesn't have to be a dreadful chore. I admit, at times friction charts and complicated formulas must be used to figure the proper sizing of a system. However, this is almost never the case in a residence, and commercial jobs are usually designed by engineers and architects, not plumbers. Let me show you some of the simple techniques that I use when sizing systems.

Pipe Sizing

Sizing pipe for a drainage system is not difficult. To size pipe for drainage, there are a few benchmark numbers you must know, but you don't have to memorize them. Your code book contains charts and tables that provide the benchmarks. All you must know is how to interpret and use the information found there.

The size of a drainage pipe is determined by using various factors, the first of which is the *drainage load*. This refers to the volume of drainage that the pipe will be responsible for carrying. In the code book, you will find ratings that assign a fixture-unit value to various plumbing fixtures (Table 13.1). For example, a residential toilet has a fixture-unit value of 4. A bathtub's fixture-unit value is 2.

Using the ratings given in the code book, you can quickly assess the drainage load for the system you are designing. Since plumbing fix-

TABLE 13.1   Fixture-Unit Ratings in Zone 3

Bathtub	2
Shower	2
Residential toilet	4
Lavatory	1
Kitchen sink	2
Dishwasher	2
Clothes washer	3
Laundry tub	2

tures require traps, you must also determine what size traps are needed for particular fixtures. Again, you don't need a math degree to do this. In fact, the code book specifies the trap sizes required for most common plumbing fixtures.

Your code book lists trap size requirements for specific fixtures (Tables 13.2, 13.3, and 13.4). For example, referring to the ratings in the code book, you will find that a bathtub requires a $1\frac{1}{2}$-in trap. A lavatory may have a $1\frac{1}{4}$-in trap. The list describes the trap needs for all common plumbing fixtures. Trap sizes are not given for toilets, since toilets have integral traps.

When necessary, you can determine a fixture's drainage unit value by the size of its trap. A $1\frac{1}{4}$-in trap, the smallest trap allowed, has a fixture-unit rating of 1. A $1\frac{1}{2}$-in trap has a fixture-unit rating of 2. A 2-in trap has a rating of 3 fixture units. A 3-in trap has a fixture-unit rating of 5, and a 4-in trap has a fixture-unit rating of 6. This information can be found in your code book (Tables. 13.5, 13.6, and 13.7) and may be applied to a fixture not specifically given a rating in the

TABLE 13.2   Recommended Trap Sizes for Zone 1

Type of fixture	Trap size, in
Bathtub	$1\frac{1}{2}$
Shower	2
Residential toilet	Integral
Lavatory	$1\frac{1}{4}$
Bidet	$1\frac{1}{2}$
Laundry tub	$1\frac{1}{2}$
Washing machine standpipe	2
Floor drain	2
Kitchen sink	$1\frac{1}{2}$
Dishwasher	$1\frac{1}{2}$
Drinking fountain	$1\frac{1}{4}$
Public toilet	Integral

TABLE 13.3 Recommended Trap Sizes for Zone 2

Type of fixture	Trap size, in
Bathtub	$1\frac{1}{2}$
Shower	2
Residential toilet	Integral
Lavatory	$1\frac{1}{4}$
Bidet	$1\frac{1}{2}$
Laundry tub	$1\frac{1}{2}$
Washing machine standpipe	2
Floor drain	2
Kitchen sink	$1\frac{1}{2}$
Dishwasher	$1\frac{1}{2}$
Drinking fountain	1
Public toilet	Integral

TABLE 13.4 Recommended Trap Sizes for Zone 3

Type of fixture	Trap size, in
Bathtub	$1\frac{1}{2}$
Shower	2
Residential toilet	Integral
Lavatory	$1\frac{1}{4}$
Bidet	$1\frac{1}{4}$
Laundry tub	$1\frac{1}{2}$
Washing machine standpipe	2
Floor drain	2
Kitchen sink	$1\frac{1}{2}$
Dishwasher	$1\frac{1}{2}$
Drinking fountain	$1\frac{1}{4}$
Public toilet	Integral
Urinal	2

TABLE 13.5 Fixture-Unit Requirements
on Trap Sizes in Zone 1

Trap size, in	Fixture units
$1\frac{1}{4}$	1
$1\frac{1}{2}$	3
2	4
3	6
4	8

TABLE 13.6   Fixture-Unit Requirements
on Trap Sizes in Zone 2

Trap size, in	Fixture units
$1\frac{1}{4}$	1
$1\frac{1}{2}$	2
2	3
3	5
4	6

TABLE 13.7   Fixture-Unit Requirements
on Trap Sizes

Trap size, in	Fixture units
$1\frac{1}{4}$	1
$1\frac{1}{2}$	2
2	3
3	5
4	6

code book. Remember, the three major plumbing codes are similar, but they are not identical. Also, plumbing codes change from time to time. Consult your local code for current, correct information.

Determining the fixture-unit value of a pump does require a little math, but it's simple. Assign 2 fixture units for every gallon per minute (gpm) of flow. For example, a pump with a flow rate of 30 gpm has a fixture-unit rating of 60. In zone 3, it is more generous. In zone 3, for every $7\frac{1}{2}$ gpm, 1 fixture unit is assigned. With the same pump, producing 30 gpm, in zone 3 the fixture-unit rating is 4. That's quite different from the ratings in zones 1 and 2.

To size drainpipe, you must also consider the type of drain and the amount of fall of the pipe. For example, the sizing for a sewer is done a little differently from the sizing for a vertical stack. A pipe with a $\frac{1}{4}$-in fall is rated differently from the same pipe with a $\frac{1}{8}$-in fall.

Sizing Building Drains and Sewers

In building drains and sewers (Table 13.8), the same factors determine the proper pipe size. To size these types of pipes, you must know the total number of drainage fixture units entering the pipe and the amount of fall placed on the pipe. The amount of fall is the distance the pipe drops in each foot of travel. A normal grade is generally $\frac{1}{4}$ in/ft, but the fall could be more or less.

TABLE 13.8   Building Drain Size for Zone 3

Pipe size, in	Pipe grade, in/ft	Maximum number of fixture units
2	$\frac{1}{4}$	21
3	$\frac{1}{4}$	42*
4	$\frac{1}{4}$	216

Note: No more than two toilets may be installed on a 3-in building drain.

When you refer to the code book, you will find information, probably a table, to help you size building drains and sewers. Let's take a look at how a building drain for a typical house is sized in zone 3.

Sizing Example

Our sample house has 2½ bathrooms, a kitchen, and a laundry room. To size the building drain for this house, we must determine the total fixture-unit load that could be placed on the building drain. To do this, we start by listing all the plumbing fixtures producing a drainage load. In this house we have the following fixtures:

One bathtub	One shower
Three toilets	Three lavatories
One kitchen sink	One dishwasher
One clothes washer	One laundry tub

Using a chart in the code book, we can determine the number of drainage fixture units assigned to each of these fixtures. When we add all the fixture units, the total load is 28 fixture units. It is always best to allow a little extra in the fixture-unit load so that the pipe will be in no danger of becoming overloaded. The next step is to look at a different chart in the code book to determine the size of the building drain.

The building drain will be installed with a ¼-in fall. Looking at the chart, we see that we can use a 3-in pipe for the building drain, based on the number of fixture units; but there is a footnote below the chart. It indicates that a 3-in pipe may not carry the discharge of more than two toilets, and our test house has three toilets. So we have to move up to a 4-in pipe.

Suppose the test house had only two toilets. What then? If we eliminate one of the toilets, the fixture load drops to 24. According to the table, we could use a 2½-in pipe, but the building drain must be at least a 3-in pipe, to connect to the toilets. A fixture's drain may enter a pipe the same size as the fixture drain or a pipe that is larger, but it may never be reduced to a smaller size, except with a 4-in by 3-in closet bend.

So, with two toilets, our sample house could have a building drain and sewer with a 3-in diameter. But should we run a 3-in pipe or a 4-in pipe? In a highly competitive bidding situation, 3-in pipe would probably win the coin toss. It would be less expensive to install a 3-in drain, and we would be more likely to win the bid on the job. However, when feasible, it would be better to use a 4-in drain. This allows the homeowner to add another toilet sometime in the future. If we install a 3-in sewer, the homeowner cannot add a toilet without replacing the sewer with 4-in pipe.

Horizontal Branches

Horizontal branches are the pipes branching off from a stack, to accept the discharge from fixture drains. These normally leave the stack as horizontal pipe, but they may turn to a vertical position, while retaining the name *horizontal branch.*

The procedure for sizing a horizontal branch (Table 13.9) is similar to that used to size a building drain or sewer, but the ratings are different. The code book contains the benchmarks you need to size, but let me give you some examples.

The number of fixture units allowed on a horizontal branch is determined by the pipe size and the pitch. All the following examples are based on a pitch of $\frac{1}{4}$ in/ft. A 2-in pipe can accommodate up to 6 fixture units, except in zone 1, where it can have 8 fixture units. A 3-in pipe can handle 20 fixture units, but not more than two toilets. In zone 1, a 3-in pipe is allowed up to 35 fixture units and up to three toilets. A $1\frac{1}{2}$-in pipe carries 3 fixture units, unless you are in zone 1.

TABLE 13.9   Example of Sizing Horizontal Branch in Zone 2

Pipe size, in	Maximum number of fixture units on horizontal branch
$1\frac{1}{4}$	1
$1\frac{1}{2}$	3
2	6
3	20*
4	160
6	620

Note: Not more than two toilets may be connected to a single 3-in horizontal branch. Any branch connecting with a toilet must have a minimum diameter of 3 in.

Note: Table does not represent branches of the building drain, and other restrictions apply under battery-venting conditions.

In zone 1, only $1\frac{1}{2}$-in pipe carries 2 fixture units, and they may not be from sinks, dishwashers, or urinals. A 4-in pipe takes up to 160 fixture units, except in zone 1, where it takes up to 216 units.

Stack Sizing

Stack sizing is not too different from the other sizing exercises we have studied. When you size a stack (Tables 13.10 to 13.13), you base your decision on the total number of fixture units carried by the stack

TABLE 13.10   Stack Sizing for Zone 2

Pipe size, in	Fixture-unit discharge on stack from a branch	Total fixture units allowed on stack
$1\frac{1}{2}$	3	4
2	6	10
3	20*	30*
4	160	240

Note: No more than two toilets may be placed on a 3-in branch, and no more than six toilets may be connected to a 3-in stack.

TABLE 13.11   Sizing of Tall Stacks in Zone 2

(Stacks with More than Three Branch Intervals)

Pipe size, in	Fixture-unit discharge on stack from a branch	Total fixture units allowed on stack
$1\frac{1}{2}$	2	8
2	6	24
3	16*	60*
4	90	500

Note: No more than two toilets may be placed on a 3-in branch, and no more than six toilets may be connected to a 3-in stack.

TABLE 13.12   Stack Sizing for Zone 3

Pipe size, in	Fixture-unit discharge on stack from a branch	Total fixture units allowed on stack
$1\frac{1}{2}$	2	4
2	6	10
3	20*	48*
4	90	240

Note: No more than two toilets may be placed on a 3-in branch, and no more than six toilets may be connected to a 3-in stack.

TABLE 13.13   Sizing of Tall Stacks in Zone 3

Pipe size, in	Fixture-unit discharge on stack from a branch	Total fixture units allowed on stack
(Stacks with More than Three Branch Intervals)		
$1\frac{1}{2}$	2	8
2	6	24
3	20*	72*
4	90	500

Note: No more than two toilets may be placed on a 3-in branch, and no more than six toilets may be connected to a 3-in stack.

and the amount of discharge into branch intervals. This may sound complicated, but it isn't.

When you look at a chart for stack sizing, note that there are three columns. The first is for the pipe size, the second represents the discharge of a branch interval, and the third shows the ratings for the total fixture-unit load on a stack. The table is based on a stack with no more than three branch intervals.

To size the stack, first you must determine the fixture load entering the stack at each branch interval. Let's look at an example. We will size a stack that has two branch intervals. The lower branch has a half-bath and a kitchen on it. Using the ratings from zones 2 and 3, we see that the total fixture-unit count for this branch is 6. This is determined by the table listing fixture-unit ratings for various fixtures.

The second stack has a full bathroom group on it. The total fixture-unit count on this branch, using sizing from zones 2 and 3, is 6, if we use a bathroom group rating, or 7, if we count each fixture individually. I would use the larger of the two numbers, for a total of 7.

Next, we look at the sizing table that contains horizontal listings for a 3-in pipe. We know the stack must have a minimum size of 3 in to accommodate the toilets. As we look across the table, we see that each 3-in branch may carry up to 20 fixture units. Well, the first branch has 6 fixture units, and the second branch has 7 fixture units, so both branches are within their limits.

The total from both branches is 13 fixture units. Continuing to look across the table, we see that the stack can accommodate up to 48 fixture units in zone 3 and up to 30 fixture units in zone 2. Obviously, a 3-in stack is adequate. If the fixture-unit loads had exceeded the numbers in either column, the pipe size would have to be increased.

In sizing a stack, the developed length of the stack may comprise different sizes of pipe. For example, at the top of the stack, the pipe size may be 3 in, but at the bottom of the stack the pipe size may be 4 in. The reason is that as you get to the lower portion of the stack, the total number of fixture units placed on the stack is greater.

Sizing Storm Water Drainage Systems

When you wish to size a storm water drainage system (Table 13.14), you must have some benchmark information to work with. One consideration is the amount of pitch of a horizontal pipe. Another factor is the number of square feet of surface area that the system will be required to drain. You will also need data on the rainfall rates in the area.

When you use the code book to size a storm water drain system, you should have access to all the key elements required to size the job, except possibly the local rainfall amounts. You can obtain rainfall data from the state or county offices. Your code book contains a table to use in the sizing calculations.

Sizing a horizontal storm drain or sewer

The first step in sizing a storm drain or sewer is to list the known data. How much pitch will the pipe have? In this example, my pipe has $\frac{1}{4}$ in/ft pitch.

What else do I know? Well, I know the system is going to be located in Portland, Maine. Portland's rainfall is rated at 2.4 inches per hour (in/h). This rating assumes a 1-h storm that is only likely to occur once every 100 years. Now I have two factors to size the system.

I also know that the surface area to be drained is 15,000 ft²; this includes the roof and parking area. I've got three of the elements needed to get this job done. But how do I use the numbers in a sizing chart? Well, there are a couple of ways to ease the burden. When you are working with a standard table, like the ones found in most code books, you must adapt the information to suit local conditions. For example, if a standardized table is based on 1 in/h of rainfall and my

TABLE 13.14 Example of a Horizontal Storm Water Sizing Table

Pipe grade, in/ft	Pipe size, in	Gallons per minute (gpm)	Surface area, ft²
$\frac{1}{4}$	3	48	4,640
$\frac{1}{4}$	4	110	10,600
$\frac{1}{4}$	6	314	18,880
$\frac{1}{4}$	8	677	65,200

Note: These figures are based on a rainfall with a maximum rate of 1 in/h of rain, for 1 full hour, and occurring once every 100 years.

location has 2.4 in/h of rainfall, I must adjust the table. But this is not difficult; trust me.

When I want to adjust a table based on a 1-in rainfall to meet my local needs, I divide the drainage area in the table by the rainfall amount. For example, if the standard chart shows that an area of 10,000 ft^2 requires a 4-in pipe, I can change the value in the table by dividing the rainfall amount, 2.4, into the surface area, 10,000 ft^2.

Dividing 10,000 by 2.4, I get 4167. All of a sudden, I have solved the mystery of computing storm water piping needs. With this simple conversion, I know that if my surface area is 4167 ft^2, I need a 4-in pipe. But my surface area is 15,000 ft^2, so what size pipe do I need? Well, I know it has to be larger than 4 in. So I look at the conversion chart and find the appropriate surface area. A 15,000-ft^2 surface area requires a storm water drain with a diameter of 8 in.

Now, let's recap this exercise. To size a horizontal storm drain or sewer, decide what pitch will be put on the pipe. Next, determine the area's rainfall for a 1-h storm, occurring each 100 years. If you live in a city, it may be listed, with its rainfall amount, in your code book. Using a standardized chart, rated for 1 in/h of rainfall, divide the surface area by a factor equal to your rainfall index—in my case, it was 2.4. This division process converts a generic table value to a customized table value, just for your area.

Once the math is done, scan the table for the surface area that most closely matches the area you have to drain. To be safe, use a number slightly higher than your projected number. It is better to have a pipe sized one size too large than one size too small. When you have found the appropriate surface area, look across the table to find the pipe size you need. See how easy that was. Well, maybe it's not easy, but it is a chore you can handle.

Sizing rain leaders and gutters

When you need to size rain leaders or downspouts, use the same procedure described above, with one exception. Use a table in the code book to size the *vertical* piping. Determine the amount of surface area your leader will drain, and use the appropriate table to find the pipe size. The conversion factors are the same.

Sizing gutters is essentially the same as sizing horizontal storm drains. You will use a different table in the code book, but the mechanics are the same.

Roof drains

Roof drains are often the starting point of a storm water drainage system. As the name implies, roof drains are located on roofs. On most roofs, the drains are equipped with strainers that protrude upward, at least 4 in, to catch leaves and other debris. Roof drains should be at least twice the size of the piping connected to them. All roofs that do not drain to hanging gutters are required to have roof drains. A minimum of two roof drains should be installed on roofs with a surface area of 10,000 ft^2 or less. If the surface area exceeds 10,000 ft^2, a minimum of four roof drains should be installed.

When a roof is used for purposes other than just shelter, the roof drains may have a strainer that is flush with the roof's surface. Roof drains obviously should be sealed to prevent water from leaking around them. The size of the roof drain is key in the flow rates designed for a storm water system. When a controlled flow from roof drains is wanted, the roof structure must be designed to accommodate it.

More sizing information

If a combined storm drain and sewer arrangement is approved, it must be sized properly. This requires converting fixture-unit loads to drainage surface area. For example, 256 fixture units are treated as 1000 ft^2 of surface area. Each additional fixture unit in excess of 256 is assigned a value of $3\frac{9}{10}$ ft^2. In sizing for continuous flow, each gallon per minute (gpm) is rated as 96 ft^2 of drainage area.

Sizing Vents

Sizing an individual vent is easy. The vent must be at least one-half the size of the drain it serves, but it may not have a diameter of less than $1\frac{1}{4}$ in. For example, a vent for a 3-in drain could, in most cases, have a diameter of $1\frac{1}{2}$ in. A vent for a $1\frac{1}{2}$-in drain may not have a diameter of less than $1\frac{1}{4}$ in (Tables 13.15 and 13.16).

Relief vents

Relief vents are used in conjunction with other vents. Their purpose is to provide additional air to the drainage system, when the primary vent is too far from the fixture. The relief vent must be at least one-half the size of the pipe it is venting. For example, if a relief vent is venting a 3-in pipe, the relief vent must have a $1\frac{1}{2}$-in, or larger, diameter.

Circuit vents

Circuit vents are used with a battery of plumbing fixtures. Circuit vents are normally installed just before the last fixture of the battery. Then, the circuit vent is extended upward to open air or is tied into

TABLE 13.15   Vent Sizing for Zone 3

(For Use with Individual, Branch, and Circuit Vents for Horizontal Drainpipes)

Drainpipe size, in	Drainpipe grade, in/ft	Vent pipe size, in	Maximum developed length of vent pipe, ft
$1\frac{1}{2}$	$\frac{1}{4}$	$1\frac{1}{4}$	Unlimited
$1\frac{1}{2}$	$\frac{1}{4}$	$1\frac{1}{2}$	Unlimited
2	$\frac{1}{4}$	$1\frac{1}{4}$	290
2	$\frac{1}{4}$	$1\frac{1}{2}$	Unlimited
3	$\frac{1}{4}$	$1\frac{1}{2}$	97
3	$\frac{1}{4}$	2	420
3	$\frac{1}{4}$	3	Unlimited
4	$\frac{1}{4}$	2	98
4	$\frac{1}{4}$	3	Unlimited
4	$\frac{1}{4}$	4	Unlimited

TABLE 13.16   Vent Sizing for Zone 3

(For Use with Vent Stacks and Stack Vents)

Drainpipe size, in	Fixture-unit load on drainpipe	Vent pipe size, in	Maximum developed length of vent pipe, ft
$1\frac{1}{2}$	8	$1\frac{1}{4}$	50
$1\frac{1}{2}$	8	$1\frac{1}{2}$	150
$1\frac{1}{2}$	10	$1\frac{1}{4}$	. 30
$1\frac{1}{2}$	10	$1\frac{1}{2}$	100
2	12	$1\frac{1}{2}$	75
2	12	2	200
2	20	$1\frac{1}{2}$	50
2	20	2	150
3	10	$1\frac{1}{2}$	42
3	10	2	150
3	10	3	1040
3	21	$1\frac{1}{2}$	32
3	21	2	110
3	21	3	810
3	102	$1\frac{1}{2}$	25
3	102	2	86
3	102	3	620
4	43	2	35
4	43	3	250
4	43	4	980
4	540	2	21
4	540	3	150
4	540	4	580

another vent that extends to the outside. Circuit vents may tie into stack vents or vent stacks. When sizing a circuit vent, you must account for its developed length. But in any event, the diameter of a circuit vent must be at least one-half the size of the drain it is serving.

Vent Sizing Using Developed Length

What effect does the length of the vent have on the vent size? The *developed length*—the total linear footage of pipe making up the vent—is used in conjunction with factors provided in code books to determine vent sizes. To size circuit vents, branch vents, and individual vents for horizontal drains, you must use this method of sizing.

The factors used in sizing a vent, based on developed length, are the grade of the drainage pipe, the size of the drainage pipe, the developed length of the vent, and those allowed by local code requirements. Let's try a few examples of sizing a vent by using this method.

First, assume the drain we are venting is a 3-in pipe with a $\frac{1}{4}$ in/ft grade. This sizing exercise is for zone 3 requirements. If you have a code book for zone 3, inspect the sizing charts in it. Note the number listed under the $1\frac{1}{2}$-in vent column. It is 97. This means that a 3-in drain running horizontally with a $\frac{1}{4}$ in/ft grade can be vented with a $1\frac{1}{2}$-in vent that has a developed length of 97 ft. It would be rare to extend a vent anywhere near 97 ft, but if your vent needed to exceed this distance, you could use a larger vent. A 2-in vent allows you to extend the vent for a total length of 420 ft. A vent larger than 2 in allows you to extend the vent indefinitely.

Second, still in zone 3, assume the drain is a 4-in pipe with a $\frac{1}{4}$ in/ft grade. In this case you could not use a $1\frac{1}{2}$-in vent. Remember, the vent must be at least one-half the size of the drain it is venting. A 2-in vent allows a developed vent length of 98 ft, and a 3-in vent allows the vent to extend to an unlimited length. As you can see, this type of sizing is not difficult.

Now, let's size a vent according to zone 1 rules. In zone 1, vent sizing is based on the vent's length and the number of fixture units on the vent. To size a vent for a lavatory, you need to know how many fixture units the lavatory represents. Lavatories are rated as 1 fixture unit. Using a table in the code book, we find that a vent serving 1 fixture unit can have a diameter of $1\frac{1}{4}$ in and may extend for 45 ft. A bathtub, rated at 2 fixture units, requires a $1\frac{1}{2}$-in vent. The bathtub vent could run for 60 ft.

Branch vents

Branch vents are vents extending horizontally and connecting multiple vents together. They are sized according to the developed-length method, just as in the examples above. A branch vent or individual vent that is the same size as the drain it serves has an unlimited developed length. Be advised, in zones 2 and 3, different tables and ratings are used to size various types of vents; in zone 1, the same rating and table are used for all normal venting situations.

Vent stacks

A *vent stack* is a pipe used only for the purpose of venting. Vent stacks extend upward from the drainage piping to open air, outside a building. Vent stacks are used as connection points for other vents, such as branch vents. A vent stack is a primary vent that accepts the connection of other vents and vents an entire system. Vent stacks run vertically and are sized a little differently.

The basic procedure for sizing a vent stack is similar to that used with branch vents, but there are some differences. You must know the size of the soil stack, the number of fixture units carried by the soil stack, and the developed length of the vent stack. With this information and the regulations of the local plumbing code, you can size your vent stack. Let's work on an example.

Assume the system has a soil stack with a diameter of 4 in. This stack is loaded with 43 fixture units. The vent stack will have a developed length of 50 ft. What size pipe do we use for the vent stack? Looking at the table, we see that a 2-in pipe, used as a vent for the described soil stack, allows a developed length of 35 ft. The vent will have a developed length of 50 ft, so we can rule out 2-in pipe. In the column for $2\frac{1}{2}$-in pipe, the rating is up to 85 ft. Since the vent is going only 50 ft, we could use a $2\frac{1}{2}$-in pipe. However, since $2\frac{1}{2}$-in pipe is not common, we would probably use a 3-in pipe. This same sizing method is used to compute the size of stack vents.

Stack vents

Stack vents are really two pipes in one. The lower portion of the pipe is a soil pipe, and the upper portion is a vent. This is the type of primary vent most often found in residential plumbing. Stack vents are sized by the same methods as those used for vent stacks.

Common vents

Common vents are single vents that vent multiple traps. They are allowed only when the fixtures being served by the single vent are on the same floor. In zone 1, the drains of fixtures being vented with a common vent enter the drainage system at the same level. Normally, not more than two traps can share a common vent, but there is an exception in zone 3. In zone 3, traps of up to three lavatories can be vented with a single common vent. Common vents are sized by the same techniques as those applied to individual vents.

Wet vents

Wet vents (Tables 13.17 and 13.18) are pipes that serve as a vent for one fixture and a drain for another. Wet vents, once you know how to use them, can save you a lot of money and time. By effectively using wet vents, you can reduce the amount of pipe, fittings, and labor required to vent a bathroom group, or two.

The sizing of wet vents is based on fixture units. The size of the pipe is determined by the number of fixture units it may be required to carry. A 3-in wet vent can handle 12 fixture units. A 2-in wet vent is rated for 4 fixture units, and a $1\frac{1}{2}$-in wet vent is allowed only 1 fixture unit. It is acceptable to wet vent two bathroom groups, six fixtures, with a single vent, but the bathroom groups must be on the same floor. In zone 2, provisions are made for wet-venting bathrooms on different floors. In zone 1, the approach to wet vents is different.

In zone 2 there are additional regulations that pertain to wet vents: The horizontal branch connecting to the drainage stack must enter at a level equal to, or below, the water closet drain. However, the branch may connect to the drainage at the closet bend. When there are two bathroom groups, the wet vent must have a minimum diameter of 2 in.

Kitchen sinks and washing machines may not be drained into a 2-

TABLE 13.17   **Sizing a Vent Stack for Wet Vents in Zone 2**

Wet-vented fixtures	Vent stack size requirements, in
1–2 Bathtubs or showers	2
3–5 Bathtubs or showers	$2\frac{1}{2}$
6–9 Bathtubs or showers	3
10–16 Bathtubs or showers	4

TABLE 13.18   **Sizing a Wet Stack Vent in Zone 2**

Pipe size of stack, in	Fixture-unit load on stack	Maximum length of stack, ft
2	4	30
3	24	50
4	50	100
6	100	300

in combination waste-and-vent. Water closets and urinals are restricted on vertical combination waste-and-vent systems.

As for the permissibility allowance of wet venting on different levels in zone 2, here are the facts. Wet vents must have at least a 2-in diameter. Water closets that are not located on the highest floor must be back-vented. If, however, the wet vent is connected directly to the closet bend with a 45° bend, the toilet being connected is not required to be back-vented, even if it is on a lower floor.

In zone 1 wet vents are limited to vertical piping. These vertical pipes are restricted to receiving only the waste from fixtures with fixture-unit ratings of 2 or less and that vent no more than four fixtures. Wet vents must be one pipe size larger than normally required, but they must never be smaller than 2 in in diameter.

Crown vents

A *crown vent* is a vent that extends upward from a trap or trap arm. Crown vent traps are not allowed. Crown vents are normally used on trap arms, but even then, they are not common. The vent must be on the trap arm, and it must be behind the trap by a distance equal to twice the pipe size. For example, on a $1\frac{1}{2}$ in trap, the crown vent has to be 3 in behind the trap, on the trap arm.

Vents for Sumps and Sewer Pumps

When sumps and sewer pumps are used to store and remove sanitary waste, the sump must be vented. In zones 1 and 2, these vents are treated about the same as vents installed on gravity systems.

If you will be installing a pneumatic sewer ejector, you need to run the sump vent to outside air, without tying it into the vents for the standard sanitary plumbing system. This ruling on pneumatic pumps applies to all three zones. If the sump will be equipped with a regular sewer pump, you may tie the vent from the sump back into the main venting system for the other sanitary plumbing.

In zone 3 there are some additional rules. The following is an outline of the requirements in zone 3: Sump vents may not be smaller than $1\frac{1}{4}$-in pipe. The size requirements for sump vents are determined by the discharge of the pump. For example, a sewer pump capable of producing 20 gpm could have its sump vented for an unlimited distance with $1\frac{1}{2}$-in pipe. If the pump is capable of producing 60 gpm, a $1\frac{1}{2}$-in pipe could not have a developed length of more than 75 ft.

In most cases, a 2-in vent is used on sumps, and the distance allowed for developed length is not a problem. However, if the pump will pump more than 100 gpm, you had better take the time to do some math. The code book has the factors you need to size the vent, and the sizing is easy. Simply look for the maximum discharge capacity of the pump, and match it with a vent that allows the developed length you need.

Another consideration in sizing vents is the distance between the vent and the trap it is serving. The grade of a drainpipe can also influence the size of a system. Again, information in your local code book (Tables 13.19 to 13.21) will help you determine the proper pipe size.

TABLE 13.19   Trap-to-Vent Distances in Zone 1

Grade on drainpipe, in	Size of trap arm, in	Maximum distance between trap and vent
$\frac{1}{4}$	$1\frac{1}{4}$	2 ft 6 in
$\frac{1}{4}$	$1\frac{1}{2}$	3 ft 6 in
$\frac{1}{4}$	2	5 ft
$\frac{1}{4}$	3	6 ft
$\frac{1}{4}$	4 and larger	10 ft

TABLE 13.20   Trap-to-Vent Distances in Zone 2

Grade on drainpipe, in	Fixture's drain size, in	Trap size, in	Maximum distance between trap and vent
$\frac{1}{4}$	$1\frac{1}{4}$	$1\frac{1}{4}$	3 ft 6 in
$\frac{1}{4}$	$1\frac{1}{2}$	$1\frac{1}{4}$	5 ft
$\frac{1}{4}$	$1\frac{1}{2}$	$1\frac{1}{2}$	5 ft
$\frac{1}{4}$	2	$1\frac{1}{2}$	8 ft
$\frac{1}{4}$	2	2	6 ft
$\frac{1}{8}$	3	3	10 ft
$\frac{1}{8}$	4	4	12 ft

TABLE 13.21   Trap-to-Vent Distances in Zone 3

Grade on drainpipe, in	Fixture's drain size, in	Trap size, in	Maximum distance between trap and vent
$\frac{1}{4}$	$1\frac{1}{4}$	$1\frac{1}{4}$	3 ft 6 in
$\frac{1}{4}$	$1\frac{1}{2}$	$1\frac{1}{4}$	5 ft
$\frac{1}{4}$	$1\frac{1}{2}$	$1\frac{1}{2}$	5 ft
$\frac{1}{4}$	2	$1\frac{1}{2}$	8 ft
$\frac{1}{4}$	2	2	6 ft
$\frac{1}{8}$	3	3	10 ft
$\frac{1}{8}$	4	4	12 ft

Sizing Potable Water Systems

In this section we show how to size potable water piping. But, be advised, this procedure is not always simple, and it requires concentration. Some parts of potable water pipe sizing are not very hard. Many times, the code book contains charts and tables (Table 13.22) to help you. Some of these graphics will detail precisely what size of pipe or tubing is required. But, unfortunately, code books cannot provide concrete answers for all piping installations.

Many factors affect the sizing of potable water piping. The type of pipe used affects your findings. Some pipe materials have smaller inside diameters than others, and other pipe materials have a rougher surface or more restrictive fittings than others. Both factors affect the sizing of a water system.

When sizing a potable water system, you must be concerned with the speed of the flowing water, the quantity of water (Tables 13.23

TABLE 13.22   Recommended Minimum Sizes for Fixture Supply Pipes

Fixture	Minimum pipe size, in
Bathtub	$\frac{1}{2}$
Bidet	$\frac{3}{8}$
Dishwasher	$\frac{1}{2}$
Hose bib	$\frac{1}{2}$
Kitchen sink	$\frac{1}{2}$
Laundry tub	$\frac{1}{2}$
Lavatory	$\frac{3}{8}$
Shower	$\frac{1}{2}$
Water closet, two-piece	$\frac{3}{8}$
Water closet, one-piece	$\frac{1}{2}$

TABLE 13.23   Recommended Capacities at Fixture Supply Outlets

Fixture	Flow rate, gpm	Flow pressure, lb/in^2
Bathtub	4	8
Bidet	2	4
Dishwasher	2.75	8
Hose bib	5	8
Kitchen sink	2.5	8
Laundry tub	4	8
Lavatory	2	8
Shower	3	8
Water closet, two-piece	3	8
Water closet, one-piece	6	20

TABLE 13.24 Sample Pressure and Pipe Chart for Sizing

Size of water meter and street service, in	Size of water service and distribution pipes, in	Maximum length of water pipe, ft					
		40	60	80	100	150	200
$3/4$	$1/2$	9	8	7	6	5	4
$3/4$	$3/4$	27	23	19	17	14	11
$3/4$	1	44	40	36	33	28	23
1	1	60	47	41	36	30	25
1	$1/4$	102	87	76	67	52	44

and 13.24) needed, and the restrictive qualities of the pipe being used to convey the water. Most materials approved for potable water piping will provide a flow velocity of 5 feet per second (ft/s). The exception is galvanized-steel pipe. Galvanized-steel pipe provides a flow speed of 8 ft/s.

It may surprise you that galvanized-steel pipe has a faster flow rate. This is the result of the wall strength of galvanized-steel pipe. In softer pipes, such as copper, fast-moving water can essentially wear a hole in the pipe. These flow ratings are not etched in stone. Undoubtedly some people will argue for either a higher or a lower rating, but these ratings are used in current plumbing codes.

To use the three factors previously discussed to determine the pipe size, you must use math that you may not have seen since schooldays, and you may not have seen it then. Let me give you an example of how a typical formula might look. A common formula might look like this:

$$Y = XZ$$

where X = flow rate of water, 5 ft/s usually
Y = amount of water in pipe
Z = inside diameter of pipe

Since many plumbers refuse to do this type of math, most code books offer alternatives. The alternatives are often tables or charts that show pertinent information on the requirements for pipe sizing.

The tables or charts in a code book are likely to discuss the following: outside diameter of the pipe, inside diameter of the pipe, flow rate for the pipe, and pressure loss in the pipe over a distance of 100 ft. These charts or tables are specific to a particular type of pipe. For example, there is one table for copper pipe and another table for PB pipe.

The information supplied in a ratings table for PB pipe might look like this:

- Pipe size is $\frac{3}{4}$ in.
- Inside pipe diameter is 0.715.
- The flow rate, at 5 ft/s, is 6.26 gpm.
- Pressure lost in 100 ft of pipe is 14.98.

This type of pipe sizing is most often done by engineers, not plumbers. When you size a potable water system, start at the last fixture and work your way back to the water service.

Commercial jobs, where pipe sizing can get quite complicated, are generally sized by design experts. All that a working plumber is required to do is to install the proper pipe sizes in the proper locations and manner. For residential plumbing, where engineers are less likely to have a hand in the design, there is a rule-of-thumb method to sizing most jobs. In the average home, a $\frac{3}{4}$-in pipe is sufficient for the main artery of the water distribution system. Normally, not more than two fixtures can be served by a $\frac{1}{2}$-in pipe. With this in mind, sizing becomes simple.

Usually $\frac{3}{4}$-in pipe is run to the water heater, and it is typically used as a main water distribution pipe. When it nears the end of a run, the $\frac{3}{4}$-in pipe is reduced to $\frac{1}{2}$-in pipe, once there are no more than two fixtures to connect to. Most water service pipes will have a $\frac{3}{4}$-in diameter, with those serving homes with numerous fixtures having a 1-in diameter. This rule-of-thumb sizing will work on almost any single-family residence.

The water supplies to fixtures must meet minimum standards. These sizes are derived from local code requirements. Simply find the fixture you are sizing the supply for, and check the column heading for the proper size.

Most code requirements seem to agree that there is no definitive way to set a boilerplate formula for establishing potable water pipe sizing. Code officers can require pipe sizing to be performed by a licensed engineer. In most major plumbing systems, the pipe sizing is done by a design professional.

Code books give examples of how a system might be sized. But the examples are not meant as a code requirement. The code requires a water system to be sized properly. However, due to the complexity of the process, the books do not set firm statistics for the process. Instead, code books give parts of the puzzle, in the form of some minimum standards, but it is up to a professional designer to come up with an approved system.

Where does this leave you? Well, the sizing of a potable water system is one of the most complicated aspects of plumbing. Very few single-family homes are equipped with potable water systems designed by engineers. I have already given you a basic rule-of-thumb method for sizing small systems. Next, I will show you how to use the fixture-unit method of sizing.

The Fixture-Unit Method

The fixture-unit method is simple, and it is generally acceptable to code officers. While this method may not be perfect, it is much faster and easier to use than the velocity method. Except for the additional expense for materials, you can't go wrong by oversizing pipe. If you are in doubt about the size, go to the next larger size. Now, let's see how to size a single-family residence's potable water system by using the fixture-unit method.

Most codes assign a fixture-unit value (Table 13.25) to common plumbing fixtures. To size by the fixture-unit method, you must establish the number of fixture units to be carried by the pipe. You must also know the working pressure of the water system. Most codes provide guidelines for these two pieces of information.

For this example, the house has the following fixtures: three toilets, three lavatories, one bathtub/shower combination, one shower, one dishwasher, one kitchen sink, one laundry hookup, and two sill cocks. The water pressure serving this house is 50 psi. There is a 1-in water meter serving the house, and the water service is 60 ft long. With this information and the guidelines provided by the local code, we can do a pretty fair job of sizing the potable water system.

The first step is to establish the total number of fixture units on the system. The code regulations provide this information. There are three toilets, so that's 9 fixture units. The three lavatories add 3 fixture units. The tub/shower combination counts as 2 fixture units, and the showerhead over the bathtub doesn't count as an additional fixture. The shower has 2 fixture units. The dishwasher adds 2 fixture units, and so does the kitchen sink. The laundry hookup counts as 2

TABLE 13.25 Common Fixture-Unit Values for Water Distribution

Fixture	Hot	Cold	Total
Bathtub	3	6	8
Bidet	1.5	1.5	2
Kitchen sink	1.5	1.5	2
Laundry tub	2	2	3
Lavatory	1.5	1.5	2
Shower	3	3	4
Water closet, two-piece	0	5	5

fixture units. Each sill cock is worth 3 fixture units. This house has a total fixture-unit load of 28.

Now we have the first piece of the sizing puzzle. The next step is to determine what size pipe will allow the number of fixture units. This house has a water pressure of 50 psi. This pressure rating falls into the category allowed in the sizing chart in the code book. First, we find the proper water meter size—1-in. Note that a 1-in meter and a 1-in water service are capable of handling 60 fixture units, when the pipe is only running 40 ft. However, when the pipe stretches to 80 ft, the fixture-unit load is reduced to 41. At 200 ft, the fixture-unit rating is 25. What is it at 100 ft? At 100 ft, the allowable fixture load is 36. See, this type of sizing is not so hard.

Now, what does this tell us? Well, we know the water service is 60 ft long. Once inside the house, how far is it to the most remote fixture? In this case, the farthest fixture is 40 ft from the water service entrance. This gives a developed length of 100 ft—60 ft for the water service and 40 ft for the interior pipe. Going back to the sizing table, we see that for 100 ft of pipe, under the conditions in this example, we are allowed 36 fixture units. The house has only 28 fixture units, so the pipe is properly sized.

What if the water meter were a ¾-in meter, instead of a 1-in meter? With a ¾-in meter and a 1-in water service and main distribution pipe, we could have 33 fixture units. This would still be a suitable arrangement, since we have only 28 fixture units. Could we use a ¾-in water service and water distribution pipe with the ¾-in meter? No. With all sizes set at ¾ in, the maximum number of fixture units allowed is 17.

In this example, the piping is oversized. But if you want to be safe, follow this type of procedure. If you are required to provide a riser diagram showing the minimum pipe sizing allowed, you will have to do a little more work. Once inside a building, water distribution pipes normally extend for some distance, supplying many fixtures with water. As the distribution pipe continues on its journey, the fixture load is reduced.

For example, assume that the distribution pipe serves a full bathroom group within 10 ft of the water service. Once this group is served with water, the fixture-unit load on the remainder of the water distribution piping is reduced by 6 fixture units. As the pipe serves other fixtures, the fixture-unit load continues to decrease. So it is feasible for the water distribution pipe to become smaller as it goes along.

Let's look at that same house and see how we could use smaller pipe. Okay, we know we need a 1-in water service. Once inside the foundation, the water service becomes the water distribution pipe.

The water heater is located 5 ft from the cold water distribution pipe. The 1-in pipe extends over the water heater and supplies it with cold water. And the hot water distribution pipe originates at the water heater. Now we have two water distribution pipes to size.

When sizing the hot and cold water pipes, we could make adjustments for fixture-unit values on some fixtures. For example, a bathtub is rated as 2 fixture units. Since the bathtub rating is inclusive of both hot and cold water, obviously the demand for just the cold water pipe is less than that shown in the table. For simplicity's sake, we do not break down the fixture units to fractions or reduced amounts. We will do the example as if a bathtub required 2 fixture units of hot water and 2 fixture units of cold water. However, we could reduce the amounts listed in the table by about 25 percent to obtain the rating for each hot and cold water pipe. For example, the bathtub, when being sized for only cold water, could take on a value of $1\frac{1}{2}$ fixture units.

Now then, let's get on with the example. We are at the water heater. We run a 1-in cold water pipe overhead and drop a $\frac{3}{4}$-in pipe into the water heater. What size pipe do we bring up for the hot water? First, we count the number of fixtures that use hot water, and we assign them a fixture-unit value. The fixtures using hot water are all fixtures, except the toilets and sill cocks. The total count for hot water fixture units is 13. From the water heater, the most remote hot water fixture is 33 ft away.

What size pipe should we bring up from the water heater? If you look at the sizing table in your code book, you will find a distance and fixture-unit count that will work in this case. Look under the 40-ft column, since the distance is less than 40 ft. The first fixture-unit number in the column is 9; this won't work. The next number is 27; this will work, because it is greater than the 13 fixture units we need. Looking across the table, you will see that the minimum pipe size to start with is $\frac{3}{4}$-in pipe. Isn't it convenient that the water heater just happens to be sized for $\frac{3}{4}$-in pipe?

Okay, now we start our hot water run with $\frac{3}{4}$-in pipe. As our hot water pipe travels along the 33-ft stretch, it provides water to various fixtures. When the total fixture-unit count remaining to be served drops to less than 9, we can reduce the pipe to $\frac{1}{2}$-in pipe. We can also run the fixture branches off the main in $\frac{1}{2}$-in pipe. We can do this because the highest fixture-unit rating on any of the hot water fixtures is 2. Even with a pipe run of 200 ft we can use $\frac{1}{2}$-in pipe for up to 4 fixture units. Is this sizing starting to seem easy? Remember the rule-of-thumb sizing I gave earlier. These sizing examples show how well the rule-of-thumb method works.

With the hot water sizing done, let's look at the remainder of the cold water. There is a pipe run of less than 40 ft to the farthest cold water fixture. There is a branch near the water heater drop for a sill cock, and there is a full bathroom group within 7 ft of the water heater drop. The sill cock branch can be $\frac{1}{2}$-in pipe. The pipe going under the full bathroom group could probably be reduced to $\frac{3}{4}$-in pipe, but it would be best to run it as a 1-in pipe. However, after the bathroom group and the sill cock have been served, how many fixture units are left? There are only 19. We can now reduce to $\frac{3}{4}$-in pipe, and when the demand drops to below 9 fixture units, we can reduce to $\frac{1}{2}$-in pipe. All the fixture branches can be run with $\frac{1}{2}$-in pipe.

This is one way to size a potable water system that works without driving you crazy. Some may dispute the sizes given in these examples, saying that the pipe is oversized. But as I said earlier, when in doubt, go bigger. In reality, the cold water pipe in the last example could probably have been reduced to $\frac{3}{4}$-in pipe where the transition was made from water service to water distribution pipe. It could have almost certainly been reduced to $\frac{3}{4}$-in pipe after the water heater drop. Local codes will have their own interpretations of pipe sizing, but this method will normally serve you well.

Well, we're done with the sizing exercises. Now it's time to discuss the proper way to design a plumbing system once it has been sized.

Installing Drain-Waste-and-Vent Systems

Installing drain-waste-and-vent (DWV) systems is a bit more complex than installing water pipes. Code regulations for drains and vents are much more numerous than those for water piping. A plumber has to know what fittings can be used to turn from a horizontal run to a vertical run. It's necessary to know the minimum pipe size for piping installed underground. How far above a roof does a vent pipe have to extend before it terminates? The answer depends on where you live, and you must be in tune with your local plumbing code to avoid getting rejection slips when your work is inspected.

Code regulations are not the only thing that complicates the installation of a DWV system. Since drains and vents are much larger in diameter than most water pipes, it is harder to find open routes where the plumbing can be installed. There is also a matter of pipe grade that must be maintained with a DWV system. This doesn't apply to water pipes.

Some plumbers prefer installing DWV systems to water systems. Others find the job of installing drains and vents to be a bother. I don't dislike either type of piping. In my opinion, water pipe is much easier to work with and install, but I enjoy running DWV pipes.

There is no question that DWV systems require planning. It's pretty easy to change directions with a water pipe, but this is not always the case with a drain or vent. Thinking ahead is crucial to the success of a job. This is true of any type of plumbing, and the rule certainly applies to DWV systems. Since many drainage systems start in the dirt, that is where we will begin—no, not in the dirt, but with underground plumbing.

Underground Plumbing

Underground plumbing is frequently overlooked in books on plumbing. Nonetheless, it is instrumental to the successful operation of some plumbing systems. Underground plumbing is installed before a house is built. When the plumbing is not installed properly, a new concrete floor will have to be broken up to relocate your pipes. This is a fast way to lose your job or make an enemy out of your customer.

Since the placement of underground plumbing is critical, you must be acutely aware of how it is installed. Miscalculating a measurement in the above-grade plumbing may mean cutting out the plumbing; but the same mistake with underground plumbing will be much more troublesome to correct.

An underground plumbing system is often called *groundwork* by professional plumbers. The groundwork is routinely installed after the footings for a home are poured and before the concrete slab is installed. When installing underground plumbing, you have to note code differences for groundworks versus above-grade plumbing. For example, while it is perfectly correct to run a $1\frac{1}{2}$-in drain for an above-grade bathtub, you must run a 2-in drain if it is under concrete. Let's take a moment to look at some of the code variations encountered with interior underground plumbing.

Code considerations

Underground plumbing is quite different from above-grade plumbing. There are several rules pertaining to groundworks that do not apply to above-grade plumbing. This is true of both the DWV system and the water distribution system. The following paragraphs will highlight common differences. This information may not apply in your jurisdiction, and the information does not include all code requirements. The following code considerations are the ones I encounter most often with underground plumbing.

The differences for DWV installations are minimal, but they must be observed. Galvanized steel pipe is an ap-proved material for a DWV system when the pipe is located above grade. But galvanized drain pipes may not be installed below ground. This rule has little effect on you, since most DWV systems today are plumbed with schedule 40 plastic pipe.

The minimum pipe size allowed under concrete is 2 in. This rule could cause trouble for the unsuspecting plumber. Suppose you check the sizing charts and see that $1\frac{1}{2}$-in pipe is an approved size. You might choose to use it under concrete. If you do, you will be in violation of most codes. While $1\frac{1}{2}$-in pipe is fine above grade, it is illegal below grade.

The types of fittings allowed also vary. For work above grade, a short-turn quarter bend is allowed in changing direction from horizontal to vertical. This is not true below concrete. Long-sweep quarter bends must be used for all 90° turns under the concrete.

When drains and vents are run above grade, the pipes must be supported every 4 ft. This is done with some type of pipe hanger. When installing the pipes underground, you will not use pipe hangers. Instead, you will support the pipe on firm earth or with sand or gravel. The pipe must be supported firmly and evenly.

When the groundworks leave the foundation of a house, the pipe will have to be protected by a sleeve. The sleeve can be a piece of schedule 40 plastic pipe, but it must be two pipe sizes larger than the pipe passing through the sleeve. This rule applies to any pipe passing through the foundation or under the footing.

Rarely will any residential drains under concrete run for 50 ft, but if they do, you must install a cleanout in the pipe. Pipes with a diameter of 4 in or less must be equipped with a cleanout every 50 ft in a horizontal run. The cleanouts must extend to the finished floor grade.

As the building drain becomes a sewer, there are yet more rules to adhere to. The home's sewer must have a cleanout for every 100 ft it runs. There should also be a cleanout within 5 ft of the house where the pipe makes its exit. It can usually be either inside or outside the foundation wall. These cleanouts must extend to the finished grade. If the sewer takes a turn of more than 45°, a cleanout must be installed. Cleanouts should be the same size as the pipes they serve, and they must be installed so that they will open in the direction of flow for the drainage system.

With pipes of a 3-in or larger diameter, there must be a clear space of 18 in in front of the cleanout. On smaller pipes, the clear distance must be at least 12 in. The sewer must be supported on a firm base of earth, sand, or gravel as the sewer travels the length of the trench.

Placement

The placement of groundworks is critical in the plumbing of a building. Before you install the underground plumbing, you must do some careful planning. The first step is the laying out of the plumbing.

When the groundworks are installed, the sewer and water service may or may not already be installed. If they are already installed, your job is a little easier. The pipes will be stubbed into the foundation, ready to be connected. If these pipes are installed, you can start with them and run the underground plumbing. If the water service and sewer are not installed, you will have to locate the proper spot for them.

Refer to your blueprints for the location of the water and sewer entrance. If the pipe locations are not noted on the plans, use common sense. Determine from where the water service and sewer will be coming to the house. If you are working with municipal water and sewer, the public works department of your town or city can help you. The public works department will tell you where your connections will be made to the mains and how deep they will be. If you will be connecting to a septic system and well, find the locations for these systems.

You have two considerations in picking the location of the water service and sewer. First, it should be the most convenient location inside the home for plumbing purposes. Second, there is an exit point that will allow successful connection to the main sewer and water service outside the home. Make your decision with a priority on connecting to the mains outside the home. You can adjust the interior plumbing to work for the incoming pipes, but you may not be able to adjust the existing outside conditions.

Once you have picked a location for the sewer and water service, you are ready to lay out the remainder of the groundworks. The locations for underground pipes will be partially determined by the blueprints. The blueprints show fixture locations. These fixture locations determine where you must have pipes in place. In addition to any grade-level plumbing, you look to the plumbing on upper floors. Where there are plumbing fixtures above the level of the concrete floor, you have to rough in pipes for the drains.

Figure 14.1 Measuring to the center of a drain for a groundwork installation.

Figure 14.2   A typical trenched-in groundwork installation.

When you turn pipes up out of the concrete for drains and vents, location of the pipes can be crucial. Many of these pipes are intended to be inside of walls to be built. If pipe placement is off by even 1 in, the pipe may miss the wall location. When this happens, a pipe will stick up through the floor in the wrong place. It may be in a hall or a bedroom, but it will have to be moved. Moving the pipe will require the breaking and repairing of the new concrete floor. To avoid these problems, make careful measurements (Fig. 14.1) and check all measurements twice.

When you have decided on all the pipes you will need in the underground plumbing, you can lay out your ditches. You will normally have to dig ditches to lay in the pipes. The easiest way to mark your ditches is with lime or flour. By placing these white substances on the ground, in the path of the ditch, you can dig the ditches accurately. The ditches have to be graded to enable the proper pitch of your pipes (Fig. 14.2). The standard pitch for household plumbing is $\frac{1}{4}$ in of fall for every 1 ft the pipe travels.

Installation

With all the planning and layout done, you are ready to start the installation. The best place to start is with the sewer. The height of the exit point is dictated by the depth of the connection at the main sewer. If the sewer is not already stubbed into the foundation, you have to establish the proper depth for the sewer leaving the foundation. At times, if you take the sewer out under the footing, the drain will be too low to make the final connection at the main. Before tunneling under the footing or cutting a hole in the foundation, you must determine the proper depth of the hole. This can be done by measuring the distance from the main sewer to the foundation. If the main connection is 60 ft away, the drain will drop 15 in from the time it leaves the house until it reaches the final destination. This determination is made by dividing the distance by 4, to allow for $\frac{1}{4}$ in of fall per 1 ft.

The sewer pipe should be covered by at least 12 in of dirt where it leaves the foundation. With this fact taken into consideration, the main connection must be at least 27 in deep. This is arrived at by taking 12 in of depth needed for cover and adding it to the 15 in of drop in the pipe's grade. It is best to allow a few extra inches to ensure a good connection point.

I will start the instructions, assuming the water service and sewer have not been run to the house when you start plumbing. With the depth of the hole known for the sewer, tunnel under the footing or cut a hole in the foundation wall. Install a sleeve for the sewer that is at least two pipe sizes larger than the sewer pipe. Install a cap on one end of the pipe to be used for the sewer. Extend the capped end of the sewer pipe through the sleeve and about 5 ft beyond the foundation. You are now ready to install the groundworks for all the interior plumbing.

In almost every circumstance, the pipe used for underground DWV plumbing will be schedule 40 plastic pipe. Working with this pipe is easy. The pipe can be cut with saws or roller-type cutters. Most plumbers cut the pipe with a saw. The type of saw varies. I use a hacksaw on pipe up through a 4-in diameter. A hand saw, like the one carpenters use, is effective in cutting the pipe and is easier to use on pipe over 3 in in diameter. There are saws made especially for cutting schedule 40 plastic pipe.

It is important to cut the pipe evenly. If the pipe cut is crooked, it will not seat completely in the fitting. This can cause the joint to leak. After you make a square cut on the pipe, clear the pipe of any burrs and rough pieces of plastic. You can usually do this with your hand, but the burrs and plastic can be sharp. It is safer to use a pocketknife or similar tool to remove the rough spots. The pipe and the interior of the fitting's hub must be clean and dry. Wipe these areas with a cloth if necessary, to clean and dry the surface.

Apply an approved cleaning solution to the end of the pipe and the interior of the fitting hub. Next, apply an approved primer, if required, to the pipe and fitting hub. Now apply the solvent or glue to the pipe and hub. Insert the pipe into the hub until it is seated completely in the hub. Turn the pipe one-quarter turn to ensure a good seal. This type of pipe and glue makes a quick joint. If you have made a mistake, you will have trouble removing the pipe from the fitting. If the glue sets up for long, you will have to cut the fitting off the pipe to correct any mistakes.

Many plumbers dry-fit the joints to confirm their measurements. Dry-fitting is putting the pipe and fittings together without glue to check alignment and measurements. I have done plumbing for a long time and don't dry-fit, but many plumbers prefer to dry-fit each joint. This procedure makes the job go much more slowly and can create its own problems. If you are dry-fitting the pipe to get an accurate feel for your measurements, the pipe must seat tightly in the fitting. When this is accomplished, it can be difficult to get the pipe back out of the fitting.

While I never dry-fit, I do sometimes verify my measurements with an easier method. I place the fitting beside the pipe, with the hub positioned as it will be when glued. I then take a measurement to the center of my fitting to confirm the length of the pipe. Holding the fitting beside the pipe, instead of inserting the pipe, I don't have to fight to get the pipe out of the fitting. Once a joint is made with schedule 40 plastic pipe, it should not be moved for a minute or so. It doesn't take long for the glue to set up, and you can continue working without fear of breaking the joint.

The first fitting to install on the building drain is a cleanout. The cleanout will usually be made with a wye and an eighth bend. The pipe extending from the eighth bend must extend high enough for the cleanout to be accessible above the concrete floor. Many plumbers put this cleanout in a vertical stack that serves plumbing above the first floor. The cleanout can be stopped at the finished floor level, but it must be accessible above the concrete.

After the cleanout is installed, you can go on about your business. As you run your pipes to the appropriate locations, they must be supported on solid ground or an approved fill, such as sand or gravel. Maintain an even grade on the pipes as they are installed. The minimum grade should be $\frac{1}{4}$ in for every 1 ft that the pipe runs. Don't apply too much grade to the pipe. If the pipe falls too quickly, the drains may stop up when used. The fast grade will drain the pipes of water, but leave solids behind. These solids accumulate to form a stoppage in the pipe and fittings.

Support all the fittings installed in the system. If you are installing a 3-by-2 wye, the 2-in portion will be higher above the ground than the 3-in portion. Place dirt, sand, or gravel under these fittings to support them. Don't allow dirt, mud, or water to get in the fittings or on the ends of the pipe. A small piece of dirt is all it takes to cause a leak in the joint.

To know where to position your pipes for fixtures, vents, and stacks,

Figure 14.3 A turn-up pipe between strings that indicate wall locations.

it helps to have string up where the walls (Fig. 14.3) will be installed. You should also know what the finished floor level will be. The blueprints will show wall locations, and the general contractor, concrete subcontractor, or supervisor will be able to give you the finished floor grade.

You must know the level of the finished floor so that the pipes will not be too high and wind up above the floor. Once you have the grade level for the finished floor, mark the level on the foundation. Drive stakes into the ground, and stretch a string across the area where the plumbing is being installed. The string should be positioned at the same level as the finished floor. You can use this string as a guide to monitor the height of the pipes.

Using the blueprints, mark wall locations with stakes in the ground. You can use a single string and keep your pipes turning up through the concrete to one side of the string, or you can use two strings to simulate the actual wall. Using two strings will allow you to position the pipes in the center of the wall.

Plumbing measurements are generally given from the center of a drain or vent. If your water closet is supposed to be roughed in 12 in off the back wall, the measurement is made from the edge of the wall to the center of the drain.

As you near completion on the underground plumbing, secure all the pipes that need to stay in their present positions. The pipes of greatest concern are pipes coming up through the concrete and pipes

Figure 14.4 Copper stakes securing a groundwork turn-up pipe.

for bathtubs, showers, and floor drains. The best way to secure these pipes is with stakes. The stakes should not be made of wood. In many regions, termites may come to the house for the wood under the floor. Use steel or copper stakes. The stakes should be located on both sides of the pipe (Fig. 14.4) to ensure it is not moved by other trades.

When all the DWV pipe is installed, cap all pipes. You can use temporary plastic caps or rubber caps, but cap the pipes. You do not want foreign objects to get into the drains and clog them. Where you have a tub or a shower to be installed on the concrete floor, you will need a trap box (Fig. 14.5). This is just a box to keep concrete away from the pipe and to allow the installation of a trap when the bathing unit is set. Put a spacer cap over the pipe turned up for a toilet (Fig. 14.6) to be set on the concrete floor. This spacer will keep concrete away from the pipe so that you can install a closet flange when the time comes to set the toilet. Make sure all pipes are secured before you consider the job finished (Fig. 14.7).

Figure 14.5 A tub box.

Figure 14.6 A spacer cap on a turn-up pipe for a water closet.

Aboveground Rough-ins

When you are installing aboveground rough-ins, you don't have to dig any ditches. This is good. But you do have to drill holes. Once a building is framed and under roof, you can start the installation of aboveground drains and vents. If the building has underground plumbing, you will be tying into the pipes you installed earlier. Not all houses have groundworks, so the framing stage might be your first visit to a job for installing pipes.

Houses without groundworks still require water services and sewers. Refer to the information given on these two components of a

Figure 14.7 A water pipe is secured to a copper stake in a groundwork setup. *Note:* Insulation will be added to protect copper from concrete.

plumbing system in earlier paragraphs. Make sure that you don't set yourself up with a building sewer that will be too high to connect to a main sewer or septic tank (Figs. 14.8 to 14.10). When you are comfortable with your exit locations, you can begin to lay out and install your DWV rough-in.

Pipes under 4 in in diameter	$\frac{1}{4}$ in/ft
Pipes 4 in or larger in diameter	$\frac{1}{8}$ in/ft

Figure 14.8 Minimum pitch of drainage pipe in zone 1.

Pipes under 3 in in diameter	$\frac{1}{4}$ in/ft
Pipes 3 in or larger in diameter	$\frac{1}{8}$ in/ft

Figure 14.9 Minimum pitch of drainage pipe in zone 2.

Pipes under 3 in in diameter	$\frac{1}{4}$ in
Pipes 3 to 6 in in diameter	$\frac{1}{8}$ in/ft
Pipes 8 in or larger in diameter	$\frac{1}{16}$ in/ft

Figure 14.10 Minimum pitch of drainage pipe in zone 3.

Hole sizes

Hole sizes are important. Some codes require that the hole size be kept to a minimum to reduce the effect of fire spreading through a home. If you drill oversized holes, they can act as chimneys for fires in a building. Choose a drill bit that is just slightly larger than the pipe you will be installing. For 2-in pipe, a standard drill bit size is $2\frac{9}{16}$ in.

When a hole is cut for a shower drain, the hole size is determined by the shower drain. Keep the hole as small as possible, but large enough to allow the shower drain to fit in it. For a tub waste, the standard hole is 15 in long and 4 in wide. This allows adequate space for the installation of the tub waste. In some jurisdictions, after the tub waste is installed, the hole must be covered with sheet metal to eliminate the risk of a draft during a house fire.

Running pipe

When your design is made and your holes are open, you are ready to run pipe. This is where you must use your knowledge of the plumbing code. There are many rules pertaining to DWV systems. If you purchase a code book, you will see pages upon pages of rules and regulations. The task seems overwhelming with all the rules to follow. If you don't lose your cool, the job is not all that difficult. By following a few rules and basic plumbing principles, you can install DWV systems with ease.

One of the most common mistakes made in drainage piping is failure to remember to allow for the pipe's grade. When you draw the design on paper, you are not likely to think of the size of the floor joists and the grade you will need for the pipes. Since many houses have their plumbing concealed by a ceiling, it is important to keep the pipes above the ceiling level. Not many people want an ugly drain pipe hanging below the ceiling in their formal dining room. These small details of planning will make a noticeable difference in the outcome of the job.

Once you know the starting and ending points of your drains and vents, you can project the space needed for adequate grade. Generally the grade is set at $\frac{1}{4}$ in/ft for drains and vents. Drains fall downward, toward the final destination. Vents pitch upward, toward the roof of the house. With a 12-ft piece of pipe, the low end will be 3 in lower than the high end. When you are drilling through floor joists, most holes will be kept at least $1\frac{1}{2}$ in above the bottom and below the top of the joist. If you follow this rule of thumb, a 2-by-8 joist will have $4\frac{1}{2}$ in for you to work with.

A 2-by-8 has a planed width of $7\frac{1}{2}$ in. From $7\frac{1}{2}$ in you deduct 3 in for the top and bottom margin. This leaves $4\frac{1}{2}$ in for you to work with. What does all this mean to you? It means that with a 2-by-8 joist system, you can run a 2-in pipe for about 10 ft before you are in trouble. The pipe diameter takes up 2 in of the remaining $4\frac{1}{2}$ in. This leaves $2\frac{1}{2}$ in to manipulate for grade. At $\frac{1}{4}$ in/ft, you can run 10 ft with the 2-in pipe. With $1\frac{1}{2}$-in pipe, you could go 12 ft. And 3-in pipe will be restricted to a distance of about 6 ft.

Under extreme conditions, you can run farther by drilling closer to the top or the bottom of the floor joist. Before you drill any closer than 1 in to either edge, consult the carpenters. They will probably have to install headers or a small piece of steel to strengthen the joists. Many experienced plumbers have trouble with running out of space for pipes. They don't look ahead and do the math before drilling the holes. After drilling several joists, they realize they cannot get where they want to go. This results in a change in layout and a bunch of joists with holes drilled in them that cannot be used. Plan your path methodically, and you will not have these embarrassing problems.

When your holes have been drilled, running the pipe will be easy. Depending on the code you are following, every fixture must have a vent. Except for jurisdictions using a combination-waste-and-vent code, you must provide a vent for every fixture. This does not mean every fixture must have an individual back-vent. Most codes allow the use of wet vents. Using wet vents will save you time and money.

Wet vents

Wet vents (Figs. 14.11 and 14.12) are pipes that serve two purposes. They are a drain for one fixture and a vent for another. Toilets are often wet-vented with a lavatory. This involves placing a fitting within a prescribed distance from the toilet that serves as a drain for the lavatory. As the drain proceeds to the lavatory, it will turn into a dry vent after it extends above the trap arm. Exact distances and specifications are set forth in local plumbing codes.

Dry vents

Many of your fixtures will be vented with dry vents (Fig. 14.13). These are vents that do not receive the drainage discharge of a fixture. Since the pipes carry only air, they are called *dry vents*. There are many types of dry vents: common vents (Fig. 14.14), individual vents (Fig. 14.15), circuit vents (Fig. 14.16), vent stacks, and relief vents (Fig. 14.17), to name a few. Don't let all these vents scare you.

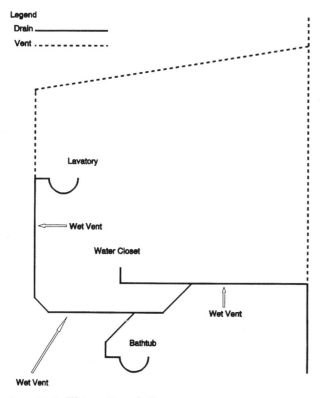

Legend

Drain ————————

Vent . – . – . – . – . – . .

Lavatory

Wet Vent

Water Closet

Wet Vent

Bathtub

Wet Vent

Figure 14.11   Wet-venting a bathroom group.

In plumbing an average house, venting the drains does not have to be complicated.

If you don't understand wet vents, you can run all dry vents. This may cost a little more in material, but it can make the job easier for you to understand. Remember, every fixture needs a vent. How you vent the fixture is up to you, but as long as a legal vent is installed, you will be okay. You must install at least one 3-in vent that will penetrate the roof of the home. You can have more than one 3-in vent, but you must have *at least* one.

After you have your mandatory 3-in vent, most bathroom groups can be vented with a 2-in vent. The majority of individual fixtures can be vented with 1½-in pipe. Most secondary vents can be tied into the main 3-in vent before the main vent leaves the attic. In a standard application, the average house will have one 3-in vent going through the roof and one 1½-in vent, which serves the kitchen sink, going through the roof on its own.

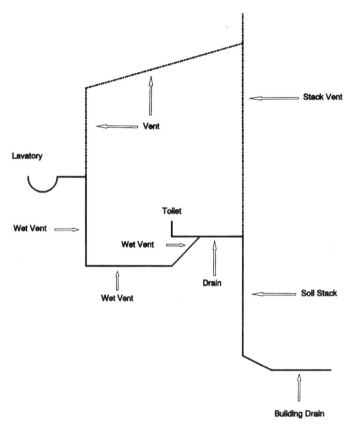

Figure 14.12   Wet-venting a toilet with a lavatory.

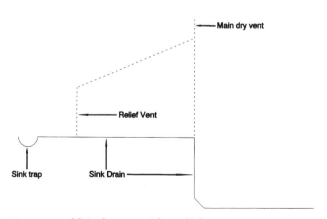

Figure 14.13   Main dry vent with a relief vent.

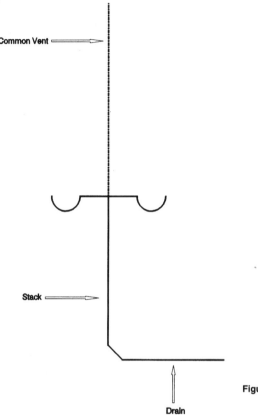

Common Vent

Stack

Drain

Figure 14.14 Common vent.

If you wanted to, you could vent each fixture with an individual vent and take them all through the roof. This would be a waste of time and money, but it would be in compliance with the plumbing code. It is more logical to tie most of the smaller vents back into the main vent, before it exits the roof. Before you can change the direction of a vent, it must be at least 6 in above the flood-level rim of the fixture it is serving.

Yoke vents (Fig. 14.18) are not common in residential plumbing, but branch vents (Fig. 14.19), stack vents (Fig. 14.20), and vent stacks (Fig. 14.21) are. Even island vents (Fig. 14.22) are used in some residential jobs. You should become familiar with the various types of vents so that you can install efficient systems.

When you start to install the drains, you must pay attention to pipe size, pitch, pipe support, and the fittings used. The sizing charts in your local plumbing code will help you to identify the proper pipe sizes. If you grade all your pipes with the standard $\frac{1}{4}$ in/ft (Fig.

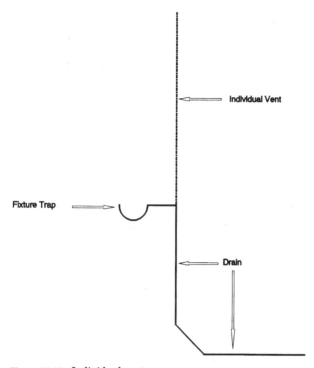

Figure 14.15   Individual vent.

14.23), you won't have a problem there. Support all the horizontal drains at 4-ft intervals or in compliance with your local code requirements (Tables 14.1 through 14.6). The last thing to learn about is the use of various fittings.

When each fixture in the system is vented, use P-traps for tubs, showers, sinks, lavatories, and washing machine drains. When the pipes rise vertically, use a sanitary tee to make the connection between the drain, vent, and trap arm. Use long-sweep quarter bends for horizontal changes in direction. You can use short-turn quarter bends above grade when the pipe is changing from horizontal to vertical; but if you want to keep it simple, use long-sweep quarter bends for all 90° turns.

Use wyes with eighth bends to create a stack or branch that changes from horizontal to vertical. Remember to install cleanouts near the base of each stack and at the end of horizontal runs, when feasible. Keep turns in the piping to a minimum. The fewer turns the pipe makes now, the fewer problems you will have with drain stoppages later. Never tie a vent into a stack below a fixture if the stack receives that fixture's drainage. If the pipes could be hit by a nail or

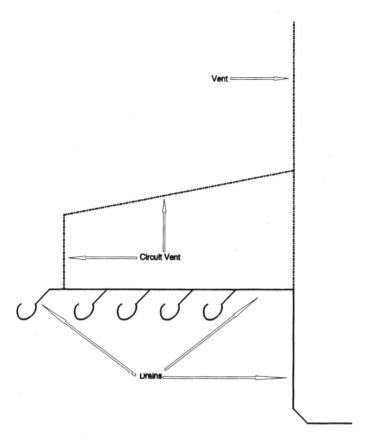

Figure 14.16 Circuit vent.

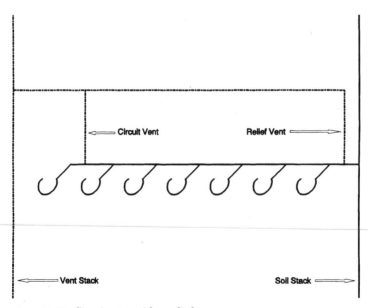

Figure 14.17 Circuit vent with a relief vent.

14.18

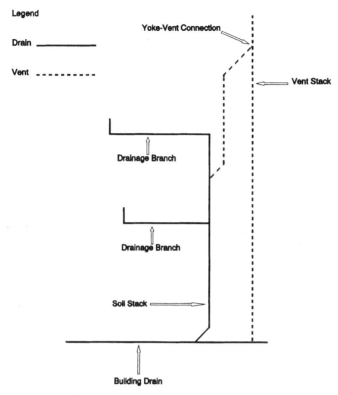

Figure 14.18 Yoke vent.

drywall screw, install a metal plate to protect the pipe. These nail plates are nailed or driven onto studs and floor joists to prevent damage to pipes.

Keep your vents within the maximum distance allowed from the fixture. For a 1½-in drain, the distance between the fixture's trap and the vent cannot exceed 5 ft. For 2-in pipe the distance is 8 ft for a 1½-in trap and 6 ft for a 2-in trap. A 3-in pipe has a maximum distance of 10 ft. With 4-in pipe, you can go 12 ft. If the fixture must be beyond these limits from the stack vent, install a relief vent.

A *relief vent* is a dry vent that comes off the horizontal drain as it goes to the fixture. The relief vent rises up at least 6 in above the flood-level rim and ties back into the stack vent.

If you are faced with a sink in an island counter, you have some creative plumbing to do for the vent. Since the sink is in an island, there will be no walls to conceal a normal vent. Under these circumstances you must use an *island vent*.

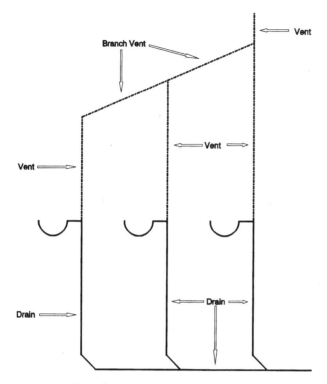

Figure 14.19 Branch vent.

When your vents penetrate a roof, they must extend at least 12 in above it. In some areas, the extension requirement is 2 ft. If the roof is used for any purpose other than weather protection, the vent must rise 7 ft above the roof. Be careful taking vents through the roof when windows, doors, or ventilating openings are present. Vents must be 10 ft from these openings or 3 ft above them (Fig. 14.24).

When you are plumbing the drain for a washing machine, keep this in mind: The standpipe from the trap must be at least 18 but not more than 30 in high. The piping must be accessible for clearing stoppages.

Fixture placement

Fixture placement is normally shown on all blueprints. Code requirements dictate elements of locating fixtures. Rough-in measurements vary from fixture to fixture and from manufacturer to manufacturer. To be safe, you should obtain rough-in books from your supplier for each type and brand of fixture you will be roughing in.

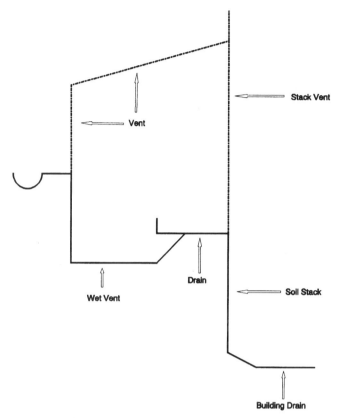

Figure 14.20 Stack vent.

Standard toilets will rough in with the center of the drain 12 in from the back wall. This measurement of 12 in is figured from the finished wall. If you are measuring from a stud wall, allow for the thickness of drywall or whatever the finished wall will be. Most plumbers rough in the toilet $12\frac{1}{2}$ in from the finished wall. The extra $\frac{1}{2}$ in gives you a little breathing room if conditions are not exactly as planned. From the front rim of the toilet, there must be a clear space of 18 in between the toilet's rim and another fixture.

From the center of the toilet's drain, there must be 15 in of clearance on both sides. So you need a minimum width of 30 in to install a toilet. If you are plumbing a half-bath, the room must have a minimum width of 30 in and a minimum depth of 5 ft. These same clearance measurements apply to bidets.

If you plan to install a tub waste using slip nuts, you will need an access panel to gain access to the tub waste. If you do not want an

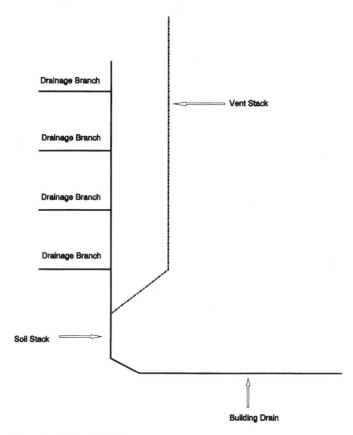

Figure 14.21 Vent stack.

access panel, use a tub waste with solvent-welded joints. Many people object to access panels in their hall or bedroom; avoid slip-nut connections, and you can avoid access panels.

If you take the time to learn your local plumbing code and are willing to think ahead in your planning, DWV systems will not give you many problems. Plumbers who are intimidated by cryptic code books can refer to Chap. 5 and to one of my other books, *National Plumbing Codes Handbook,* also published by McGraw-Hill, for help in sizing and other code-related issues. The book is as easy to understand as this one is, and it offers guidance for interpreting all three of the major plumbing codes.

Legend

Vent ┈┈┈┈┈┈

Drain ─────────

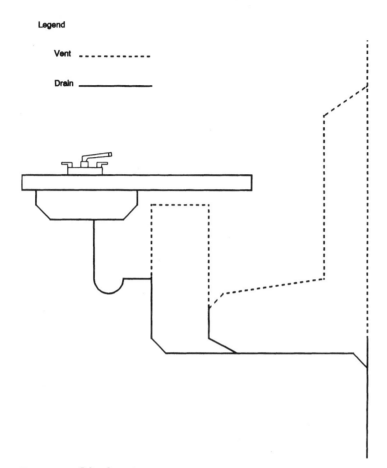

Figure 14.22 Island vent.

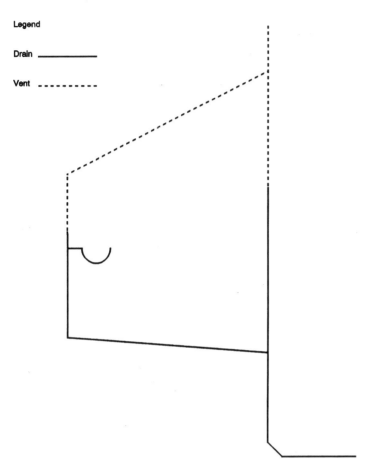

Figure 14.23 Graded-vent connection.

TABLE 14.1 **Horizontal Pipe Support Intervals in Zone 1**

Type of vent pipe	Maximum distance between supports, ft
ABS	4
Cast iron	At each pipe joint*
Galvanized	12
Copper ($1\frac{1}{2}$-in and smaller)	6
PVC	4
Copper (2-in and larger)	10

*Note: Cast-iron pipe must be supported at each joint, but supports may not be more than 10 ft apart.

TABLE 14.2 Horizontal Pipe Support Intervals in Zone 2

Type of vent pipe	Maximum distance between supports, ft
ABS	4
Cast iron	At each pipe joint
Galvanized	12
PVC	4
Copper (2-in and larger)	10
Copper ($1\frac{1}{2}$-in and smaller)	6

TABLE 14.3 Horizontal Pipe Support Intervals in Zone 3

Type of vent pipe	Maximum distance between supports, ft
Lead	Continuous
Cast iron	5
Galvanized	12
Copper tube ($1\frac{1}{4}$ in)	6
Copper tube ($1\frac{1}{2}$ in and larger)	10
ABS	4
PVC	4
Brass	10
Aluminum	10

TABLE 14.4 Vertical Pipe Support Intervals in Zone 1

Type of drainage pipe	Maximum distance between supports
Lead	4 ft
Cast iron	At each story
Galvanized	At least every other story
Copper	At each story*
PVC	Not mentioned
ABS	Not mentioned

*Support intervals may not exceed 10 ft.
Note: All stacks must be supported at their bases.

TABLE 14.5 Vertical Pipe Support Intervals in Zone 2

Type of vent pipe	Maximum distance between supports
Lead	4 ft
Cast iron	At each story*
Galvanized	At each story†
Copper ($1\frac{1}{4}$ in)	4 ft
Copper ($1\frac{1}{2}$ in and larger)	At each story
PVC ($1\frac{1}{2}$ in and smaller)	4 ft
PVC (2 in and larger)	At each story
ABS ($1\frac{1}{2}$ in and smaller)	4 ft
ABS (2 in and larger)	At each story

*Support intervals may not exceed 15 ft.
†Support intervals may not exceed 30 ft.
Note: All stacks must be supported at their bases.

TABLE 14.6 **Vertical Pipe Support Intervals in Zone 3**

Type of vent pipe	Maximum distance between supports, ft
Lead	4
Cast iron	15
Galvanized	15
Copper tubing	10
ABS	4
PVC	4
Brass	10
Aluminum	15

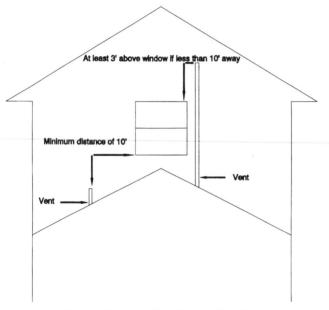

Figure 14.24 Layout of vent position to operable windows.

15

Installing
Water Distribution
Systems

Installing water distribution systems is usually easier than installing a drain-waste-and-vent (DWV) system. For one thing, the pipes being installed with a water system are smaller. Also, the requirements for pipe grading are less stringent for water systems. Code requirements for water systems are easier for most plumbers to understand. When all elements are considered, a residential water system is pretty simple. Commercial jobs are more complex, but piping diagrams for commercial jobs make the installations easy if you are good at following directions.

Sizing water systems can be tricky. Most residential jobs are fairly simple, but sometimes a plumber can run into some complicated sizing situations. Since most residential water systems are designed by plumbers, not by engineers and architects, residential plumbers have to know how to interpret the plumbing code for sizing requirements.

Sizing is not the only part of installing a water system that can get tricky. It may be hard to decide what pipe will work best. The cost-effective routing of pipes in a layout is important to the plumber's profit picture. There's a lot to learn about water systems, so let's get started.

Potable Water

For what purposes do we require potable water? Potable water is needed for drinking, cooking, bathing, and the preparation of food and medicine. Potable water is also used for other activities, but those listed are the uses dealt with in the plumbing code. The plumbing of fixtures used for any of the above purposes must be done so that only potable water is accessible to them.

When must hot water be provided? Hot water is required in buildings where people work and in all permanent residences. What about cold water? Cold water must be supplied to every building that contains plumbing fixtures and is used for human occupancy.

How much water pressure is required? The water pressure for a water distribution system must be high enough to provide proper flow rates at each of the fixtures. Flow rates for various fixtures are determined by referring to the tables or text in your local code.

Generally, 40 pounds per square inch (psi) is considered adequate water pressure. If the incoming pressure is 80 psi or more, the pressure must be controlled with a pressure-reducing valve. This valve is installed between the water service and the main water distribution pipe. The device allows the water pressure to be kept at a lower value. There are, of course, exceptions to most rules; check your local code for the requirements in your jurisdiction.

On the subject of code requirements, note that three major plumbing codes are used in the United States. I have these codes broken down by location, and I refer to them as zones 1, 2, and 3 (Figs. 15.1 to 15.3).

Figure 15.1   States in zone 1.

Washington
Oregon
California
Nevada
Idaho
Montana
Wyoming
North Dakota
South Dakota
Minnesota
Iowa
Nebraska
Kansas
Utah
Arizona
Colorado
New Mexico
Indiana
Parts of Texas

Figure 15.2 States in zone 2.

Alabama
Arkansas
Louisiana
Tennessee
North Carolina
Mississippi
Georgia
Florida
South Carolina
Parts of Texas
Parts of Maryland
Parts of Delaware
Parts of Oklahoma
Parts of West Virginia

Figure 15.3 States in zone 3.

Virginia
Kentucky
Missouri
Illinois
Michigan
Ohio
Pennsylvania
New York
Connecticut
Massachusetts
Vermont
New Hampshire
Rhode Island
New Jersey
Parts of Delaware
Parts of West Virginia
Parts of Maine
Parts of Maryland
Parts of Oklahoma

Water conservation

Water conservation is an issue that most codes address. The codes
restrict the flow rates of certain fixtures, such as showers, sinks, and
lavatories. An unmodified shower head can normally produce a flow
of 5 gallons per minute (gpm). This flow rate is often reduced to 3
gpm with the insertion of a water-saver device. The device is a small
disk with holes in it.

Other water-saver regulations apply to buildings with public rest rooms. It is frequently required that all public-use lavatories be equipped with self-closing faucets. These faucets have restricted flow rates, and they will shut themselves off after use.

Urinals should have a flow rate of not more than $1\frac{1}{2}$ gallons (gal) per flush. Toilets generally must not use more than 4 gal of water during the flush cycle.

Antiscald devices

Antiscald devices are required on some plumbing fixtures. Different codes require these valves on different types of fixtures, and sometimes the maximum water temperature varies. Antiscald devices are valves or faucets that are specially equipped to prevent burns due to hot water. These devices come in different configurations, but they all have the same goal: to prevent scalding.

Whenever a gang shower is installed, such as in a school gym, antiscald shower valves should be installed. Some codes require antiscald valves on residential showers, but others don't. The maximum hot water temperature in some codes is 110°F. Other codes set it at 120°F.

Choosing the Type of Pipe and Fittings to Be Used

The three primary types of pipe used in above-grade water systems are cross-linked polyethylene (PEX), copper, and chlorinated polyvinyl chloride (CPVC). All these pipes provide adequate service as water distribution carriers. Which pipe you choose will be largely a matter of personal preference. Many homeowners opt for CPVC because they do not have to solder the joints. Soldering seems difficult to someone who has no experience with it. It seems easier to glue plastic pipe than to solder copper joints.

What pipe will you use? Before you answer this question, consider the differences in installation procedures for each type of pipe. The first consideration might be the cost of materials. Although plastic pipes are cheaper than copper pipes, plastic fittings are often more expensive than copper ones. At the end of plumbing an average home, there will not be a major difference in the material costs. The primary considerations are the durability of the pipe and the ease of installation.

CPVC Pipe

CPVC pipe has been used as potable water pipe for many years. The fittings for the pipe are installed with a solvent weld. You do not need soldering skills or equipment with CPVC. The pipe has some flexibility and is easy to snake through floor joists. It is suitable for hot and cold water applications and can be cut with a hacksaw. It is easy to install.

Professional plumbers I know, and have known, do not show great affection for CPVC. To make a waterproof joint, the pipe and fittings must be properly prepared. This preparation is very much like the procedures used with PVC drainage pipe. A cleaning solution must be applied to the pipe and fittings. Then a primer is applied to both the pipe and the fittings. Joints are made with a solvent or glue.

Going through all the steps of making the joint is a slow process. Since professionals seek to complete their jobs as quickly as possible, CPVC is not an ideal choice. Not only do you lose time with all the preparations, but also the joints cannot be disturbed for some time. If they are jarred before they have cured, the joints may suffer from voids that will leak. The cure time for the joint will vary depending upon the temperature.

The time lost waiting for the joints to set up is a drawback for professionals. When you work with polybutylene or copper, the time spent waiting to work with new joints is greatly reduced. This reason alone is enough to cause professionals to use a different pipe, but it is not the only reason professionals choose other pipes.

Even after a CPVC joint is made, it is not extremely strong. Any stress on the joint can cause it to break loose and leak. CPVC becomes brittle in cold weather. If the pipe is dropped on a hard surface, it can develop small cracks. These cracks often go unnoticed until the pipe is installed and tested. When the pipe is tested and leaks, it must be replaced. This means more lost time for the professional. Copper and polybutylene are not subject to these same cracks when dropped.

If the pipe or fitting has any water on it, the glue may not make a solid joint. With polybutylene, water on the pipe has no effect on the integrity of the joint. Water on the outside of copper pipe will turn to steam and normally does not cause a leak. The simple act of hanging the pipe is even a factor in making a choice. When you hang the pipe, you will probably use a hammer to drive the hanger into place. If the hammer slips and hits your pipe, CPVC may shatter. Copper may dent, but it will not shatter. Polybutylene will bounce right back into shape from a hammer blow. These may be small differences, but they add up to stack the deck against CPVC for professionals.

There is one advantage that CPVC has over copper for professional plumbing applications. It is not adversely affected by acidic water. When potable water is being provided by a well, acid in the water can cause copper pipe to leak. The acid causes pinhole leaks in copper after some time. Since CPVC is plastic, it is not affected by the acid.

Polybutylene Pipe and PEX Tubing

If you are going to install PB or PEX pipe, learn how to do it correctly. These types of pipe are not like polyethylene. Special fittings and clamps should be used to ensure good joints. PEX and polybutylene slide over ridged insert fittings (I prefer copper fittings) and are held in place with special clamps. These clamps are not adjustable, stainless-steel clamps, like those used on polyethylene pipe. The clamps used on PEX and polybutylene are solid metal and are installed with a special crimping tool. There are many advantages to working with these plastic tubings.

PEX and polybutylene come coiled in a roll. The pipe is extremely flexible and can be run very much as electric wire is. After holes are drilled, the pipe can be pulled through the holes in long lengths. The pipe is approved for hot and cold water and can be cut with a hacksaw, although special cutters do a neater job. In cold climates, PEX and polybutylene offer another advantage: they can expand a great deal before splitting during freezing conditions. Acidic water will not eat holes in PEX or PB pipe.

When houses are plumbed with PEX or polybutylene, the piping design is often different from that used when the house is plumbed with a rigid pipe. CPVC and copper installations incorporate the use of a main water pipe and several branches to feed individual fixtures. With polybutylene or PEX, most plumbers run individual lengths of pipe from each fixture to a common manifold. By installing the pipe in this way, there are no joints concealed in the walls or ceilings. This reduces the likelihood of a leak that will be difficult to access when the house is completed.

Basically, the water service or a main water pipe is run to the manifold location. The manifold receives its water from the water service or main and distributes it through the individual pipes. This type of installation requires more pipe than traditional installations, but polybutylene pipe is cheap. It is the fittings that are expensive, but in a manifold installation, very few fittings are used.

Manifold systems are convenient for service plumbers and homeowners. The fact that PEX and PB pipe can be installed with very few concealed joints is an attractive attribute.

Copper Pipe

Copper has long been the leader in water distribution systems. While copper is expensive and has some drawbacks, it remains the most common type of pipe in potable water systems. Many homeowners specify copper in their building plans, and old-school plumbers swear by it. As a seasoned plumber, I like copper and use it frequently. However, I seem to be using more and more PB pipe as people become more accepting of it. Copper is a proved performer with a solid track record.

When soldering joints on potable water pipes, you must use a low-lead solder. In the old days, 50/50 solder was used, but not today. Buy lead-free solder for your joints. The old 50/50 solder is still available for making connections on nonpotable pipes, but don't use it on potable water systems.

Copper is easy to cut when you use roller cutters (Fig. 15.4). To cut the pipe, simply place the pipe between the cutting wheel and the rollers, tighten the handle, and turn the cutter. As you rotate the cutter around the pipe, tighten the handle to maintain steady pressure. After a few turns, the pipe will be cut. Be careful of jagged pieces of copper after you cut the pipe. Be especially careful when you sand the pipe to clean it. If the pipe does not cut evenly, there may be sharp, jagged edges protruding from the edge of the pipe. These copper shards can cut you or become embedded in your skin.

Copper pipe is durable and is approved for hot and cold water pip-

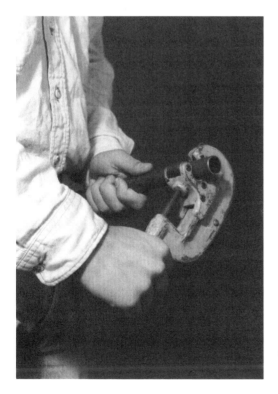

Figure 15.4 Roller cutters for cutting copper tubing and pipe.

ing. Once you know how to solder, copper is easy to work with. The pipe installs quickly and produces a neat-looking job. Since the pipe is rigid, it can be installed to allow the water system to drain. This is a factor in seasonal cottages and other circumstances where the water is turned off for the winter.

The Pipe Decision

Which pipe to use is up to you. The average person would probably choose polybutylene. For a professional, it would be a toss-up between polybutylene and copper. The two factors that I would consider are the working conditions and type of plumbing being installed. Personally, I would not use CPVC. Consider all your options and make your choice based on those factors you feel strongly about.

Materials approved for water distribution

There are a number of materials approved for water distribution, but only a few are used frequently in new plumbing systems. Probably copper is still the number-one choice as a water distribution material, with PB pipe second and CPVC a distant third.

Materials approved for use in water distribution systems may vary slightly from code to code. The following information is valid in many jurisdictions, but remember to always check your local code before installing plumbing.

Galvanized steel pipe

Galvanized steel pipe is an approved water distribution pipe. It has been used for a very long time, and you will still find it in many older plumbing systems. It is not, however, used much in modern plumbing.

This pipe is heavy and requires threaded joints. The pipe tends to rust and deteriorate more quickly than other approved water distribution pipes. There are few, if any, occasions when galvanized steel pipe is used for modern plumbing installations.

Brass pipe

Brass pipe is approved for water distribution, but like galvanized steel pipe, it is rarely used.

Copper pipe

We talked about the use of copper as a water distribution pipe; it is, of course, approved for that application. There are, however, different types of copper. By types, I mean ratings of thickness. The two types most commonly used for water distribution are type L and type M. Type K copper is also fine for water distribution, but it is more expensive and usually the extra-thick pipe is not needed. Type M copper, the thinnest of the three, is not approved in all areas for water distribution. For years type M was the standard, but now type L is required by some administrative authorities.

CPVC pipe

Chlorinated polyvinyl chloride, or CPVC as it is commonly called, is another approved material that we have discussed. You may find this pipe ideal, but I don't favor it. Try it and see what you think, but in experience I do better with other types of piping.

Sizing

The task of sizing water pipes strikes fear into the hearts of many plumbers. They take one look at the friction-loss charts in their code books and see some math formulas that read like a foreign language, and they run the other way. Sizing water pipes can be intimidating.

There are two approaches to sizing water pipe: the engineer's approach and the plumber's approach. Large jobs typically are required to be designed by an engineer or other suitable professionals, other than plumbers. Most plumbing codes will not allow plumbers to design complex water distribution systems. For most plumbers, myself included, this is a blessing.

Your local code book will provide information on how you can use formulas, friction charts, and other complicated aids to size water pipes; but it will also give you some easier ways around the problem.

Your code book will make clear what size pipe is needed for all common plumbing fixtures. Let me give you some idea of what you will find. One of the major codes allows fixture supplies with diameters of $3/8$ inch (in) to be used for the following fixtures: bidets, drinking fountains, single showers, lavatories, and water closets—with the exception of one-piece toilets. If you increase the fixture supply size to $1/2$ in, you can feed these fixtures: bathtubs, domestic dishwashers, hose bibs, residential kitchen sinks, laundry trays, service sinks, flush tank urinals, wall hydrants, and one-piece toilets.

By increasing the fixture size to $3/4$ in, you can connect to commercial kitchen sinks, flushing-rim sinks, and flush valve urinals. Stepping up to a 1-in pipe, you can serve toilets that use flush valves.

Now wasn't that easy? All you have to do is look at your code book to quickly see the minimum size requirements for your water fixture supplies. Up to the first 30 in of a section of pipe connecting to a fixture is considered a fixture supply. Once the supply is more than 30 in away from the fixture, it is considered a water distribution pipe. If you want to know the required flow rates and pressures for various fixtures, simply refer to your code book. All this information is available to you in an easy-to-understand format.

The next step is not quite as easy, but it still is not difficult. There are several methods for sizing water pipe. Most code books disclaim any responsibility for the examples they give on sizing. The examples are just that—examples, not carved-in-stone procedures. One of the easiest ways to size water pipe is with the use of the fixture-unit method, and that is the one we concentrate on. Let me show you how it works.

The fixture-unit method

The fixture-unit method of sizing water distribution pipes is pretty easy to understand. You will need to know the total number of fixture units that will be placed on your pipes. This information can be obtained from listings in your code book.

The total developed length of your water pipes must be known. If you are on the job, you can measure this distance with a tape measure. When you are working from blueprints, a scale rule will give you the numbers you need. The measurement begins at the location of the water meter.

You will also need to know the water pressure on the system. If you are not controlling the water pressure, as you might with a pressure-reducing valve or a well system, you can call the municipal water department to obtain the pressure rating on the water main.

To pinpoint the accuracy of your sizing, find out the difference in elevation between the water meter and the highest plumbing fixture. However, by being generous with your pipe sizes, you can get by without this information—most of the time.

The fixture units assigned to fixtures for the purpose of sizing water pipe will normally include both the hot and cold water demand of the fixture. The ratings for fixtures used by the public will be different from those fixtures installed for private use. All the information you need for this method of sizing is easily obtained from your local code book.

Now, let's size the water pipe for a small house. The house has one full bathroom and a kitchen. It also has a laundry hookup and one hose bib. We need to know what fixture-unit (FU) ratings to assign to these fixtures. A quick look at a code book might give us the following ratings:

Bathtub: 2 FU   Lavatory: 1 FU

Toilet: 3 FU   Sink: 2 FU

Hose bib: 3 FU   Laundry hookup: 2 FU

Okay, now we know the total number of fixture units. The working pressure on this water system is 50 psi. The total developed length of piping, from the water meter to the most remote fixture, is 95 feet (ft).

We now have all we really need to size water pipe, with a little help from a table in our code book. When we look for the sizing table in the code book, we see different tables for different pressure ratings. We choose the table that matches our rating of 50 psi.

There will be columns of numbers to identify fixture units and developed lengths of pipe. We will see sizes for building water supplies and branches. Even the sizes for water meters will be available. All we have to do is put our plan into action.

Our water meter for this example is a $\frac{3}{4}$-in meter. First we have to size the water service. How many fixture units will be on the water service? When we count all the fixture units, the total is 13. The pipe has to run 95 ft, so what size does it have to be? Looking at the sizing table in your code book, you will quickly see that the required size is $\frac{3}{4}$-in water service. I was able to determine this just by running my finger down and across the sizing table.

The sizing table is based on the length of the pipe run, the water pressure, and the size of the water meter. Knowing those three variables, you can size pipe quickly and easily. The same cross-referencing that was done to size the water service also works for sizing water distribution pipes.

For example, assume that the most remote fixture in the house is 40 ft from the end of the water service. When I look at my table, I see that a $\frac{3}{4}$-in pipe is required to begin the water distribution system. As the main $\frac{3}{4}$-in pipe reaches fixtures and lowers the number of fixture units remaining in the run, the pipe size can be reduced. As a rule of thumb, when there are only two fixtures remaining to be served on a residential water main, the pipe size can be reduced to a $\frac{1}{2}$-in pipe.

Let's say the $\frac{3}{4}$-in pipe is running down the center of the home. The kitchen sink is perpendicular to the main water line and lies 22 ft away. The sink has a fixture-unit demand of 2. Checking my sizing table, I see that the pipe branching off the main to serve the sink can be a $\frac{1}{2}$-in pipe.

Once you sit down with your code book and work with the fixture-unit method of pipe sizing, you will see how easy it is. Now, we are done with our sizing exercise, and we are going to move on to some techniques used to install water distribution pipes.

Installing the Potable Water System

Regardless of how the connections are made, CPVC and copper systems are installed according to the same principles. Polybutylene systems could be installed along the same lines as copper, but a manifold installation makes more sense with polybutylene. PB pipe is available in semirigid lengths, in addition to coils.

Is a water service part of a water distribution system? It is part of the water system, but it is not normally considered to be a part of a water distribution system. This can be confusing, so let me explain.

A water service does convey potable water to a building, but it is not considered a water distribution pipe. Water services fall into a category all their own. Water distribution pipes are the pipes found inside a building. In water services the majority of the pipe length is outside a building's foundation. They are usually buried in the ground. Water distribution pipes are installed within the foundation of the building and don't normally run underground or outside a foundation. However, they can, as in the case of a one-story building where the water distribution pipes are buried under a concrete floor. In any event, water services and water distribution pipes are dealt with differently by the plumbing codes.

Why are water distribution pipes considered different from water services? Water distribution pipes are dealt with differently because they serve different needs. Water distribution pipes pick up where water services leave off. The pipes distributing water to plumbing fixtures do not have to meet the same standards as underground water services.

For one thing, the water pressure on a water distribution pipe is often less than that of a water service. Most homes have water pressure ranges between 40 and 60 psi. A water service from a city main could easily have an internal pressure of 80 psi, or more.

Water services are also buried in the ground. Water distribution pipes rarely are. Underground installation can affect the types of pipes that are approved for use. Another big difference between water service pipes and water distribution pipes is the temperature of the water they contain. Water services deliver cold water to a building. Water distribution pipes normally distribute both hot and cold water. Since most codes require the same type of pipe to be used for the cold water lines as for hot water pipes, the temperature ranges can limit the applications of some types of materials. This is a significant factor when you are choosing a water distribution material.

Valves

Many valves are required in a legal water distribution system. The types of valves used and the places where they are required are mandated by the code. For example, a full-open valve is required near the origination of a water service. In the case of a city water hookup, the valve is supposed to be installed near the curb of the street. Water services extending from a well are not required to have such a valve at their origination. A full-open valve is a valve like a gate valve or ball valve. When these valves are open, water has the full diameter of the valve to flow through, rather than a restricted opening, as in valves with washers and seats.

A full-open valve is required on the discharge side of every water meter. The main water distribution pipe must be fitted with a full-open valve where it meets the water service. In nonresidential buildings, a full-open valve must be installed at the origination of every riser pipe. In these nonresidential buildings, a full-open valve is required at the top of all drop-feed pipes.

If a water supply pipe is feeding a water tank, such as a well pressure tank, a full-open valve is required. All water supplies to water heaters must be equipped with full-open valves. If a building contains multiple dwellings, the water supplies for each dwelling must be equipped with full-open valves. Remember that not all codes have the same requirements. Check your local code for applicable restrictions.

The valves we are about to discuss are not required to be full-open valves. The valves used in the following locations are usually either stop-and-waste valves or supply-stop valves. Valves are required on all supply pipes feeding sill cocks and hose bibs. If a water supply is serving an appliance, such as an ice maker or dishwasher, a valve is required.

Some types of buildings are required to have individual cutoffs on the supplies to all plumbing fixtures. Other types of buildings are not. It is generally standard procedure to install cutoffs on all water supplies, with the possible exception of the pipes feeding tub and shower valves.

When a valve is required by the code, that valve must be accessible. It may be concealed by a door that opens or some other means of accessible concealment, but it must be accessible. Stop-and-waste valves are normally not approved for use below ground.

Backflow Protection

Backflow protection for water systems has become a major issue in the plumbing trade. The forms of protection required involve such issues as cross-connections, backflow, and vacuums. If backflow preventers are not installed, entire water mains and water supplies can be at risk of contamination.

Many types of cross-connections are possible. For example, hot water may pass into cold water pipes through a cross-connection. This shouldn't contaminate the system, but it can create a health risk. Someone who turns on what is believed to be cold water and discovers that it's hot water could be burned. Normally, this type of cross-connection is more of a nuisance than a hazard, but it does have the potential to cause injury.

A much more serious type of cross-connection might occur in a commercial photography processing plant. What would happen if the contents of a pipe conveying photography chemicals were allowed to mix with the potable water system? The results could be quite serious, indeed.

Cross-connections are normally created by accident, but sometimes they are designed. If a cross-connection of some sort is to be installed, it must be protected with the proper devices to avoid unwanted mixing. Few, if any, codes will allow a cross-connection between a private water source, such as a well, and a municipal water supply. Check your local code requirements for backflow prevention devices.

An air gap

An *air gap* is the open airspace between the flood-level rim of a fixture and the bottom of a potable water outlet. For example, usually there is an air gap between the outlet of a spout on a bathtub and the flood-level rim of the tub.

By having such an air gap, nonpotable water, such as dirty bath water, cannot be sucked back into the plumbing system through the potable water outlet. If the tub spout were mounted so that it was submerged in the tub water, the dirty water could be pulled into the water pipes if a vacuum formed in the plumbing system.

The amount of air gap required on various fixtures is determined by the plumbing code. For example, in a table or in the text of your code, you might find that the tub spout opening must not be closer than 2 in to the flood-level rim. The same rule for lavatories might require the spout of the faucet to remain at least 1 in above the flood-level rim.

There is another type of air gap used in plumbing. This fitting accepts the waste from a dishwasher for conveyance to the sanitary drainage system. The device allows the water draining out of the dishwasher to pass through open airspace on its way to the drainage system. Since the water passes through open air, it is impossible for wastewater from the drainage system to be siphoned back into the dishwasher.

Backflow preventers

Backflow preventers are devices that prohibit the contents of a pipe from flowing in a direction opposite its intended flow. These protection devices come in many shapes and sizes, and they cost from a few dollars to thousands of dollars. Backflow preventers must be installed in accessible locations.

When a potable water pipe is connected to a boiler to provide a supply of water, the potable water pipe must be equipped with a backflow preventer. If the boiler has chemicals in it, such as antifreeze, the connection must be made through a backflow preventer of an air gap that works on the principle of reduced pressure.

Anytime a connection is subject to back pressure, the connection must be protected with a reduced-pressure type of backflow preventer.

Check valves

Check valves have an integral section, something of a flap, that allows liquids to flow freely in one direction but not at all in the opposite direction. Check valves are required on potable water supply pipes that are connected to automatic fire sprinkler systems and standpipes.

Vacuum breakers

Vacuum breakers are installed to prevent a vacuum from being formed in a plumbing system that could result in contamination. Some of these devices mount in line on water distribution pipes, and others screw onto the threads of sill cocks and laundry tray faucets.

When vacuum breakers are required, the critical level of the device is usually required to be set at least 6 in above the flood-level rim of the fixture. What is the critical level? The *critical level* is the point at which the vacuum breaker would be submerged, prior to backflow.

Normally, any type of fixture that is equipped with threads for a garden hose must be fitted with a vacuum breaker. Some sill cocks are available with integral vacuum breakers. For the ones that are not, small vacuum breakers that screw onto the hose threads are available. After the devices are secured on the threads, a setscrew is turned into the side of the device. The screw is tightened until its head breaks off. This ensures that the safety device cannot be removed easily.

Since vacuum breakers have an opening that leads to the potable

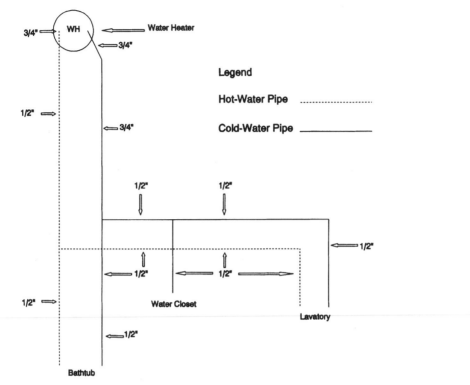

Figure 15.5 Water pipe riser diagram.

water system, they cannot be installed where toxic fumes or vapors may enter the plumbing system through the openings. One example of a place where they could not be installed is under a range hood.

The starting point for your potable water system will be where the water service enters the house (see Figs. 15.5 and 15.6). Connect your main cold water pipe to the water service. In most places you will be required to install a backflow preventer on the water service pipe. If your water service is polyethylene pipe, it must be converted to a pipe rated for hot and cold water applications. Polyethylene cannot handle the high temperature of hot water. The code requires you to use the same type of pipe for both hot and cold water. Since polyethylene cannot be used for hot water, it cannot be used as an interior water distribution pipe in a house that has hot water.

You can convert the polyethylene pipe to the pipe of your choice with an insert adapter. The insert portion fits inside the polyethylene pipe and is held in place by stainless-steel clamps. The other end of the fitting is capable of accepting the type of pipe you are using for

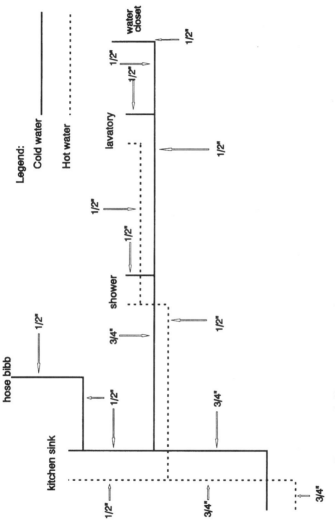

Figure 15.6 Diagram of efficient water supply.

water distribution. In most cases the adapter is threaded on the conversion end. The threads allow you to mate any type of female adapter to the insert-by-male adapter.

Once the conversion is made, install a gate valve on the water distribution pipe. When required, install a backflow preventer following the gate valve. It is a good idea to install another gate valve after the backflow preventer. By isolating the backflow preventer with the two gate valves, you can cut off the water on both sides of the backflow preventer. This option will be appreciated if you must repair or replace the backflow preventer.

As you bring pipe out of the second gate valve, consider installing a sill cock near the rising water main. If the location is suitable, you will use a minimum amount of pipe for the sill cock. The main water distribution pipe for most houses will be a $\frac{3}{4}$-in pipe. If you have already designed your water distribution layout, install the pipe accordingly. If you haven't made a design, do it now.

The $\frac{3}{4}$-in pipe should go to the inlet of the water heater, undiminished in size. You may choose to branch off the main to supply fixtures with water as the main goes to the water heater. This is acceptable and economical.

Water pipe will normally be installed in the floor or ceiling joists so that they can be hidden from plain view. When possible, keep the pipe at least 2 in from the top or the bottom of the joists. Keep in mind that your local building code will prohibit you from drilling, cutting, or notching joists close to their edges. Keeping pipes near the middle of studs and joists reduces the risk of the pipes being punctured by nails or screws. If the pipe passes through a stud or joist where it might be punctured, protect it with a nail plate.

Proper pipe support is always an issue that all plumbers must be aware of. Copper tubing with a diameter of $1\frac{1}{4}$ in or less should be supported at intervals not to exceed 6 ft. Larger copper can run 10 ft between supports. These recommendations are for pipes installed horizontally. If the pipes are run vertically, both sizes should be supported at intervals of no more than 10 ft.

If your material of choice is CPVC, it should be supported every 3 ft, whether installed vertically or horizontally. PEX and polybutylene pipe that is installed vertically must be supported at intervals of 4 ft. When installed horizontally, PB pipe should have a support every 32 in.

Avoid placing water pipes in outside walls and areas that will not be heated, such as garages and attics. The main delivery pipe for hot water will originate at the water heater. When you are piping the water heater, the cold water pipe should be equipped with a gate valve before it enters the water heater. Many places require the installation of a vacuum breaker at the water heater. These devices prohibit backsiphoning of the contents of the water heater into the cold water pipes.

Do not put a cutoff valve on the hot water pipe leaving the water heater. If a valve is installed on the hot water side, it could become a safety hazard. Should the valve become closed, the water heater could build excessive pressure. The combination of a closed valve and a faulty relief valve could cause a water heater to explode.

During the rough-in, you will not normally be installing valves on various fixtures. Most fixtures get their valves during the finish plumbing stage. If you install your sill cocks or hose bibs during the rough-in, install stop-and-waste valves in the pipe supplying water to the sill cock or hose bib. Stop-and-waste valves have an arrow on the side of the valve body. Install the valve so that the arrow is pointing in the direction in which the water flows to the fixture.

Rough-in numbers

Refer to a rough-in book (available from plumbing suppliers for various brands of fixtures) to establish the proper location for your water pipes to serve the fixtures. The examples given here are common rough-in numbers, but your fixtures may require a different rough-in. The water supplies for sinks and lavatories are usually placed 21 in above the subfloor. Kitchen water pipes are normally set 8 in apart, with the centerpoint being the center of the sink. Lavatory supplies are set 4 in apart, with the center being the center of the lavatory bowl.

Most toilet supplies will rough in at 6 in above the subfloor and 6 in to the left of the closet flange, as you face the back wall. Shower heads are routinely placed 6 ft 6 in above the subfloor. They should be centered above the bathing unit. Tub spouts are centered over the tub's drain and roughed in 4 in above the tub's flood-level rim. The faucet for a bathtub is commonly set 12 in above the tub and centered on the tub's drain. Shower faucets are generally set 4 ft above the subfloor and centered in the shower wall. If you ask your supplier for a rough-in book, the supplier should be able to give you exact information for roughing in your fixtures.

Secure all pipes near the rough-in for fixtures. When you rough in for a shower head, use a wing ell. These ells have 0.5-in female threads to accept the threads of a shower arm. The wing ell has an ear on each side that allows you to screw the ell to a piece of backing in the wall. Securing the ell is convenient for later installation of the shower arm. Where necessary, install backing in the walls to which all water pipes can be attached. It is very important to have all pipes firmly secured.

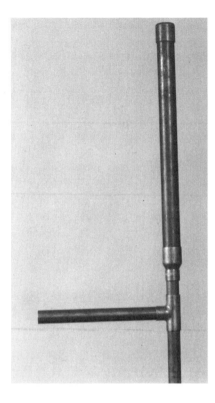

Figure 15.7 Air chamber.

If you are soldering joints, remember to open all valves and faucets before soldering them. Failure to do this could damage the washers and internal parts once heat is applied. Consider installing air chambers to reduce the risk of noisy pipes later. The air chamber can be made from the same pipe as you use in the rest of the house. Air chambers should be at least one pipe size larger than the pipes they serve. A standard air chamber will be about 12 in tall (Fig. 15.7).

Installing cutoff valves in the pipes feeding the bathtub or shower is optional. Most codes do not require the installation of these valves, but the valves come in handy when you have to work on the tub or shower faucets in later years. Stop-and-waste valves are typically used for this application. Be sure to install the valves where you will have easy access to them after the house is finished.

Avoid extremely long, straight runs of water pipe. If you are running the pipe a long way, install offsets in it. The offsets reduce the risk of annoying water hammer later. A little thought during the rough-in can save you a lot of trouble in the future.

A Manifold System

The rules for plumbing a water distribution system are basically the same for all approved materials. But when polybutylene is used, often it makes more sense to use a manifold than to run a main with many branches. One of the most desirable effects of this type of installation is the lack of concealed joints. All the concealed piping is run in solid lengths, without fittings.

To plumb a manifold system, run the water main to the manifold. The manifold can be purchased from a plumbing supplier. It will come with cutoff valves for the hot and cold water pipes. The cold water comes into one end of the manifold and goes out the other to the water heater. The hot water comes into the manifold from the water heater. The manifold is divided into hot and cold sections. The pipe from each fixture connects to the manifold. When there is a demand for water at a fixture, it is satisfied with the water from the manifold. Once you have both the hot and cold supply to the manifold, all you have to do is run individual pipes to each fixture.

You can snake coiled polybutylene pipe through a house in much the same way as you would electric wires. By installing the water distribution system in this way, you save many hours of labor. You also eliminate concealed joints, and you can cut off any fixture with the valves on the manifold. This type of system is efficient, economical, and the way of the future.

Tips for Underground Water Pipes

To close this chapter, let me give you a few tips for underground water pipes. When you are ready to run water pipe below grade, there are a few rules you must obey. Pipe used for the water service must have a minimum working pressure of 160 psi at 73.4°F. If the water pressure in your area exceeds 160 psi, the pipe must be rated to handle the highest pressure available to the pipe. If the water service is a plastic pipe, it must terminate within 5 ft of its point of entry into the home. If the water service pipe is run in the same trench as the sewer, special installation procedures are required (refer to your local plumbing code).

When running the water service and the sewer in a common trench, you must keep the bottom of the water service at least 12 in above the top of the sewer. The water service must be placed on a solid shelf of firm material to one side of the trench. The sewer and the water service should be separated by undisturbed or compacted earth. The water service must be buried at a proper depth to protect it from freezing temperatures. The depth required will vary from state to state. Check with the code enforcement office for the proper depth in your area.

Copper pipe run under concrete should be type L or type K copper. Where copper pipe will come into direct contact with concrete, it should be sleeved to protect the copper. Concrete can cause a damaging reaction when in direct contact with copper. The pipe may also vibrate during use and wear a hole in itself by rubbing against rough concrete. All water pipe should be installed so as to prevent abrasive surfaces from coming into contact with the pipe.

Rules for underground piping are different from those for above-ground systems. Consult your local plumbing code to identify differences which affect plumbers in your area.

16

Well Systems

Plumbers in rural locations work with well systems regularly. They install them, and they service them. City-based plumbers have little experience with well systems. Even plumbers in the country don't always have a lot of experience with well systems. In some regions, well drillers control most of the pump business. But if you work in an area where water wells and pumps are used, you should be prepared for the service calls you may get.

There is good money to be made in well systems. Whether you are installing new systems or fixing problems with existing pumps, the pay can be lucrative. How much do you have to know? It depends on what your plans are. It takes less knowledge to install new well systems than it does to troubleshoot and repair them.

Most pump makers offer booklets to professional plumbers (through plumbing suppliers) that offer troubleshooting tips and techniques. These publications are invaluable when you are in the field with a pump system that you are not familiar with. Ask your supplier for booklets on all the brands of pumps used in your area. You will probably have to visit several suppliers to obtain literature on all the various types of pumps. If suppliers can't help you, contact the pump makers directly. They will be happy to send you all sorts of information on their products. When you combine specific manufacturer information with the data in this chapter, you will be a formidable force in the installation and repair of well systems.

Shallow-Well Jet Pumps

Shallow-well pumps are used with wells where the lowest water level will not be more than 25 ft below the pump. Many factors influence the type and size of pump that an installation will require. The first consideration is the height to which water will be pumped. If a pump has to pump water higher than 25 ft, a shallow-well pump is not a viable choice. Shallow-well pumps are not intended to lift water more than 25 ft. This limitation is due to the way that a shallow-well pump works.

These pumps work on a suction principle. The pump sucks the water up the pipe and into the home. With a perfect vacuum at sea level, a shallow-well pump may be able to lift water to 30 ft. This maximum lift is not recommended and is rarely achieved. If the pump has to lift water higher than 25 ft, investigate other types of pumps. When conditions allow the use of a shallow-well jet pump, this section tells you how to install it. Talk to your pump dealer or refer to the manufacturer's recommendations for proper sizing of the pump.

A suction pump

When you install a suction pump, the single pipe from the well to the pump usually has a diameter of $1\frac{1}{4}$ in. A standard well pipe material is polyethylene, rated for 160 psi. Make sure that the suction pipe is not coiled or in any condition that may cause it to trap air as it is being installed. If the pipe holds an air pocket, priming the pump can be quite difficult. In most cases a foot valve is installed on the end of the pipe that is submerged in the well.

Screw a male insert adapter into the foot valve. Place two stainless-steel clamps over the well pipe, and slide the insert fitting into the pipe. Tighten the clamps to secure the pipe to the insert fitting. When you lower the pipe and foot valve into the well, don't let the foot valve sit on the bottom of the well. If the suction pipe is too close to the bottom of the well, it may suck sand, sediment, or gravel into the foot valve. If this happens, the pipe cannot pull water from the well.

When the pipe reaches the upper portion of the well, it usually takes a 90° turn to exit the well casing. This turn is made with an insert-type elbow. Always use two clamps to hold the pipe to its fittings. When the pipe leaves the well, it should be buried underground. Run the well pipe through a sleeve so that the well casing will not chafe the pipe during use and wear a hole in it. The pipe must be deep enough that it will not freeze in the winter. This depth will vary from state to state. Your local plumbing inspector can tell you to what depth you must bury the water supply pipe.

When you place the pipe in the trench, be careful not to lay it on sharp rocks or other objects that might wear a hole in it. Backfill the trench with clean fill dirt. If you dump rocks and cluttered fill on the pipe, it can be crimped or cut. When you bring the pipe into the home, run it through a sleeve as it comes through or under the foundation. The sleeve should be two pipe sizes larger than the water supply pipe.

Once the pipe is inside the home, you may have to convert it to copper, CPVC, PB, or one of the other approved materials. When PE pipe is used as a water service, it must be converted to some other type of pipe within 5 ft of entry into the home, assuming that the house will have both hot and cold water. If you convert the pipe, the conversion will typically be done with a male insert adapter. The water supply pipe should run directly to the pump. The foot valve acts as a strainer and as a check valve. When there is a foot valve in the well, there is no need for a check valve at the pump.

The incoming pipe will attach to the pump at the inlet opening with a male adapter. At the outlet opening, install a short nipple and a tee fitting. At the top of the tee, install reducing bushings and a pressure gauge. From the center outlet of the tee, the pipe will run to another tee fitting. A gate valve should be installed in this section of pipe, near the pump. At the next tee, the center outlet is piped to a pressure tank. From the end outlet of the tee, the pipe will run to yet another tee fitting. At this tee, the center outlet becomes the main cold water pipe for the house. Another gate valve should be installed in the pipe feeding the water distribution system. On the end outlet of the tee, install a pressure relief valve. All these tee fittings should be in a close proximity to the pressure tank.

The pump is equipped with a control box that requires electric wiring. This job should be done by a licensed electrician. If you are an electrician, you know how to do the job. If you are not an electrician, do *not* attempt to wire the controls.

Priming the pump

The pump has a removable plug in the top of it, to allow the priming of the pump. Remove the plug, and pour water into the priming hole. Continue this process until the water is standing in the pump and visible at the hole. Apply pipe dope to the plug, and screw it back into the pump. When you turn on the pump, you should have water pressure. If you don't, continue the priming process until the pump is pumping water. This can be a time-consuming process; don't give up.

Setting the water pressure

Once the pump is pumping water and filling the pressure tank, setting the water pressure is a priority. When the tank is filled, the pressure gauge should read between 40 and 60 psi. The pump's controls are preset at cut-on and cutoff intervals. These settings regulate when the pump cuts on and off. Typically, a pump cuts on when the tank pressure drops below 20 psi. The pump cuts off when the tank pressure reaches 40 psi.

If your customer prefers a higher water pressure, the pressure switch can be altered to deliver higher pressure. You might have the controls set to cut on at 40 psi and off at 60 psi. These settings are adjusted inside the pressure switch, around electric wires. There is possible danger of electrocution when you are making these adjustments. To avoid problems, cut off the power to the pressure switch while you are making the adjustments.

The adjustments are made by turning a nut that sits on top of a spring in the control box. You will see a coiled spring, compressed with a retaining nut. By moving this nut up and down the threaded shaft, you can alter the cut-on and cutoff intervals. Refer to the manufacturer's recommendations for precise settings.

Deep-Well Jet Pumps

When the water level is more than 25 ft below the pump, you have to use either a deep-well jet pump or a submersible pump. In today's plumbing applications, submersible pumps are normally used in deep wells. However, deep-well jet pumps will get the job done.

Deep-well jet pumps resemble shallow-well pumps. They are installed above ground and are piped in a similar manner to shallow-well pumps. The noticeable difference is the number of pipes going into the well. A shallow-well pump has only one pipe. Deep-well jet pumps have two pipes. The operating principles of the two types of pumps differ. Shallow-well pumps suck water up from the well. Deep-well jet pumps push water down one pipe and suck water up the other; this is why there are two pipes on deep-well jet pumps.

The only major difference between a shallow-well pump and a deep-well pump is the number of pipes used in the installation and a pressure control valve. Deep-well pumps still use a foot valve. A jet body fitting is submerged in the well and attached to both pipes and the foot valve. The pressure pipe connects to the jet assembly first. The foot valve hangs below the pressure pipe. There is a molded fitting on the jet body for the suction line to connect to. With this jet body, both pipes are allowed to connect in a natural and efficient manner.

Deep-well jet pumps exert pressure *down* the pressure pipe; with water pushed through the jet assembly, it makes it possible for the suction pipe to pull water *up* from the deep well. From the suction pipe, water is brought into the pump and is distributed to the potable water system. At the head of a deep-well jet pump, you will see two openings for the pipes to connect to. The larger opening is for the suction pipe, and the smaller opening is for the pressure pipe. The suction pipe usually has a diameter of $1\frac{1}{4}$ in. The pressure pipe typically has a diameter of 1 in.

The piping from the pump to the pressure tank needs a pressure control valve installed in it. This valve ensures a minimum operating pressure for the jet assembly. Shallow-well pumps are not required to have a pressure control valve. Once the pressure control valve is installed, the remainder of the piping is done in the same manner as for a shallow-well pump.

Submersible Pumps

Submersible pumps are very different from jet pumps. Jet pumps are installed outside the well. Submersible pumps are installed in the well, submerged in the water. Jet pumps use suction pipes. Submersible pumps have only one pipe, and they push water up the pipe, from the well. Jet pumps use a foot valve; submersible pumps don't. Submersible pumps are much more efficient than jet pumps; they are also easier to install. Under the same conditions, a 0.5-horsepower (hp) submersible pump can produce nearly 300 gal more water than a 0.5-hp jet pump. With so many advantages, it is almost foolish to use a jet pump, when you could use a submersible pump. The one exception occurs when you are installing a pump for a shallow well. Then a jet pump makes the most sense.

Installing a submersible pump requires different techniques from those used with a jet pump. Since submersible pumps are installed in the well, electric wires must run down the well to the pump. Before you install a submersible pump, consult a licensed electrician about the wiring needs of the pump. Some plumbers do this wiring themselves, but many jurisdictions require the connections to be made by a licensed electrician.

You will need a hole in the well casing to install a pitless adapter. The pitless adapter provides a watertight seal in the well casing for the well pipe to feed the water service. When you buy the pitless adapter, it should be packaged with instructions on what size hole is needed in the well casing.

Cut a hole in the well casing with a cutting torch or a hole saw. The pitless adapter attaches to the well casing and seals the hole. On the inside of the well casing, there is a tee fitting on the pitless adapter. This is where the well pipe is attached. This tee fitting is designed to allow you to make all pump and pipe connections above ground. After all the connections are made, lower the pump and pipe into the well, and the tee fitting slides into a groove on the pitless adapter.

To make up the pump and pipe connections, you need to know how deep the well is. The well driller should provide you with the depth and rate of recovery for the well. Once you know the depth, cut a piece of plastic well pipe to the desired length. The pump should hang at least 10 ft above the bottom of the well and at least 10 ft below the lowest expected water level.

Apply pipe dope to a male-insert adapter, and screw it into the pump. This fitting is normally made of brass. Slide a torque arrester over the end of the pipe. Next, slide two stainless-steel clamps over the pipe. Place the pipe over the insert adapter, and tighten the two clamps. Compress the torque arrester to a size slightly smaller than the well casing, and secure it to the pipe. The torque arrester absorbs thrust and vibrations from the pump and helps to keep the pump centered in the casing.

Slide the torque stops down the pipe from the end opposite the pump. Space the torque stops at routine intervals along the pipe to prevent the pipe and wires from scraping against the casing during operation. Secure the electric wiring to the well pipe at regular intervals to eliminate slack in the wires. Apply pipe dope to a brass male-insert adapter, and screw it into the bottom of the tee fitting for the pitless adapter. Slide two stainless-steel clamps over the open end of the pipe, and push the pipe onto the insert adapter. After tightening the clamps, you are ready to lower the pump into the well.

Before you lower the pump, it is a good idea to tie a safety rope onto the pump. After the pump is installed, this rope will be tied at the top of the casing to prevent the pump from being lost in the well, if the pipe becomes disconnected from the pump. Next, screw a piece of pipe or an adapter into the top of the tee fitting for the pitless adapter. Most plumbers use a rigid piece of steel pipe for this purpose.

Once you have a pipe extending up from the top of the tee fitting, lower the whole assembly into the well casing. This job is easier if you have someone to help you. Be careful not to scrape the electric wires on the well casing as you lower the pump. If the insulation on the wires is damaged, the pump may not work. Holding the assembly by the pipe extending from the top of the pitless tee, guide the pitless adapter into the groove of the adapter in the well casing. When the adapter is in the groove, push down to seat it in the mounting bracket. This concludes the well part of the installation.

Attach the water service pipe to the pitless adapter on the outside of the casing. You can do this with a male-insert adapter. Run the pipe to the house in the same way as described for jet pumps. Once the pipe is inside the house, install a union on the water pipe. The next fitting should be a gate valve, followed by a check valve. From the check valve, the pipe should run to a tank tee.

The tank tee is a device that screws into the pressure tank and allows the installation of all related parts. The switch box, pressure gauge, and boiler drain can all be installed on the tank tee. When the pipe comes to the tank tee, the water is dispersed to the pressure tank, the drain valve, and the water main. Where the water main leaves the tank tee, you should install a tee to accommodate a pressure relief valve. After this tee, you can install a gate valve and continue piping to the water distribution system. All that is left is to test the system; you do not have to prime a submersible pump.

Pump Problems

Pump problems are not uncommon. If you offer your services as a plumber who troubleshoots and repairs pumps, you have to be prepared for whatever you might find. I've already told you that pump makers often offer troubleshooting booklets to professionals. Take advantage of this. Many problems are the same from pump to pump, regardless of brand, but it is always best to have information specific to the particular type of pump that you are working with.

I can't describe every detail for all the various types of pumps being used. But I can give you some solid troubleshooting tips that will often apply to all types of pumps. Servicing pumps is sometimes a matter of trial-and-error testing, but if you have the right guidelines to follow, you can shorten the time you spend trying to identify a problem. Let's start with jet pumps.

Troubleshooting Jet Pumps

When a jet pump will not run

Circuit breakers and fuses.   When a pump won't run (Table 16.1), there are five likely causes of the malfunction. First check the fuse or circuit breaker that controls the pump's electric circuit. If the fuse is blown or the breaker is tripped, the pump cannot run. When you check the electric panel and find these conditions, replace the blown fuse or reset the tripped breaker. In future references to fuses or circuit breakers, I will only refer to circuit breakers. If a home has fuses, when I say to reset the circuit breaker, replace the blown fuse.

Damaged or loose wiring.   If the circuit breaker is not tripped, inspect all the electric wiring that affects the pump. A broken or loose wire may be preventing the pump from running. Remember, you are dealing with electricity and must use appropriate caution to avoid electric shocks.

TABLE 16.1   Jet Potable-Water Pumps

Symptoms	Probable cause
Won't start	No electric power Wrong voltage Bad pressure switch Bad electrical connection Bad motor Motor contacts open Motor shaft seized
Runs, but produces no water	Needs to be primed Foot valve is above water level in well Strainer clogged Suction leak
Starts and stops too often	Leak in piping Bad pressure switch Bad air control valve Waterlogged pressure tank Leak in pressure tank
Low water pressure in pressure tank	Strainer on foot valve partially blocked Leak in piping Bad air charger Worn impeller hub Lift demand too much for pump
Pump does not cut off when working pressure is obtained	Pressure switch bad Pressure switch needs adjusting Blockage in piping

Pressure switches. Pressure switches can become defective. It is also possible that the switch is out of proper adjustment. To correct this problem, adjust or replace the pressure switch. Remove the cover of a pressure switch; you will see two nuts sitting on top of springs. The nut on the short spring is set at the factory and should not need any adjustment. However, if you want a pump to cut off at a higher pressure, turn this nut clockwise. To have the pump cut off at a lower pressure, turn the nut counterclockwise. The nut on the taller spring controls the cut-on/cutoff cycle of your pump. To make the pump cut on at a higher pressure, turn the nut clockwise. If you want the pump to cut on at a lower pressure, turn the nut counterclockwise.

Stopped-up nipples. If the tubing or nipples on the pressure switch become restricted, the pump may not run. If you suspect that this is the problem, take apart the pipe and fittings and clean or replace them. In a new installation, this problem is unlikely, but it is possible.

Overloaded motors. If the motor becomes overloaded, protection contacts will remain open. This problem will solve itself. The contacts will close automatically after a short time.

Seized pumps. If a pump becomes mechanically bound, it will not run. To correct this problem, first remove the end cap from the pump. Now you should be able to turn the motor shaft with your hands. If the shaft rotates freely, reassemble the pump and test it. The problem should be corrected.

Voltage problems. If a pump is not receiving proper voltage, it will not run. This problem calls for direct contact with the electrical system. Because of the danger involved in working with electricity, I cannot help you with this aspect of troubleshooting. If you have eliminated all the nonelectrical problems, call in a professional electrician to fix the pump.

Defective motors. Obviously, defective motors can prevent pumps from running. To determine if a pump's motor is bad, you must do extensive electrical troubleshooting. Check the ground, capacitor, switch, overload protector, and winding continuity. If you are experienced in working with electrical systems, you already know how to do this. If you are not, I am afraid to give you instructions on such a dangerous job. One wrong move could deliver a potent electric shock. Again, if the problem is electrical, call in a professional electrician.

Pump runs, but produces no water

Loss of the pump's prime. If a pump loses its prime, the pump will run without producing water. With a shallow-well pump, remove the priming plug and pour water into the pump. When the water level stands static at the opening, apply pipe dope to the priming plug and replace it. Start the pump, and you should get water pressure. With a deep-well jet pump, trying to prime the pump through the priming hole is a losing proposition.

With a two-pipe system, disconnect the pipes from the pump. Pour water down each pipe until they are both full of water. When water has filled both pipes and is holding its level, reconnect the pipes to the pump. Prime the pump through the priming hole. Start the pump, and you should get water. If you don't, the problem may be in the pressure control valve.

Pressure control valves. In a two-pipe system, the pressure control valve may need to be adjusted when a pump runs without producing water. When the pressure control valve is set too high, the air volume control may not function. When the pressure control valve is set too low, the pump may not cut off. To reduce the water pressure, turn the adjusting screw on the pressure-reducing valve counterclockwise. To increase pressure, turn the adjusting screw clockwise.

Lack of water. If a well pipe is not below the water level, the pump cannot produce water. With shallow wells, it is not uncommon for the water levels to drop during hot, dry months. If a pump is running but is not producing water, check the water level in the well. You can do this with a roll of string and a weight. Tie the weight on the end of the string, and lower it into the well. When the weight hits bottom, withdraw the string. The end of the string should be wet. Measure the distance along the wet string. This will tell you how many feet of water are in the well. This information, along with the records of how long the well pipe is, will tell you if the end of the pipe is submerged in water.

Clogged foot valves. If a foot valve is too close to the bottom of the well, it may become clogged with mud, sand, or sediment. It is not uncommon for shallow wells to fill in with sediment. What starts out as a 30-ft well may become a 25-ft well as the earth settles and shifts. Even if a foot valve was originally set at an appropriate height, it may now be hanging too low in the well. Sometimes, shaking the well pipe will clear the foot valve of its restriction. In other cases, you have to pull the pipe out of the well and clean or replace the foot valve. If the foot valve is pulling up sediment from the bottom of the well, shorten the well pipe.

Malfunctioning valves. Foot valves act as a check valve. If the foot valve or check valve is stuck in the closed position, you will not get any water from the pump. Pull the well line and inspect the check valve or foot valve. If the valve is stuck, replace it.

Defective air volume control. It is possible that the diaphragm in an air volume control has a hole in it. By disconnecting the tubing and plugging the connection in the pump, you can determine whether there is a hole in the diaphragm of the air volume control. If plugging the hole corrects the problem, you must replace the air volume control.

Failing jet assemblies. If the jet assembly fails, the pump will not produce water. Inspect the jet assembly for possible obstructions. If you find an obstruction, try to clear it. If you cannot clean the assembly, replace it.

Suction leaks. A common ailment of shallow-well systems is a leak in the suction pipe. To check for this problem, you must pressurize the system and inspect it for leaks. If you find a leak, plug it and you will have water again.

Pump will not cut off

If a pump builds pressure, but does not cut off, there may be a bad pressure switch. Before you replace the switch, there are a few things you should check. Check all the tubing, nipples, and fittings associated with the switch. If they are obstructed, clear them and test the pump. If the problem persists, check that the pressure switch is set properly. Do this by checking the cut-on and cutoff controls discussed earlier. If all else fails, replace the pressure switch.

Pump doesn't get enough pressure

A lack of water pressure can be caused by several problems. If a pump is trying to lift water too high, the pump may not be able to handle the job. This requires checking the rating of the pump to be sure it is capable of performing under the given conditions. If there are leaks in the piping, the pump will not be able to build adequate pressure. Check all piping to confirm that there are no leaks. If the jet or the foot valve is partially obstructed, the water pressure will suffer. Check these items to be sure they are free to operate. If they are blocked, clear the obstructions and test the pump.

When an air volume control is on the blink, you may not be able to build suitable pressure. Test the control as described above, and replace it if necessary. On older pumps, the impeller hub or guide vane bore may be worn. If this is the case, call a pump expert to verify your diagnosis. If either of these parts is worn, it must be replaced.

Pump cycles too often

Leaks in piping can cause a pump to cut on and off more frequently than it should. When the system is pressurized, check all piping for leaks. If any are found, repair them. If the pressure switch is not working correctly, the pump may run randomly. Follow the directions given earlier to troubleshoot the pressure switch. If necessary, replace the switch. The problem could be in the suction lift of the system. If water floods the pump through the suction line, you must control the water flow with a partially closed valve. Also there may not be enough vacuum on the suction line. During troubleshooting, check out the air volume control. Perhaps this valve is defective. The instructions given earlier explain how to check the air volume control.

Pressure tanks. The most likely cause of pumps that cycle too often is a faulty pressure tank. If the tank has a leak in it, it cannot hold pressure. In old tanks, this is a common problem. If you find a leak in the tank, replace the tank. If the air pressure in the pressure tank is not right, the pump will cut on frequently. The pump may cut on every time water to a faucet is turned on. This condition should not be ignored; it could burn out the motor of the pump.

Modern captive air tanks come with a factory precharge of air.

These tanks use a diaphragm to control the air volume. Older tanks do not have these diaphragms. Older tanks commonly become waterlogged. When the tank is waterlogged, it does not contain enough air and does contain too much water.

Air pressure. To check the air pressure in a tank, cut off the electric power to the pump. Drain the tank until the pressure gauge reads zero. A rule of thumb for proper air charge is that the air pressure should be 2 psi less than the cut-on pressure of the pressure switch. For example, if the pressure switch is set to cut on at 40 psi, the air in the tank should be set at 38 psi. You can check the air pressure by putting a tire gauge on the air valve of the tank.

If the air pressure is too low, pump air into the tank. You can do this with a bicycle pump or an air compressor. Monitor the pressure as you fill the tank with air. When the air pressure is at the proper setting, turn the electric power back on to the pump. When the water floods the tank, the pressure gauge should have a normal reading. The system should work properly if the air pressure remains at the proper setting.

If the pressure tank is old, it may have a small leak that allows the air to escape. It is possible that a new tank is defective. If this procedure does not work and keep the system working, consider replacing the pressure tank.

Submersible Pumps

Pump won't start

Circuit breakers. The first item to check when a submersible pump won't start (Table 16.2) is the circuit breaker for the pump. If the breaker is tripped, the pump cannot run. If you check the electric panel and find the breaker tripped, reset the tripped breaker.

Voltage problems. If the circuit breaker is not tripped, the voltage to the pump must be checked. This inspection can be done quickly with a voltage meter.

Pump starts but does not run

Submersible pumps that start but do not run should be checked in the following manner:

Fuses. Check all fuses or circuit breakers to see that they are not blown or tripped. If improper fuses are installed, the pump may start but not run.

TABLE 16.2   Submersible Potable Water Pumps

Symptom	Probable cause
Won't start	No electric power
	Wrong voltage
	Bad pressure switch
	Bad electrical connection
Starts, but shuts off fast	Circuit breaker or fuse inadequate
	Wrong voltage
	Bad control box
	Bad electrical connections
	Bad pressure switch
	Pipe blockage
	Pump seized
	Control box too hot
Runs, but does not produce water or produces only a small quantity	Check valve stuck in closed position
	Check valve installed backward
	Bad electric wiring
	Wrong voltage
	Pump sitting above water in well
	Leak in piping
	Bad pump or motor
	Broken pump shaft
	Clogged strainer
	Jammed impeller
Low water pressure in pressure tank	Pressure switch needs adjusting
	Bad pump
	Leak in piping
	Wrong voltage
Pump runs too often	Check valve stuck open
	Pressure tank waterlogged and needs air injected
	Pressure switch needs adjusting
	Leak in piping
	Wrong-size pressure tank

Voltage.   Incorrect voltage can allow the pump to start, but not allow it to continue running.

Seized pump.   If a pump has seized with sand or gravel, pull the pump and clean it. When the pump has seized, a high amperage reading is found from tests with a meter.

Damaged wiring.   If the electric wiring to the pump is damaged, the pump may not continue to run. This condition can be tested with a meter by reading the resistance. The insulation on the wires may have been damaged when the pump was lowered into the well. A spliced connection may have become open or short-circuited. When the problem is located, the wiring must be repaired or replaced.

Pressure switch. Your problem may be caused by a defective pressure switch. Check the points on the pressure switch, and replace it if the points are bad.

Pipe obstructions. If the tubing or nipples for the pressure switch are clogged, the pump may not run. Check for obstructions and clear the tubing or nipples if necessary.

Loose wires. It is possible that you are a victim of loose wires. Check the wiring at the control box and the motor. If the wires are loose, the pump may start, then stop.

Control box. If the pump has never run properly, it may have been fitted with the wrong control box. Inspect the box and confirm that it is the correct one for the pump. When the control box is exposed to high temperatures, customers may experience trouble with pumps. If a control box is located in an area where the temperature exceeds 120°F, the high temperature may be causing the problem.

Pump runs but gives no water

Clogs. If the protective screening on the pump is obstructed, water cannot enter the pump. This will cause the pump to run without providing water. When impellers are blocked by obstructions, you will experience the same problem. Pull the pump and check the strainer and impellers for clogs.

Bad motor. If the pump motor is badly worn, it may not produce water, even if it runs. You will have to pull the pump to inspect the motor. If it is worn, replace it.

Electrical problems. Improper voltage, incorrect connections, or loose connections could be the cause of your problem. You might discover these problems with an electric meter. If you are not comfortable working with electric wires, call an electrician to help you.

Leaks. If there are leaks in the piping, water may not be able to get to the house. Connect an air compressor to the pipe, and pump it up with air to test for leaks. Inspect all piping for leaks, and repair or replace the pipe as needed.

Check valves. Sometimes a check valve is installed backward. If there are problems with a new installation, check that the check valve was properly installed. It is also possible that the check valve has stuck in the closed position. Do a visual inspection, and replace or reinstall the check valve if necessary.

No water. Normally, drilled wells maintain a good water level, but at times the water level drops below the pump. If the pump is not submerged, it cannot pump water. Check the water level in the well with a string and a weight. If the water is low, you may be able to lower the pump; but don't hang the pump too close to the bottom of the well.

When a deep well is drilled, the installer should provide a recovery rate for the well. The recovery rate is stated in gallons per minute (gpm). The minimum recovery rate normally accepted is 3 gpm, and 5 gpm is a better rate. If a pump is pumping faster than the well can recover, the well may run out of water. Pumps are rated according to how many gallons per minute they pump. Compare the pump's rating to the recovery rate of the well, and you may find the cause of the trouble.

Low water pressure

If there are leaks in the piping, a pump will not be able to build adequate pressure. Check all piping to confirm that no leaks are present. Check the setting on the pressure switch. An improper setting will affect water pressure. If the pump is worn or the voltage to the pump is incorrect, you may not have strong water pressure.

Pump cycles too often

If a check valve is stuck in the open position, a pump will have to work overtime. Inspect the check valve to be sure it is operating freely. If the valve is bad, replace it. Leaks in the piping may cause a pump to cut on and off frequently. Check all piping for leaks. If any are found, repair them. If the pressure switch is not working correctly, the pump may cycle too often. Follow the directions given earlier to troubleshoot the pressure switch. If necessary, replace the switch.

The most likely cause of this problem is a faulty pressure tank. If the tank has a leak in it, it cannot hold pressure. In old tanks, this is a common problem. If you find a leak in the tank, replace the tank. If the air pressure in the pressure tank is not right, the pump will cut on frequently. The pump may cut on every time water to a faucet is turned on. This condition should not be ignored; it could burn out the pump motor.

Modern captive air tanks come with a factory precharge of air. These tanks use a diaphragm to control the air volume. Older tanks do not have these diaphragms. Older tanks commonly become waterlogged. When the tank is waterlogged, it does not contain enough air and does contain too much water.

To check a pressure tank's air pressure, cut off the electric power to the pump. Drain the tank until the pressure gauge is at zero. A rule of thumb for the proper air charge is that the air pressure should be 2 psi less than the cut-on pressure of the pressure switch. For example, if a pressure switch is set to cut on at 40 psi, air in the tank should be set at 38 psi. Check the air pressure by putting a tire gauge on the air valve of the tank.

If the air pressure is too low, pump air into the tank. You can use a bicycle pump or an air compressor. Monitor the pressure as you fill the tank with air. When the air pressure is at the proper setting, turn the electric power back on to the pump. When the water floods the tank, the pressure gauge should have a normal reading. The system should work properly if the air pressure remains at the proper setting. The pressure tank may have a small leak in it, allowing air to escape. If the tank cannot hold air, it cannot build pressure.

Well, there you have it. With the help of this chapter, you should be able to solve most problems associated with pump systems. In the next chapter we talk about septic systems. After all, if you are working with well pumps, you are probably working in an area where septic systems are common.

Troubleshooting and Repairs

17

Troubleshooting Toilets

Toilets are common plumbing fixtures, and they are not particularly complicated, but troubleshooting them can be a frustrating job. They are so simple, in some ways, that when there is a problem, it can be difficult to find. Plumbers with years of service experience have seen most everything a toilet can offer in terms of complications, but a lot of plumbers don't think about flush holes, condensation, and similar hidden defects that can drive a troubleshooter crazy. During my 20-year career, I've seen some mighty fine plumbers baffled by troublesome toilets, myself included.

Troubleshooting is largely a matter of logic, common sense, and systematic deduction of probable causes. As simple as the methodical act of troubleshooting is, it still gives a lot of plumbers plenty of misery. This chapter is going to help you avoid the hours of aggravation I have endured over the years.

In theory, there should be only a limited number of problems that may contribute to a toilet's malfunction. Theory is a wonderful thing, but it does not always prove itself out. I have found toilets with crumpled beer cans in their traps, and I've seen good plumbers replace toilets only to discover that the problem they were faced with wasn't the toilet at all. If you're lucky, you will never come face to face with a problem you can't solve, but the odds are good that if you stay in the trade long enough, you will get stumped by a stubborn toilet problem.

As we move through this chapter, you will learn aspects of toilet trouble that you may never have encountered. The benefit of learning what to anticipate, with the help of this book, will reduce your exposure to embarrassment and possible financial losses on the job. To stress this point, let me give you two real-life examples of how good, experienced plumbers were left dumbfounded by toilets. As we go through these examples, see if you can solve the problem before I give you the answer.

The Commercial Toilet That Couldn't Flush

This first example is about the commercial toilet that couldn't flush properly. The fixture was a wall-hung unit (Figs. 17.1 and 17.2), and it was installed beside another wall-hung toilet that flushed correctly. Both toilets shared the same drain and vent.

A call came in that one of the toilets was not flushing properly. I dispatched an experienced service plumber to investigate the problem. Shortly after arriving on the job, my plumber called in to say the drain was partially plugged and would need to be snaked out. We were busy that day, so I directed the on-site plumber, who had not been able to clear the stoppage with his small drain cleaner, to another call and told him to inform the property owner that a drain-cleaning crew would be there soon.

When the next plumber arrived with the big drain-cleaning

PLACIDUS
Siphon Jet Action
Wall Hung Closet

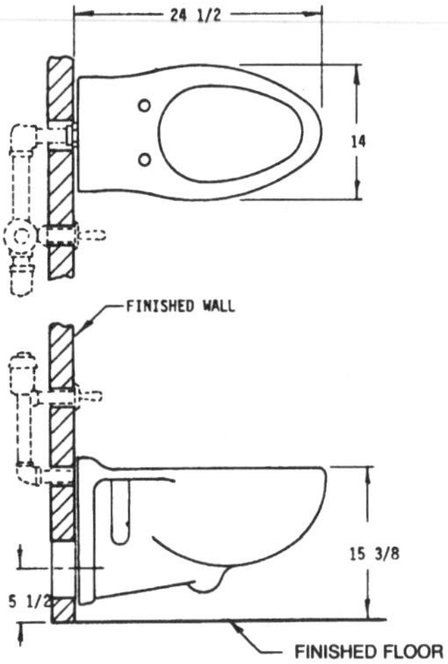

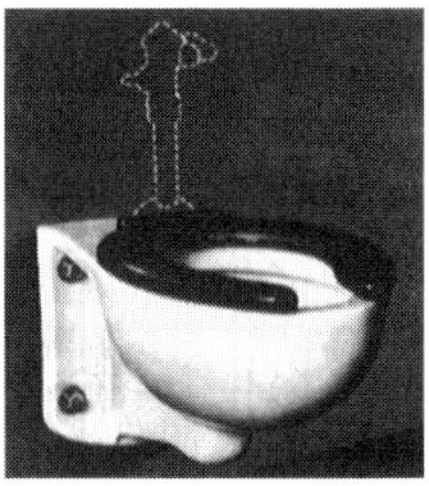

Placidus 3-449

Bowl: 3-449 **Placidus** water economy direct fed siphon jet, elongated rim, whirlpool action, 1½" back spud wall hung bowl.

Valve:

Bowl.

Support:

Note: Operation – For efficient operation of the bowl, a minimum flowing water pressure of 15 P.S.I. is required at the valve.

Figure 17.1 Wall-hung water closet with concealed flush valve. (*Courtesy of Crane Plumbing.*)

PLACIDUS CLOSET
VITREOUS CHINA WATER ECONOMY
SIPHON JET WHIRPOOL ACTION
CONCEALED FLUSH VALVE
ELONGATED RIM

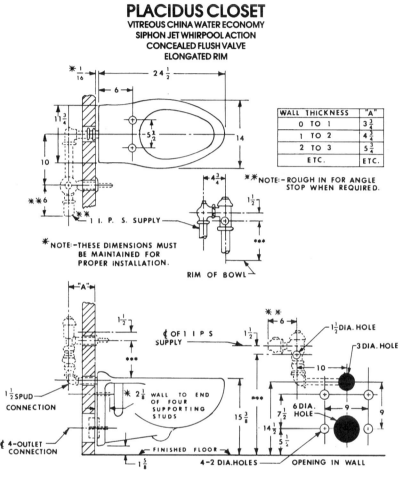

WALL THICKNESS	"A"
0 TO 1	$3\frac{3}{4}$
1 TO 2	$4\frac{3}{4}$
2 TO 3	$5\frac{3}{4}$
ETC.	ETC.

NOTE: Roughing-in dimensions for closet supporting unit to be obtained direct from manufacturer. Check with manufacturer for proper outlet coupling extension beyond finished wall.

***Variable-check with manufacturer of flush valve for proper roughing-in dimension.

Figure 17.2 Cut-away and rough-in details for a wall-hung water closet with concealed flush valve. (*Courtesy of Crane Plumbing.*)

machine, he took the initiative to investigate the problem prior to snaking the drain. In less than 5 minutes he found that the first plumber was mistaken in his diagnosis of the problem. How did he come to this conclusion so quickly?

It was simple logic that led him to the first elimination in the troubleshooting process. The first plumber had been shown the defective toilet by an employee of the retail building. Acting only on the designated toilet, the first plumber had never checked the neighboring toilet. When my second mechanic arrived, he flushed both toilets, and the one on the upstream side of the problem toilet flushed correctly. This indicated that there couldn't be any obstruction in the primary drain pipe or vent. In the short time it took to flush the second toilet, the troubleshooter had ruled out snaking the main lines. This saved the plumber time and the property owner money. There was, however, still the problem with the one toilet that would only swirl and drain very slowly when flushed. This called for further troubleshooting.

The plumber ran his closet auger through the toilet bowl and met with no resistance. He then removed the toilet from its bracket to inspect the drain outlet. Using a mirror, the plumber inspected all parts of the toilet's trap and flushing mechanisms; he found no problems. He then turned his attention to the drain arm.

With nothing being visibly wrong, the plumber connected a garden hose to a silcock and ran a full stream of water into the pipe's drain outlet. The water flowed perfectly, proving their was no stoppage in the line.

With all these steps taken, the plumber was convinced the problem was with the toilet fixture, and that it would have to be replaced. He obtained authorization from the customer to replace the toilet and did so.

Once the new toilet was installed, the plumber tested his new installation. It was then that he realized he had made a mistake in his diagnosis. The new toilet acted the same as the one the plumber had replaced.

Baffled, the plumber used his radio to call the shop and discussed the problem with the service manager. The service manager believed the problem was a plugged vent. This assumption put the plumber on the roof with a snake to clear the stoppage in the vent. Some time later the plumber was back on the radio with the service manager. Snaking the vent had done no good in solving the problem.

The service manager came to me and asked for my opinion. He filled me in on all the steps that had been taken to resolve the problem. After listening to what had been done, I had some questions of my own. You are about to find out what the real cause of the problem was. Have you figured it out yet?

I asked the service manager how long the toilet had been flushing slowly. He didn't know. I suggested that the plumber check the fitting at the drain outlet to be sure that it was an approved, long-turn fitting and not a sanitary tee. The service manager gave me a funny look and left my office shaking his head.

Well, it turned out that I solved the problem without ever leaving my office and without ever seeing the toilet. After talking with the store owner, my plumber discovered that the toilet had never worked properly, but that the owner had just put up with it until he had called us.

Upon further inspection, the plumber found my guess of a sanitary tee being used improperly to be correct. He cut out the fitting and replaced it with a long-turn fitting. When the toilet was reinstalled, it worked like a charm.

In this case, two experienced plumbers and a seasoned service manager had overlooked a very simple explanation. The first plumber didn't check enough options before making a judgment call. The second plumber did a good job of troubleshooting, but his lack of experience prevented him from thinking about the carrier fitting. As for the service manager, he should have figured the problem out but didn't.

How was I able to solve the problem so quickly and easily? I used troubleshooting information supplied by the second plumber to eliminate possible causes. When I thought about all that had been done, the carrier fitting was the only option left to consider.

If I had been the plumber on the job, I would have followed basically the same steps the second plumber did, but I would have inspected the carrier fitting when flushing the line with the water hose. This would have saved time and the purchase of an unneeded new toilet.

If you determined the cause of the problem on your own, congratulations. If you didn't, you now have one more valuable piece of experience in your head to help you if you run across a similar situation. Now, let's look at another example and see how well you do in guessing the cause of the problem.

A Residential Toilet

This true story is about a residential toilet (Figs. 17.3 and 17.4) that would not flush with enough force on a consistent basis. Sometimes the toilet would evacuate its contents, and sometimes it wouldn't.

The plumber who responded to the call for service had years of experience in residential service plumbing. His skills were good, and his dedication to perfection was superb. After checking the toilet, he found that it often swirled nearly to the flood rim before gradually subsiding. The first thing the plumber checked was the water level in the flush tank; it was okay. Then he ran a closet auger through the trap and met with no resistance. Next he checked the other plumbing fixtures in the house. They all drained properly, ruling out a full septic tank or clogged sewer. It was then that he removed the toilet from its flange and did a visual inspection of the toilet bowl and the drain opening under the closet flange. He found no problems evident.

Confused, the plumber called me. I asked if the wax seal was intact or if it may have been obstructing the drain in the toilet bowl. The wax was fine (Fig. 17.5). I asked if he had checked the flush holes for

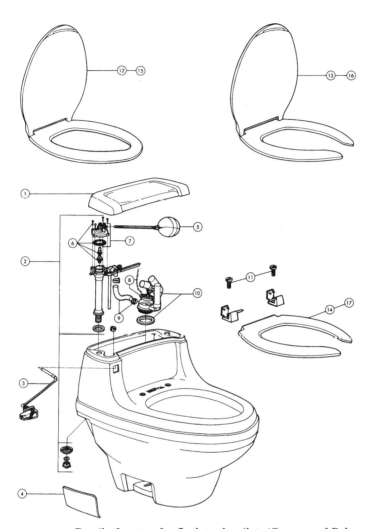

Figure 17.3 Detail of parts of a flush-tank toilet. (*Courtesy of Delta Faucet Company.*)

mineral deposits. He hadn't, and that seemed a logical next step. The plumber hung up, and I went back to my office work.

A little later I heard the plumber talking to the service manager on the radio. The homeowner had authorized a new toilet, and my plumber was on his way to get one. Later in the day, my phone rang again. This time my plumber was calling to explain that he had installed a new toilet, but that the problem persisted. I agreed to go to the job and see what I could do.

When I arrived, the plumber showed me the old toilet and the prob-

MODELS

A

KEY	PART DESCRIPTION	FINISH	ORDER NUMBER	4000	4000 OF	4000 OFC	4001	4002	4007	4100	4100 OF	4100 OFC	4100 OF VRL	4100 OFC VRL	4101	4102	4106	4107	4108	4109
1.	Tank Lid	White	RP8083	●	●	●				●	●	●	●	●						
		Bone	RP8084				●								●					
		Parchment	RP8085					●								●				
		Antique Red	RP8574														●			
		French Vanilla	RP9791						●									●		
		Cerulean Blue	RP12072																●	
		Americana Brown	RP12073																	●
2.	Ballcock Diverter Assy (Complete)		RP8076	●	●	●	●	●	●	●	●	●	●	●	●	●	●	●	●	●
3.	Trip Lever Assy (Complete)	Chrome	RP9469	●	●	●	●	●	●	●	●	●	●	●	●	●	●	●	●	●
4.	Mounting Bolt Covers (2)	White	RP8087	●	●	●				●	●	●	●	●						
		Bone	RP8088				●								●					
		Parchment	RP8089					●								●				
		Antique Red	RP8575														●			
		French Vanilla	RP9792						●									●		
		Cerulean Blue	RP12074																●	
		Americana Brown	RP12075																	●
5.	Arm & Float		RP8077	●	●	●	●	●	●	●	●	●	●	●	●	●	●	●	●	●
6.	Bonnet Screw, Diaphram, Plunger Assy. & Valve Seat		RP8124	●	●	●	●	●	●	●	●	●	●	●	●	●	●	●	●	●
7.	Complete Bonnet Assy. & Diaphram		RP8125	●	●	●	●	●	●	●	●	●	●	●	●	●	●	●	●	●
8.	Flapper & Chain		RP8091	●	●	●	●	●	●	●	●	●	●	●	●	●	●	●	●	●
9.	Hose & Clamps		RP8090	●	●	●	●	●	●	●	●	●	●	●	●	●	●	●	●	●
10.	Flush Valve Assy. (Complete)		RP8092	●	●	●	●	●	●	●	●	●	●	●	●	●	●	●	●	●
11.	Seat Bolts (2)		RP8075	●	●	●	●	●	●	●	●	●	●	●	●	●	●	●	●	●
12.	Seat-Regular*	White	RP8072	●																
		Bone	RP8073				●													
		Parchment	RP8074					●												
		French Vanilla	RP9498						●											
13.	Seat-Regular OFC w/Cover*	White	RP8164			●														
14.	Seat-Regular OF without Cover*	White	RP8435		●															
15.	Seat-Elongated*	White	RP9489							●										
		Bone	RP9490												●					
		Parchment	RP9491													●				
		Antique Red	RP9492														●			
		French Vanilla	RP9493															●		
		Cerulean Blue	RP12070																●	
		Americana Brown	RP12071																	●
16.	Seat-Elongated OFC w/Cover	White	RP9799									●		●						
17.	Seat-Elongated OF without Cover	White	RP9800								●		●							

Figure 17.4 Description of parts shown in Fig. 17.3. (*Courtesy of Delta Faucet Company.*)

Figure 17.5 Standard installation of closet bolts and wax ring on a closet flange.

lem he was experiencing with the new toilet. He took me into the basement to show me that he had run a snake through the cleanout under the toilet with no success. As soon as I entered the basement, I knew what the problem was. Have you got any idea at this point what was causing the trouble?

My experienced plumber had been standing right in front of the problem and had never noticed it. The vertical riser from the building drain came out of a sanitary tee and extended upward for about 4 feet. Now have you figured it out?

The vertical drop was so great that when the toilet was flushed, the water would hit the back of the tee and splash back up the riser. This back pressure would not allow the toilet bowl to empty properly, and toilet tissue was being left in the riser because the water was dropping too quickly. The vertical riser from the building drain to the closet flange was much taller than the plumbing code allows, and it was the primary cause of the problem.

We cut out the illegal piping and installed a wye and an eighth-bend to shorten the vertical drop and to give a smoother transition from the toilet to the building drain. While my plumber didn't argue with me on this decision, neither did he think my repair idea was going to work; he was wrong. After the new piping was in place, we tested the toilet over and over again. It flushed perfectly every time.

This was another case where a new toilet was bought unnecessarily. Fortunately, the homeowner was not angry, and she wanted to keep the new toilet. In both this case and the previous one, I reduced the labor charges to reflect what the customers would have been charged if the plumbers had been better troubleshooters. As a business owner, I don't like to give money away, but there was no way I could charge those people for my plumbers' mistakes.

What is the moral of these two stories? It is that knowing proper troubleshooting techniques will save you time, money, and embarrassment. Both of the plumbers in the above examples were experienced journeyman plumbers, but they couldn't find the solution to their on-the-job problem. Now that you've seen how easy it is for experienced plumbers to overlook obvious causes of problems, let's delve into the troubleshooting procedures for toilets in a step-by-step, problem-by-problem manner.

Flush-Tank Toilets That Run

Toilets that run constantly are not very difficult to troubleshoot. There are a number of possible causes for this type of problem, but they are easy to identify and eliminate. Let's look at each of the possibilities individually.

The ballcock

The ballcock is a likely suspect when it comes to a toilet that won't stop filling itself with water. There are two basic types of ballcocks. One style uses a horizontal float rod and a float ball. The other type has its float on the vertical shaft of the ballcock. Both types work on the same principle. When the volume of water in the flush tank is sufficient to raise the float of a ballcock to a predetermined point, the valve closes, and the water is cut off. This is a very simple principle, and one that is easy to check. Let's look first at the type of ballcock that uses a float rod.

Float-rod ballcocks. Float-rod ballcocks are the most common type found in toilets. The ballcock rises vertically in the flush tank, and the float rod extends from the top of the valve into the flush tank. One end of the rod is attached to the ballcock, and the other end has a float ball screwed onto it. The ball may be made of styrofoam, or it may be a hollow, plastic ball.

If you are inspecting the refill tube from the ballcock and find water running through it, you can assume your running toilet is the victim of a faulty ballcock. It is, however, possible that the ballcock is leaking water into the tank from some other point. To test for this, you should first cut off the water supply to the toilet.

With the water cut off, flush the toilet. Once the tank is drained, hold the float of the ballcock in an extreme upward position and turn the water supply to the toilet back on. No water should come from the ballcock. If the ballcock produces water under these conditions, it would be a good idea to replace it.

While most ballcocks can be repaired, they are so inexpensive that you are usually doing yourself, and your customer, a favor to replace them. If you replace a diaphragm and leave the job, you may be called back to correct a similar problem with the old ballcock. This aggravates customers, and it is common to have to do the work without charge under a warranty basis. Unpaid callbacks are bad for your boss or your business.

If the ballcock works properly in the dry test, release the float and observe the filling action of the ballcock. Does the float ball rub against the side of the flush tank? If it does, you need to bend the float rod to allow the ball freedom from obstructions.

After the tank is filled, does water run over the overflow tube? If it does, the float rod needs to be bent downward. This will cause the valve to cut off sooner, resulting in the solution to your problem.

Should you have a persistent problem in adjusting the float rod, check the float ball. Float balls sometimes become waterlogged (filled with water). The weight of the retained water can be enough to cause the ballcock to malfunction. All that is required to correct this situation is the simple replacement of the float ball.

Vertical ballcocks.   Vertical ballcocks (Fig. 17.6) have their floats attached to the vertical shaft of the valve. There is no float rod to contend with, and the float is not likely to take on excess water.

You can check a vertical ballcock with the same methods described earlier. If the float needs to be adjusted, there is normally a small chrome device that allows you to control the position of the float. By squeezing the small piece of metal between your thumb and forefinger, you can slide it up and down to adjust the position of the float.

Many of these types of ballcocks are designed to be adjusted for different heights in flush tanks. This requires turning the entire shaft, and the ballcock must be removed from the tank to accomplish the task. It is unlikely that you will need to go to this extreme but be aware that the option exists.

If the nature of the problem is a toilet that runs continuously, the

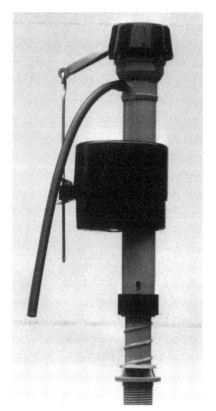

Figure 17.6   Vertical ballcock.

solution most likely lies in the ballcock. If the problem is a ballcock that cycles on and off at various intervals, the problem is more likely to be in one of the other elements of the toilet.

The overflow tube

The overflow tube in a flush tank can cause the toilet's water supply to run endlessly. Many inexperienced plumbers overlook ways that are possible for this to occur. It is not uncommon for service plumbers to look only at the top of the overflow tube. They assume that if the water level is not above the edge of the tube, the tube couldn't be the problem. This is not always true.

Many new toilets have plastic overflow tubes, but older toilets are equipped with brass tubes. These tubes are screwed into the flush valve and offer two distinct possibilities for hard-to-find leaks.

The thin threads of a metal overflow tube can disintegrate over time. If a hole develops in the tube or around the threads, water can leak into the toilet bowl. Since this leak is below the water surface in the flush tank, it can be difficult to find.

The plumbing in houses served by well water is frequently affected by minerals and acid found in the water. It is possible for acidic water to eat holes in any portion of a metal overflow tube. Water seeping into these holes and running into the toilet bowl can create a problem that may seem to be the ballcock's fault.

With either of the above situations, a ballcock will have to cut on periodically to replace the water lost through the overflow tube. If you believe the overflow tube is leaking, replace it.

Flush valves

Many new flush valves are made from plastic materials, but there are still a great number of metal flush valves in use. Just as overflow tubes can be affected by various water conditions, so can flush valves.

Metal flush valves can become pitted with depressions or holes. These pitted areas create a void between the flush valve's surface and the tank ball or flapper that seals it. A strong magnifying glass and a good flashlight make these conditions easier to find.

To check for a pitted flush valve, you must first cut off the water supply to the toilet and drain the flush tank. If you find pitted areas, you have two choices. You can use sandpaper to smooth out the pitted surface, or you can replace the entire flush valve assembly. Sanding the surface area will normally provide some temporary success, but the best way to ensure long-term satisfaction is replacement, preferably with a plastic flush valve.

Tank ball or flapper

The tank ball or flapper can be the cause of a running toilet. As time passes, these devices can become worn to a point where they no longer seal the opening in the flush valve. If this happens, water will leak into the toilet bowl, and the ballcock with be forced to cut on at periodic intervals. Inspect the seating surface of the tank ball or flapper and replace it if the rubber looks worn or damaged.

The lift rods and guide for a tank ball can also contribute to a running toilet. If these items work their way out of adjustment, the tank ball cannot seat properly. It is not uncommon for the lift-rod guide to become misplaced. If this happens, you can reset its adjustment, and the problem should be resolved.

In the case of flappers, the tank handle or chain can prevent the device from seating properly. By flushing the toilet several times and observing the movement of the flapper, you will be able to detect any problem with its alignment.

If the tank handle rubs the side of the flush tank, it may be holding the flapper up, allowing water to run through the flush valve when it shouldn't be. A chain that is too short or tangled can have the same affect on the flapper's ability to seal the flush opening.

Commercial Flush-Valve Toilets That Run

Commercial flush-valve toilets (Figs 17.7 and 17.8) that run require you to disassemble and inspect the flushing mechanism. Since the toilets don't use flush tanks, they don't have ballcocks, flappers, ank balls, lift wires, refill tubes, overflow tubes, or the tank-type flush valves found in flush-tank toilets.

To troubleshoot these toilets, you must be familiar with the inner workings of commercial-grade flush valves. Because of the washers, diaphragms, and numerous other possible culprits in defective flush valves of this nature, it is wise to install a replacement kit of all the essential elements to solve your problem. These kits are readily available for all major brands of flush valves, and they are the fastest, surest way of remedying the defect. By replacing all the small parts, you leave little chance of a recurring problem.

Toiloto That Fluch Slowly

Toilets that flush slowly can be devilish to troubleshoot. You saw in the two examples I gave early in the chapter how even experienced plumbers can be fooled by slow-moving toilets. There are, to be sure, many possible causes for a lazy flush. Let's take some time now to

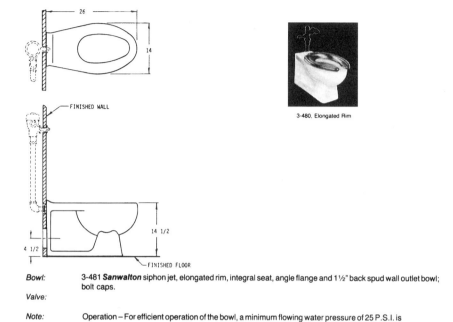

3-480, Elongated Rim

Bowl: 3-481 **Sanwalton** siphon jet, elongated rim, integral seat, angle flange and 1½" back spud wall outlet bowl; bolt caps.

Valve:

Note: Operation – For efficient operation of the bowl, a minimum flowing water pressure of 25 P.S.I. is required at the valve.

Figure 17.7   Floor-mounted flush-valve-type water closet with concealed flush valve. (*Courtesy of Crane Plumbing.*)

explore these possibilities and identify ways to find and correct the causes of lackluster flushes.

Commercial flush valves

Commercial flush valves can be responsible for slow flushes. If the valves don't release an adequate amount of water, the toilet bowl cannot be cleansed properly.

If a commercial flush valve is not producing the desired amount of water, check the control stop. Remove the chrome cover plate and use a screwdriver to make sure the valve is open to its proper position.

Assuming the valve is open and the flush valve is still not working properly, you will have to inspect the inner workings for defects or obstructions. Check the water-saver device to make sure the disks are working properly. If you find no obvious obstruction in the working parts of the valve, remove the flush connection and check for obstructions in the spud hole.

SANWALTON

VITREOUS CHINA
SIPHON JET
CONCEALED FLUSH VALVE
ELONGATED RIM, INTEGRAL SEAT
WALL OUTLET BOWL

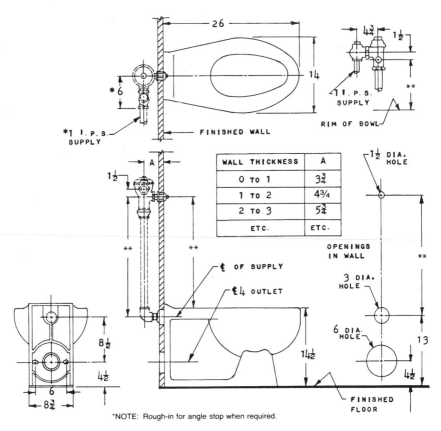

WALL THICKNESS	A
0 TO 1	$3\frac{3}{4}$
1 TO 2	4¾
2 TO 3	$5\frac{3}{4}$
ETC.	ETC.

*NOTE: Rough-in for angle stop when required.

Figure 17.8 Cut-away and rough-in details of a floor-mounted flush-valve-type water closet with concealed flush valve. (*Courtesy of Crane Plumbing.*)

Flush tank

When working on a toilet with a flush tank, check the volume of water in the tank. Sometimes people reduce the amount of water used in a flush tank to conserve water and save money; this can result in a slow flush. If the water level is low, adjust the ballcock to bring the water up to the fill line. When there are water-saving devices placed in the tank (such as bricks), remove them and make sure the tank fills to the fill line.

Once you have a full tank of water, flush the toilet. It is a good idea to put toilet tissue in the bowl before flushing. This simulates an actual flush better than an empty bowl would. If the paper is not removed quickly and entirely from the bowl, you still have some troubleshooting work to do.

Malfunctioning flappers and tank balls

Malfunctioning flappers and tank balls can cause the water supply to a toilet bowl to cut off prematurely. If the flapper or ball drops too quickly, the flush hole will be sealed before enough water has passed to cleanse the toilet bowl. This problem can usually be corrected by adjusting or replacing the flapper or tank ball assembly. It may require repositioning the toilet handle, chain, lift wires, or lift-wire guide.

As an example, a chain that is too long may not pull a flapper far enough up so that it stays suspended long enough to properly flush the bowl. By shortening the chain, the flapper is pulled farther up, and the water rushes out of the tank to flush the bowl. This is a very simple procedure, but it is one many professional plumbers fail to notice.

Flush holes

I have had experienced plumbers on my payroll who didn't know what flush holes were. Flush holes are the small holes around the inner flush rim of a toilet. They can't be seen by looking in a toilet bowl unless a mirror is used or you are able to get your head below the rim and look up.

These holes can become obstructed with mineral deposits and inhibit the normal flushing of a toilet. If you flush a toilet and don't see numerous streams of water running down the inside of the bowl, there is a good chance the flush holes are plugged up.

The flush holes are designed to clean the sides of the bowl, and if they aren't able to function, the toilet may flush slowly. When these holes become blocked, a sturdy piece of wire, such as a coat hanger, is all that is needed to clear them. Pushing the end of the wire into the holes will normally break up the mineral deposits and return the toilet to good working condition. Occasionally, the deposits will be too large to remove, and the bowl will have to be replaced. Problems with flush holes occur most often with toilets served by well water, but they are worth checking under any conditions.

Obstructed traps

Obstructed traps can certainly account for a slow flush from a toilet. Traps can become blocked by any number of foreign objects, some of which a closet auger will pass right by without clearing. Strange objects in traps probably account for most of the mysterious malfunctions of slow-flushing toilets. Since this condition is so prevalent and so often misunderstood, let's take a few moments to look at some of what my plumbers and I have removed from the traps of water closets. The details of what objects we have discovered and how they nearly avoided detection will prepare you for your own future surprises.

A beer can.   On one occasion I found a beer can in the trap of a toilet. When you consider the size of an opening in the bowl of a toilet, it is hard to believe a beer can could find its way inside, but I have seen the evidence, and it can happen. The can had been crumpled and crushed, but it was all in one piece.

The beer can was found early in my plumbing career, and its presence in the trap had me befuddled. I had augered the trap, pulled the toilet, snaked the drain, and all but given up on fixing the fixture. I assumed the toilet must be replaced. I was about to call the owner of the property to get authorization to replace the toilet when I decided to take one more look at the toilet.

With the toilet laying on its side, I looked into the drain from the base of the bowl. I couldn't see any reason for a problem. Then I ran my closet auger through the trap and saw it come out clean. It was not until I went to set the toilet back upright that I noticed anything.

When I picked the toilet up I heard something rattle inside it. I removed the flush tank and shook the bowl. There was clearly something inside it that didn't belong there. I fished around in the trap with a wire coat hanger and could feel something, but I couldn't retrieve it.

While I was working on the toilet the owner of the property came in. I explained to him what I had found and asked how he wanted to proceed. He knew the bowl would have to be replaced, and we were both curious about what was making the noise. The owner asked me to break the bowl open, and I did. That's when we found the beer can.

The property owner assumed that one of his rental tenants must have caused the problem. I installed a new toilet bowl and went on about my business, a little smarter for my experience.

The beer can, in its crushed condition, was acting as a baffle. It would allow water to pass through the trap, but it would catch solids and tissue paper, causing the toilet to drain slowly, if at all. Even the closet auger would slide right past the can without interference. As you can imagine, this type of problem is hard to identify.

A rubber ball.   One of my plumbers was faced with an unusual situation when he encountered a rubber ball in the trap of a toilet. In his case, the closet auger was meeting resistance that it could not get past. Water would drain slowly from the toilet, but solids would cause overflowing conditions. This was another occasion when the cause of the problem was not identified until the toilet bowl was broken open.

Other strange encounters. Some of the other strange encounters with foreign objects in toilet traps have accounted for a rather unique collection of plumber's experience. I have found all types of toys in toilet traps. Wash rags, hair curlers, shampoo tops, makeup containers, and a host of other items have all served to make toilets flush slowly. Most of these items were never dislodged with a closet auger. They were found either by breaking the bowls open or with a mirror. A long-handled inspection mirror and a flashlight can make your trap troubleshooting much more effective.

Wax obstructions

Wax obstructions can contribute to early aging in plumbers. When a toilet is set, the wax ring sometimes spreads out too far over the drain opening. If the wax protrudes from the flange and hangs out over the drain, it can catch a lot of toilet tissue. In time, this buildup can make emptying the toilet bowl a slow process.

A wax obstruction will usually show up on the end of a closet auger. If you put your auger through the trap and it comes back with wax on the end of it, be prepared to pull the toilet.

When you remove the toilet from the flange, you will probably find a disfigured wax seal. You have to scrape off the old wax, install a new ring, and reset the toilet. If the toilet was not set down on the flange straight and evenly, or if the bowl was twisted during installation, there is a chance for a wax obstruction.

The bad thing about this type of problem is that it doesn't show up right away. The toilet can be tested and work perfectly, only to begin giving trouble a few weeks down the road.

Partially blocked drainage pipes

Partially blocked drainage pipes can cause toilets to drain slowly. If the drain lines have any type of buildup in them, like tree roots or grease, the full flow of the pipe will be restricted. This condition can go unnoticed with other fixtures in the house, but the volume of water produced from a flush may expose the problem.

The small amount of water going down the drain from a sink or bathtub may not be noticeably slow. It may be that the drain only seems to act up when a toilet is flushed. This can be a difficult problem to put your finger on. When everything else in the house seems to be draining correctly, it would lead you to believe the toilet is at fault. There is, however, an easy way to check for such a problem.

When you are faced with a possible drain blockage, remove the toilet from its flange. Take a 5-gallon bucket of water and pour it down the drain at the closet flange. Follow that water with another bucket of water as quickly as possible. If the drain can take 10 gallons of water quickly, the odds are the problem is in the toilet. If, however, the water gurgles and backs up, check the drain and vent system.

Illegal drainage piping

Illegal drainage piping is yet another cause of slow-flushing toilets. If a pipe doesn't have the proper amount of fall on it, the toilet draining into the pipe may flush slowly. As you saw in one of the earlier examples, a long, vertical drop can cause a toilet to malfunction. If you have access to the piping under a toilet, check it out for illegal fittings, improper grade, or excessive riser heights.

Toilets That Overflow

It is easy to pinpoint trouble in toilets that overflow. The problem is either in the toilet's trap or in the drainage piping. A closet auger is a fast way to check out the trap. If the auger goes through smoothly, the problem is probably downstream.

If the contents of the drainage piping are backing up into other fixtures, such as a bathtub, you can automatically rule out the toilet trap; you know you have a blockage in the drainage pipe.

If you remove the toilet from its flange and find standing water in the pipe, you know you are dealing with an obstructed drainage system. With houses that use septic systems, it would be wise to check the septic tank before investing a lot of time and effort in snaking the drain. Homeowners often forget to have their septic tanks pumped, and that can result in a complete backup of sewage in the house.

Another consideration before you begin to snake drains is to inquire about the presence of any type of sewage pump. If the toilet is connected to a sump pit with a pump in it, the pump may be failing. If after checking all of these possibilities you haven't found a reason for the backup, you will have to snake the drain.

Toilets That Condensate

Toilets that condensate can fool plumbers. If the tank is condensating in a hard-to-see area, the water developing slowly on the bathroom floor may appear to be coming from the base of the toilet. This would indicate a faulty wax seal, when in reality, it is only condensation.

Condensation typically occurs when there is an extreme temperature variance between the water in the toilet and the room temperature. This can easily be the case for toilets supplied by well water during a hot summer.

Some homeowners dress their toilets up in cloth covers. These covers can conceal condensation and lead an inexperienced plumber on a wild goose chase. If you have unexplained water on the floor around a toilet, inspect carefully for condensation before removing the toilet from its flange.

There are a few ways to reduce the problems associated with condensation. A liner, usually made of insulating foam, can be installed on the interior of a flush tank to reduce condensation. These foam panels are glued to the tank, so the tank must be completely dry prior to their installation.

A second alternative is to install a condensate tray under the tank of the toilet to catch any water droplets. This arrangement can be ugly, and it will do nothing for a toilet bowl that is condensating.

The third, and most effective, option is to install a mixing valve on the water supply to the toilet. The mixing valve allows hot and cold water to commingle in the toilet tank, eliminating any chance of condensation.

Toilets That Fill Slowly

Toilets that fill slowly are usually in need of a new ballcock. It is sometimes possible to adjust an existing ballcock to increase its flow rate, but replacement is usually the most logical course of action.

A crimped supply tube can cause a toilet to fill slowly. Every now and then a supply tube is hit with a mop or kicked to a point that it crimps. These occasions are rare, but they do happen. Don't neglect to inspect the closet supply when troubleshooting for the reason a toilet is filling slowly.

The cutoff valve under a toilet should always be checked to prove that it is in a full-open position. It is possible that debris has gotten into the stop valve or the supply tube. To check for this, disconnect the ballcock nut and check the water pressure at the top of the closet supply tube. If it is inadequate, remove the supply tube and see what the pressure is like at the valve. If you are still experiencing low pressure, look to the water distribution piping.

If a toilet is supplied with water from a galvanized steel water pipe, there is a good chance the water distribution pipe will need to be replaced. Galvanized pipe tends to rust and clog up with age. To determine if the problem is in the distribution pipe, all you have to do is remove the stop valve and check the pressure at the end of the water pipe. Be careful not to flood the bathroom when conducting this test.

To avoid flooding, have an assistant standing by the main water cutoff to quickly turn the water on and off. You won't need much time to see how much water shoots out of the pipe. If you have to work alone, you can loosen the stop valve until water sprays out around the connection between the valve and pipe. By doing this you can tighten the connection quickly and keep the water spillage to a minimum.

Toilets That Leak around Their Bases

Toilets that leak around their bases are usually suffering from one of three problems: The wax ring needs to be replaced, the bowl is cracked, or the drainage pipe is backing up.

The most common cause of water leaking around the base of a toilet is a faulty wax ring. This condition requires the replacement of the existing wax. When the flange is too far below the finished floor level, it may be necessary to use two wax rings, one stacked on top of the other (Fig. 17.9).

A cracked bowl can allow water to seep out and imitate the symptoms of a bad wax ring. Condensation, a cracked flush tank, loose tank-to-bowl bolts, and faulty flush-hole gaskets can also cause water to gather on the floor around a toilet. To check for these conditions, all you have to do is rub all of the areas with toilet tissue. The toilet tis-

Figure 17.9 A double wax-ring seal on a closet flange.

sue will get wet if there is a leak, and you will find the problem faster than you could with just your eyesight.

When the drain for a toilet backs up, it sometimes spills out under the toilet. This buildup of water can seep out around the base and appear to be the fault of the wax seal. There is generally some evidence that discloses this type of problem to a trained eye that is watching a toilet flush. There will usually be a bubbling or a slight rise in the water as it rests in the bowl. If you suspect this is your problem, pull the toilet and pour 10 gallons of water down the drain as quickly as possible. If there is a partial blockage, the test should reveal it.

The Key to Troubleshooting

The key to effective troubleshooting is a systematic approach. If you use the information in this chapter, and the ones that follow, to eliminate possible causes one at a time, you will be able to find the root cause of your problem. Once you have identified the problem, correcting it will be much easier.

Troubleshooting Sinks and Lavatories

Have you ever had a lavatory that wouldn't hold water but that didn't show evidence of a leak anywhere? There are times when lavatories can drain themselves without a trace of water being left as evidence of a leak. If you think I'm describing a pop-up plug that is not sealing properly, you're wrong. The type of leak I'm talking about can happen even when a rubber stopper is placed in the drain opening. No matter how tightly the drain opening is sealed, the type of leak being described can happen. What kind of leak could empty a lavatory bowl full of water without so much as a drop of water to disclose it? Well, you'll find the answer a little later in the chapter. In the meantime, think about your past experiences and see if you can come up with a viable answer.

Have you ever been curled up under a sink, looking for a leak you couldn't find? Was the base of the cabinet saturated with water even though there was no evidence of a leak, no matter how often you filled and emptied the sink? This situation can be frustrating and uncomfortable physically. A plumber could test the sink in this scenario time and time again and never find the cause of the problem. Why? Because the problem is not a true plumbing problem. Oh yes, the water is there to prove a leak exists, but the leak is not from a faucet or a drainage fitting. What else could be causing the cabinet to get wet? No, it's not in the water distribution pipes, stop valves, or supply tubes. If you don't already know the answer to the question, you will find it a little later in the chapter.

Are you tired of facing questions without answers? Well, that is very much what troubleshooting is; you have questions but no answers. As a troubleshooter, it is your job to take the information you have and assess it for a suitable solution. If you don't have enough information, you must continue testing with a process of elimination to find answers.

This chapter is all about sinks and lavatories. While there are many types of sinks and lavatories in use, most of them share the same basic plumbing characteristics. There are, of course, some differences in how the various types should be treated during troubleshooting. As you go through this chapter, you will find some advice on specific types of sinks and lavatories. Most of the text, however, applies to any type of sink or lavatory.

We will look at all aspects of troubleshooting sinks and lavatories in the next several pages. Because of the vast number and complexities of faucets, they are not covered in this chapter. Faucets and valves will be explained in Chap. 24.

Kitchen Sinks

We will begin our journey with kitchen sinks since they are the most common type of sink. As you probably know, there are many, many types and styles of kitchen sinks available. Some have single bowls; some have double bowls (Fig. 18.1), and some have specialty bowls. The two most common materials for kitchen sinks to be made from are stainless steel and cast iron. Some sinks are equipped with garbage disposers, some with continuous wastes, and some with individual traps. These are only a few of the possibilities for differences in sinks, but even so, they are all about the same when it comes to troubleshooting. Basic principles apply to all sinks and can be used to solve most problems. Let's look now at some common and some not-so-common problems associated with kitchen sinks.

A wet base cabinet

Our first sample service call involves a wet base cabinet beneath a kitchen sink. Put yourself in the place of the plumber on the service-call example you are about to be given.

A homeowner calls your office and complains that the cabinet under the kitchen sink is soaked with water. You ask all the over-the-phone questions needed to determine the urgency of the call. The homeowner has never seen where the water is coming from; it is only known that something is leaking. You schedule an appointment to correct the problem.

When you arrive at the house and look under the sink, you see water stains in the base cabinet. It is obvious a significant amount of water is finding its way into the cabinet. Now it is your job to find out where the water is coming from.

Figure 18.1 Double-bowl, stainless-steel, kitchen sink. (*Courtesy of Republic sinks by UNR Home Products.*)

The kitchen sink is a stainless-steel, double-bowl model, equipped with a garbage disposer and an end-outlet continuous waste. There are an ice-maker connection and a dishwasher connection under the sink. The faucet is a single-handle model, with a spray attachment (Figs. 18.2 and 18.3).

After your initial look at the circumstances, you suspect the spray attachment is the problem. Your past experiences have proven that kitchen sprays often leak. As your first troubleshooting step, you turn on the faucet and depress the spray handle. Water comes out of the sprayer, and as you watch under the sink, there is no leak. A bit surprised, you continue your troubleshooting procedure.

You fill both bowls of the sink to capacity and then release the water. A close inspection of the drainage systems reveals no evidence of a leak. The problem is becoming more difficult to explain. The faucet isn't leaking, the drainage connections are not leaking, and the

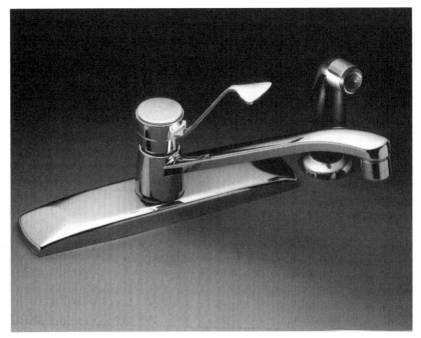

Figure 18.2   Single-handle kitchen faucet with spray. (*Courtesy of Moen, Inc.*)

spray attachment is not leaking. This leads you to a close inspection of the supply tubes, stop valves, and compression connections. After a thorough investigation, these items are not found to be defective.

As a last resort, you fill the side of the sink where the garbage disposer is connected. When the sink is full, you release the water and turn on the disposer. Still no leak. Scratching your head, you are perplexed at what could be causing the leak.

You take a paper towel and rub all the water-supply connections; the towel remains dry. When you apply the same towel test to the drainage connections there is no dampness. Many plumbers would be calling their service managers for advice at this time, but you are determined to solve the problem on your own. The only problem is, you don't know what else to check. What would you do?

Do you know where the water is coming from yet? I'll give you a hint; it has to do with caulking. Now do you know what test to conduct?

The leak in this case is not a true plumbing leak. The water supply and drainage connections are all sound. The water is running from the kitchen counter under the rim of the sink and dripping onto the base cabinet. This leak will not show up during any standard plumbing test. The only way to identify this type of problem is by putting water on the counter, around the edges of the sink.

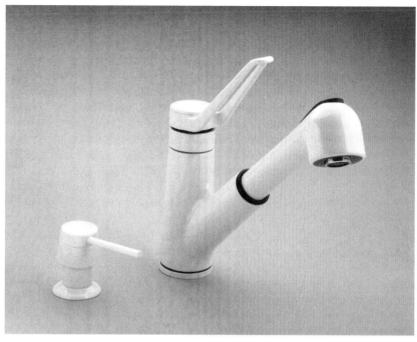

Figure 18.3 Single-handle kitchen faucet with pull-out spray. (*Courtesy of Moen, Inc.*)

When homeowners use their kitchen sinks, water frequently gets on the surrounding countertop. If the rim of the sink is not caulked properly, the water will seep under the edge of the sink and drip into the base cabinet (Fig. 18.4). A simple bead of fresh caulking will correct the problem, but many plumbers fail to find the source of the water. If the cause of a problem cannot be found, the problem cannot be fixed.

As a business owner and supervisor, I've run into this type of situa-

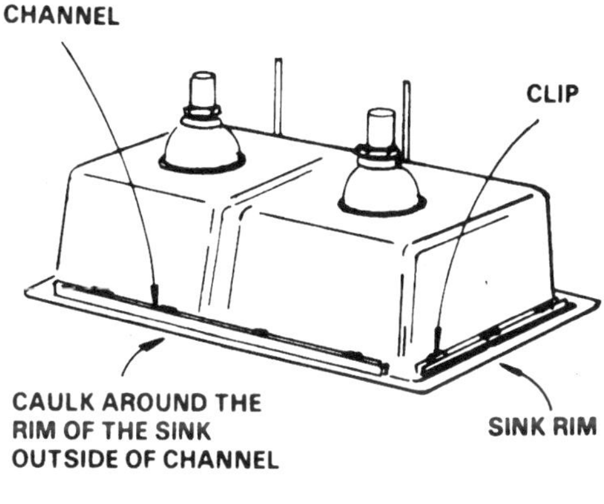

CHANNEL

CLIP

CAULK AROUND THE
RIM OF THE SINK
OUTSIDE OF CHANNEL

SINK RIM

Figure 18.4 Caulking location on a kitchen sink. (*Courtesy of Republic sinks by UNR Home Products.*)

tion often. Plumbers under my direction have been stumped by such mysterious leaks. When I've asked them if the caulking under the sink was checked, they invariably answer, "No." Once they apply the splash test to the edges of the sink, they are amazed at the volume of water that can flood into the cabinet.

Having this kind of knowledge can mean the difference between success and failure. Callbacks are expensive for all plumbing companies. If you do not positively find and correct a problem like the one in the above example, someone is going to lose money. It may be you or your boss, but someone is going to have to keep going back to the job until the situation is rectified. In addition to the financial losses, the homeowner is likely to lose confidence in the plumber and the plumbing company. A simple bead of caulking can cost you a customer.

You see, not all troubleshooting in plumbing involves what you may think of as plumbing. There are times when you have to look deeper than the common or obvious causes for leaks.

I once secured a major account from a situation similar to the one just described. A large motel was having problems in their kitchenette units. Many of the units were suffering from wet base cabinets. The owner of the motel had instructed the manager to have the leaks taken care of. The base cabinets were becoming stained from the water, and rot was going to set in if the leaks were not fixed. The manager called in one plumbing company who checked all the plumbing connections and couldn't find a cause for the leaks.

A second plumbing company was called. This company did extensive work in the leaking kitchens. The tubular fittings were all disconnected and reinstalled. New slip-nut washers were installed, new seals were placed under the basket strainers, and the supply-tube connections were all done over. The plumbing company assured the manager that the problems were resolved and would not recur. The cost of the work was extensive, but the manager assumed it was needed work and would have paid the bill, except that the leaks did recur. When the plumbing company came back, they were unable to find out where the water in the base cabinets was coming from. This obviously frustrated the motel manager. That's when she called my company.

We dispatched a service plumber to the motel. When he arrived, the manager filled him in on the last two companies who had done the previous troubleshooting. My plumber was a seasoned veteran, and as soon as he had the background information to go on, he had a good idea of what the problem was.

The manager accompanied the plumber to one of the units where there was water standing under the kitchen sink. The plumber looked under the cabinet and shined his light on the water connections; they were dry. Then he filled the single-bowl sink to capacity and released the water. None of the drainage fittings leaked.

The plumber took a towel out of his work bucket and placed it over the water in the cabinet. Then he spread paper towels out over the heavy towel and asked the manager to watch what was going to happen. The manager was confused, but she paid attention. Reaching into the wall cabinets, the plumber removed a water glass. He filled the glass partially with water and poured the water around the edges of the sink. In a matter of moments, the water was dripping down onto the paper towels. The manager was amazed.

With this problem pinpointed, the plumber and the manager moved on to the other kitchenettes that had leaks. In every case the cause of the leaks was a lack of caulking around the edges of the sinks. About an hour later, the leaks were all fixed. The manager was elated that a few tubes of silicone caulking and less than 2 hours of labor were all that was required to solve her problems. The cost for this work was a fraction of what the other plumbing company had charged, and most importantly, my plumber fixed the problem.

From that day on, the manager never called another plumbing company. The experience that my plumber had was what made it possible to fix the leaks and to secure a great account. I imagine the other plumbing companies could have figured out what the problem was if they had looked far enough, but they didn't. Their concentration on pure plumbing problems left an opportunity for my company to be a hero. By looking beyond the normal reaches of plumbing, my plumber was able to impress the manager in a way that was not forgotten.

A faucet-hole leak

Have you encountered a faucet-hole leak? This type of leak will normally appear to be a problem with the kitchen faucet or spray assembly. Water will have accumulated on the floor of the base cabinet and may be clinging to the supply tubes or spray hose. When you first see the water, it is natural to assume the problem is with the faucet.

In a way, the problem is with the faucet, but it is not a faucet defect or a problem with the faucet connections. The problem is a bad seal between the faucet base and the sink top. It could be a bad gasket or a deteriorated putty seal. In either case, water will leak under the faucet base and run through the faucet-mounting holes in the sink. To correct this problem, the faucet should be removed and a new seal of putty or a new gasket should be installed. Applying a bead of caulking around the perimeter of the faucet base is another option for solving this type of problem.

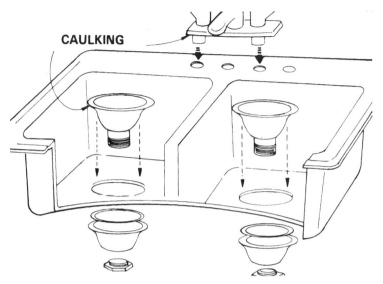

Figure 18.5 Caulking location for basket strainers. (*Courtesy of Republic sinks by UNR Home Products.*)

Basket strainers

Basket strainers can be involved in two types of kitchen-sink problems. If a sink will not hold water, the basket strainer is usually the problem. If the basket has a bad seal on it, the entire basket should be replaced. The existing drain housing can normally remain intact, but the basket insert should be replaced.

A bad seal under the lip of basket-strainer housing can allow water to seep under it and into the base cabinet. If the putty used to create this seal becomes old and brittle, leaks are likely. This type of leak will often seep out to the tailpiece and run down the tubular drainage fittings. The leak may appear to be coming from a slip nut or the rim gasket under the sink drain. To correct the problem, you must remove the basket-strainer housing and install a new seal of putty (Fig. 18.5).

Garbage disposer drains

Garbage disposer drains are very similar to basket-strainer assemblies in the ways that they may leak. If the rim of the drain collar does not have a good seal of putty under it, water will seep past it. If you find water running down the side of a disposer, dripping off the mounting bracket, or under the disposer, check the putty seal under the drain rim.

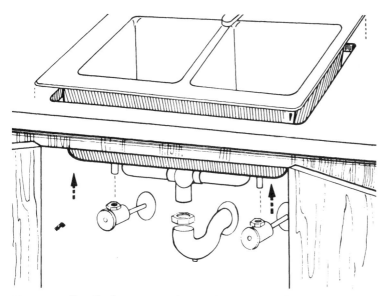

Figure 18.6 Detail of a center-outlet continuous waste for a kitchen sink. (*Courtesy of Republic sinks by UNR Home Products.*)

Continuous wastes

Continuous wastes and slip nuts account for many of the leaks under a kitchen sink (Fig. 18.6). When people store objects under a sink, they often hit and disturb the tubular drainage pipe and slip nuts. If a slip-nut washer is old and brittle, it doesn't take much stress to make it leak.

Many plumbers make a big mistake when they are looking for leaks in the drainage system of a kitchen sink. I've seen numerous good plumbers make the same mistake when they were testing out new installations and when they were looking for leaks in existing plumbing. They frequently fail to fill the sink to capacity before looking for the leaks.

Slip-nut leaks don't always occur every time water is drained past them. For example, if a kitchen faucet is turned on and the water runs directly from the faucet down the drain, there may not be enough pressure on the faulty slip-nut washer to create a drip. However, if the sink bowl is filled and all of the water is released at one time, the volume and pressure of the water can cause a significant leak at the weak joint.

Homeowners often fill their sinks to capacity to wash dishes. When they release the water, leaks can occur that a plumber will never find without filling the sink with water. This can cause a plumber to tighten the slip nuts and leave the job, satisfied that the leak is fixed. But the next time the homeowner fills the sink to do dishes, the leak will reappear. This creates a callback for the plumber, aggravation for the homeowner, and lost money for the plumbing company.

When you are looking for leaks in the continuous waste of a sink, always fill the sink to capacity and release all of the water at one time. If you are working on a double-bowl sink, fill both bowls and release them simultaneously.

Traps

Finding a leak in a trap should be done with the same procedure that is used on continuous wastes. Fill the sink to capacity and release all of the water at once.

Drainage problems

Kitchen sinks are prime targets for drainage problems. This is especially true when the sinks drain into older plumbing pipes, such as galvanized steel. Many foreign objects find their way into the drainage systems beneath kitchen sinks. Grease, however, is one of the largest contributors to drainage problems in the kitchen.

As a service plumber, I have removed rings, forks, knives, spoons, shampoo-bottle tops, bracelets, earrings, and even aerators from kitchen drains. With the crossbar strainers in modern sink drains, not nearly as many large objects fall into the drainage system as did in earlier years.

Most drainage problems in sinks are easy to troubleshoot. The trap is removed and inspected. If it is clear, the problem is normally in the trap arm or drain, and a snake will generally clear the obstruction. We are going to get into much more detail on clogged drains and vents in Chap. 30.

Bar Sinks

Bar sinks are basically just miniature versions of kitchen sinks. All of the troubleshooting information given for kitchen sinks can be applied to bar sinks.

Laundry Sinks

Laundry sinks don't offer many troubleshooting challenges. These sinks are pretty straightforward, and the problems associated with them are easily found. The same techniques used to troubleshoot kitchen sinks can be applied to laundry sinks.

Laundry tubs that are equipped with pumps, however, are another story altogether. It is not unusual for laundry tubs to be fitted with pumps to lift their contents up to a suspended drain line. These pumps and the objects that find their way into the drains of laundry tubs can create a host of problems. Let me give you a couple of examples from my past experiences.

As you probably know, most laundry-tray pumps have very small discharge lines, usually only about $\frac{3}{4}$ inch in diameter. The pumps themselves are not large, and these two factors combine to create problems. Laundry tubs are meant to be convenient, but when they are set up with pumps, they can be anything but convenient.

My first story involves a laundry pump that one of our plumbers worked on twice before I bailed him out. The plumber responded to a call for a laundry tub with a pump that would not drain. He did his troubleshooting and decided that the check valve was not working properly. This wasn't an unreasonable assumption, but it turned out to be the wrong diagnosis. The plumber replaced the check valve and tested the sink. It worked, so he left.

In less than a week, the customer called back and complained that her sink was once again unable to drain. The same plumber took the callback. He disconnected the drain line and inspected the recently installed check valve. The spring in the check valve was laced with threads of clothing, so he replaced the valve. This replacement lasted about as long as the other one had. The customer called back, in a fury, and fussed and fussed and fussed.

As a damage control effort, I responded to the complaint personally. Before going out to the job, I talked to my plumber and reviewed his previous findings. It made sense that the check valve was failing and allowing the water to run back into the sink, only to keep the pump running constantly.

When I arrived at the house, I inspected the sink and pump arrangement. There was a problem with the check valve. It was jammed open, allowing pumped water to retreat back into the sink. When I investigated further, I found the same lint and clothing threads wound up in the valve that my plumber had found on previous calls. It was obvious that the homeowner was not being conscientious in preventing foreign objects from entering the drain.

Since I had already had to eat one service call and one new check valve, I wanted to make my visit the last one that would be necessary. I questioned the homeowner about the types of things she used the laundry tub for, and she was somewhat evasive. When I asked if she ever used the basket strainer that my plumber had given her and told her to use after the first call, she assured me that she did. Somehow, I doubted that she was telling the truth.

I told the homeowner that I would personally fix her problem. This calmed her, and I went to my truck for parts. One of the parts was a new check valve, and the other was a piece of screen wire. I was sure the woman was not using a strainer in the sink, but at that particular point I couldn't prove it.

I cut a section of the screen wire and installed it between the tailpiece and the drain fitting on the sink. From up above, looking into the drain, the screen was not apparent. When I had replaced the check valve, I instructed the homeowner to always use the basket strainer. In fact, I even had her sign the service ticket that said that the warranty on the work would be voided if she failed to use the basket strainer. She wasn't happy about signing the agreement, but she did.

It hadn't been 3 days since I left the house before the complaining homeowner was shouting in my ear over the telephone. It seemed that her laundry tub was not draining again. As you might imagine, I took the call myself. When I arrived in the home, I questioned the woman about her use of the basket strainer. She assured me she had used it on every occasion. Well, guess what? She lied.

While she was standing in the laundry room, I disconnected the tailpiece and showed her the plugged-up wire that I had placed over the drain. Her eyes got big and her bottom lip started to quiver. She had been caught. I explained that the junk in the wire couldn't have gotten through the basket strainer; I even proved it to her. It wasn't long before she admitted to forgetting to use the strainer, sometimes. I put the sink back together and left the job.

We never got any more work from that customer, but then, I didn't really want her work anymore. Lint and strings can wreak havoc with the check valves in laundry-tray pump lines.

On another occasion, we had a similar problem with a laundry tray on a pump system, except this time, the pump kept seizing up. The plumber who responded to the call got the pump going, but a few weeks later the same trouble was happening all over again. A different plumber took the callback, and he found the problem quickly. What do you think might have caused the pump to jam? Would you believe gravel from an aquarium? Well, that's what it was.

The homeowner in this case washed his aquarium out in the laundry tub. When he did, some of the gravel from the bottom of the fish tank would escape and find its way into the pump. The gravel would lock up the pump and render the laundry tub useless. You can imagine my plumber's surprise when he found colored gravel in the pump.

What is the point to these stories? The point is simple: expect the unexpected. Laundry-tray pumps that attach directly to the fixture are not that great to begin with, and when you factor in negligence, they can become downright troublesome.

Service Sinks

Service sinks are no more difficult to troubleshoot than laundry sinks. While service sinks are often larger and heavier and have different types of traps than laundry sinks, the same troubleshooting principles apply. Most service sinks are equipped with cleanout plugs that allow easy inspection of the drainage system for obstructions.

Lavatories

Lavatories come in many sizes, styles, and shapes. They work on different principles than the other sinks we have discussed. The biggest difference is the presence of a pop-up assembly.

There are a few minor differences in the types of problems you may encounter with various styles of lavatories. For example, a wall-hung unit may have water splashing down behind it, but that is unlikely with a drop-in lavatory. Let's take a quick look at each type of common lavatory and explore problems that may be unique to them. When we have finished this, we will look at the troubleshooting steps for lavatories in general.

Wall-hung lavatory

A wall-hung lavatory should have a waterproof seal installed at the point where the fixture meets the wall it hangs on. If this seal is defective or nonexistent, water can run down behind the fixture and imitate a leak in the drainage or water-supply systems.

It is not uncommon for people to lean on wall-hung sinks; this can break the waterproof seal between the fixture and the wall. If the top of the sink is used to hold soap or if the users of the fixture splash water onto the top of the sink, the water can run over the unprotected back edge and find its way under the sink. If you are looking for the source of a mysterious leak with a wall-hung lavatory, check the seal between the fixture and wall.

Pedestal lavatory

A pedestal lavatory also needs a waterproof seal where its back edge meets the host wall. Since a pedestal lavatory is basically a wall-hung lavatory with a pedestal placed under it, the fixture shares the same characteristics of a wall-hung lavatory.

Drop-in lavatory

A drop-in lavatory should have a waterproof seal installed under its rim and a bead of caulking applied to the outside edges of the rim.

Silicone caulking is typically used to create the waterproof seals needed for lavatories.

When a drop-in lavatory is installed without the proper seal between it and the counter, water can run under the fixture. We talked about this same type of condition when we were on the subject of kitchen sinks. Any water that gathers on the counter may be able to slip under the lavatory rim and give the impression of a plumbing leak.

Rimmed lavatory

A rimmed lavatory should have a waterproof seal under the support rim. Since these fixtures hang below the surface of a counter, they are prime candidates for under-the-rim leaks.

Molded lavatory tops

Molded lavatory tops have integral lavatory bowls and normally are made with a backsplash. This type of lavatory is the least likely to produce leaks that are not plumbing related. The only place these units are likely to leak is around the faucet holes. All lavatories have the potential to leak through these holes if their faucets are not properly sealed and mounted.

Troubleshooting Lavatories

Troubleshooting lavatories is not very different from working with any type of sink, but the pop-up assembly does allow for some additional problem possibilities.

The slip-nut connections and traps for lavatories should be checked out with the same procedures described for kitchen sinks. Remember to always fill the bowl to capacity and release all the water at once to check for drainage leaks. It can be helpful to use toilet tissue to rub around possible leak locations. The paper will expose dampness that you might not see otherwise.

Pop-up assemblies

Pop-up assemblies (Figs. 18.7 through 18.13) are the primary difference between lavatories and sinks. If a pop-up plug is not properly adjusted, the bowl will not hold water. This problem is usually obvious, and it is normally corrected by adjusting the lift rod. There are, however, times when a lavatory bowl will lose its water even when the pop-up plug is sealing properly. How can this be? Do you remember the situation I described in the opening of this chapter? I told you

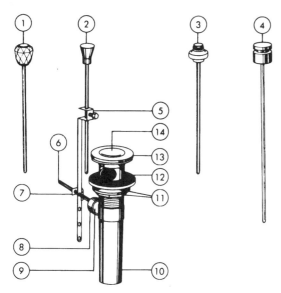

Figure 18.7 Parts detail for a lavatory pop-up assembly. (*Courtesy of Delta Faucet Company.*)

that a lavatory could lose all of its water and never show a trace of a leak. Well, you are about to find out how that can happen.

I have seen good plumbers struggle with pop-up adjustments over and over again when the pop-up plug was not the cause of the problem. Common sense tells you that if water is leaving the bowl, it must be getting past the pop-up plug. While this is usually the case, there is an exception.

If the rim of a lavatory drain is not sealed with putty, water can leak under the rim. We've talked about this in the section on kitchen sinks, but the leak is harder to find when dealing with lavatories. When the leak occurs on a kitchen sink, water is evident at some point beneath the sink. This is not the case with a lavatory.

Water that seeps under the rim of a lavatory drain winds up in the drain assembly. The water is carried down the tubing and into the trap, without leaving any evidence of a leak. The only clue available is the knowledge that the bowl is losing its water. A plumber who is not aware of this type of possibility could work for days adjusting a pop-up assembly without success in solving the puzzle.

How can you tell if the problem is a pop-up adjustment or a bad drain seal? The easiest way to make a definite determination is to remove the pop-up plug and install a rubber stopper in the drain opening. If you fill the bowl with the rubber stopper in the drain and

MODELS — A

KEY	PART DESCRIPTION	ORDER NUMBER	RP5651	RP5651 BC	RP5651 CB	RP5651 CG	RP5651 G	RP5651 SG	RP6266	RP7526	RP7526 BC	RP7526 CG	RP7530	RP10109 BC OAK	RP10109 CG ROS	RP10109 OAK	RP10109 ROS	RP12498
1.	Knob & Rod Assy.-Crystal	RP7524								●								
		RP7524 BC									●							
		RP7529										●	●					
2.	Knob & Rod Assy.-Brass	RP6146	●															
		RP6146 BC		●	●													
		RP6146 G					●											
		RP6146 SG						●										
		RP6282				●			●									
3.	Knob & Rod Assy.-Wood	RP10108 BC OAK												●				
		RP10108 CG ROS													●			
		RP10108 OAK														●		
		RP10108 ROS															●	
4.	Knob & Rod Assy.-Brass	RP12501																●
5.	Strap & Screw	RP6136	●	●	●	●	●	●	●	●	●	●	●	●	●	●	●	●
6.	Horizontal Rod w/Clip	RP6134	●	●	●	●	●	●	●	●	●	●	●	●	●	●	●	●
7.	Spring Clip	RP6144	●	●	●	●	●	●	●	●	●	●	●	●	●	●	●	●
8.	Nut	RP6132	●	●	●	●	●	●	●	●	●	●	●	●	●	●	●	●
9.	Pivot Seat & Gasket	RP6130	●	●	●	●	●	●	●	●	●	●	●	●	●	●	●	●
10.	Tailpiece	RP6128	●	●	●	●	●	●	●	●	●	●	●	●	●	●	●	●
11.	Nut & Washer	RP6140	●	●	●	●	●	●	●	●	●	●	●	●	●	●	●	●
12.	Gasket	RP6142	●	●	●	●	●	●	●	●	●	●	●	●	●	●	●	●
13.	Flange	RP6126	●		●				●							●	●	●
		RP6126 BC		●							●			●				
		RP6126 CG				●						●			●			
		RP6126 G					●											
		RP6126 SG						●										
		RP6268								●			●					
14.	Stopper	RP5648	●		●				●							●	●	●
		RP5648 BC		●							●			●				
		RP5648 CG				●						●			●			
		RP5648 G					●											
		RP5648 SG						●										
		RP6313								●			●					

Figure 18.8 Parts description for the pop-up assembly in Fig. 18.7. (*Courtesy of Delta Faucet Company.*)

still lose the water, you can bet the problem is a bad drain seal. If the rubber stopper holds the water in the bowl, you have a pop-up assembly that is out of adjustment or a bad seal on the pop-up plug. This simple test can save you hours of time and frustration.

When you do the rubber-stopper test, use only a rubber stopper. If you use a rubber disk, such as is sometimes used to seal kitchen sink drains, the disk will cover the drain rim, and you won't get accurate data.

If you find the cause of your leak to be a bad drain seal, you must disassemble the drain assembly and install a new seal of putty.

When you know the problem lies with the pop-up plug, you can

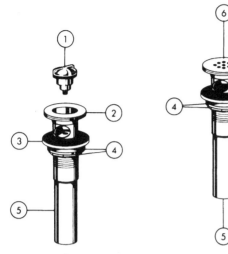

Figure 18.9 Lavatory pop-up with a rotary stopper. (*Courtesy of Delta Faucet Company.*)

Figure 18.10 Lavatory pop-up with a grid strainer. (*Courtesy of Delta Faucet Company.*)

| | | | MODELS | |
| | | | **B** | **C** |
KEY	PART DESCRIPTION	ORDER NUMBER	RP6339	RP6346
1.	Rotary Stopper	RP6354	●	
2.	Flange	RP6126	●	
3.	Gasket	RP6142	●	●
4.	Nut & Washer	RP6140	●	●
5.	Tailpiece	RP6128	●	●
6.	Grid Flange	RP6344		●

Figure 18.11 Parts description for the pop-up assemblies in Figs. 18.9 and 18.10. (*Courtesy of Delta Faucet Company.*)

adjust the lift rod to compensate for the poor fit. If the leak persists, check the seal on the pop-up plug. If the gasket is bad, either replace it or the entire pop-up plug.

The threads of a pop-up assembly can allow water to leak by if they are not coated with a sealant. This type of leak will not normally show up when water is just being run out of the faucet. To expose this type of leak, the lavatory bowl should be filled and all the water released at once. You need the volume and pressure this type of test provides to find leaks in the threads.

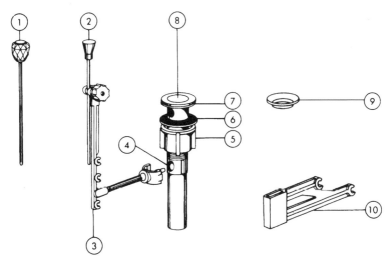

Figure 18.12 A Snap'n Pop pop-up assembly. (*Courtesy of Delta Faucet Company.*)

			MODELS	
			D	
KEY	PART DESCRIPTION	ORDER NUMBER	RP6463	RP7525
1.	Knob & Rod Assy.	RP6146	●	
2.	Knob & Rod Assy.	RP7524		●
3.	Pivot Rod & Strap Assy.	RP6477	●	●
4.	"O" Ring	RP6481	●	●
5.	Nut	RP6475	●	●
6.	Gasket	RP6142	●	●
7.	Foam Gasket	RP6876	●	●
8.	Stopper	RP5648	●	●
9.	Spacer	RP10994		
10.	Extender	RP7585		

Figure 18.13 Parts detail for pop-up assembly in Fig. 18.12. (*Courtesy of Delta Faucet Company.*)

Sometimes water will leak out around the retainer nut that holds the pivot rod of a pop-up assembly in place. This usually indicates a bad washer or the absence of a required washer. If you remove the retainer nut and pivot rod, there should be a rubber washer installed in front of the nylon ball. If the washer is missing, cut, or worn, a leak can occur.

Slow drains

Slow drains are common for lavatories. Human hair is routinely found to be the cause of these sluggish drains. Unlike most sinks, dropping a trap and inspecting it may not reveal the reason for a lavatory draining poorly.

Hair is commonly caught on the base of pop-up plugs and around pivot rods. A plumber who only inspects a trap and then opts to snake the drain of a lavatory may be wasting a lot of time and energy. The root cause of the problem may be hidden in the pop-up assembly.

Before you start snaking a slow lavatory drain, check the pivot rod and pop-up plug for hair and other obstructions. If the sink is not equipped with a pop-up plug, look closely in the drain assembly for unwanted objects, such as toothpaste container tops.

Open lavatory drains frequently collect tops from toothpaste containers and shampoo bottles. Coins and a wide variety of other unexpected items can be found in the drain assemblies of lavatories. To check for these items, remove the trap. Shine a light down the drain and do a visual inspection. The pivot rod may have to be removed in order to clear obstructions in the drain assembly.

Another possible cause for a sluggish lavatory drain is a pop-up plug that is not being lifted high enough by the lift rod. If the lift rod does not protrude high enough above the lavatory faucet, it may not provide adequate leverage to give a full opening below the pop-up seal. When this problem occurs, a minor adjustment in the height of the lift rod will correct it.

Lavatories that are served by galvanized drain pipes require frequent attention. The rough inner surface of the galvanized pipe catches hair and other substances (Fig. 18.14), causing the drain to flow slowly. Additionally, the short-turn fittings used with galvanized pipe do not result in fast-flowing drains. We will go into more details on the drains and vents of plumbing systems in Chap. 30.

Common Sense

Common sense and a systematic approach will go a long way in any type of troubleshooting. This is particularly true when working with sinks and lavatories. There are very few complex reasons for a sink or lavatory to leak.

I have given you examples of the two most common hard-to-find leaks when dealing with sinks and lavatories—a lack of putty around the drain rim and a lack of putty around the fixture rim. Aside from these two types of problems, most sink and lavatory leaks are easy to find. It is, however, important that you remember to fill the bowls to

Figure 18.14 A galvanized steel drain clogged with grease and other debris.

capacity and release all of the water at once when testing for drainage leaks. Also remember to use paper towels or toilet tissue to help locate tiny leaks. If you remember these principles and troubleshoot sinks and lavatories with a good flashlight and a systematic approach, you shouldn't have much trouble finding the causes of leaks.

We are now ready to move onto the next chapter and investigate troubleshooting techniques for bathtubs.

Troubleshooting Bathtubs

Problems with bathtubs can cause a lot of damage. If you've done much service work, you know how fast a bad bathtub can ruin a downstairs ceiling or floor. If you're new to the trade, you will soon find out how destructive the leaks associated with bathtubs can be.

Because of the nature of bathtubs, they can offer plenty of challenge to a troubleshooter. There are many ways that leaks having to do with bathtubs can avoid detection or cast suspicion on other parts of a plumbing system.

Bathtubs can produce leaks from their wastes and overflows, tub spouts, shower heads, valves, traps, and even their surrounding walls and floors. With so many possibilities to consider, troubleshooting bathtubs requires strict attention to detail and a plan of action.

Let's start this chapter with a look at a real-life situation I have encountered numerous times. The example you are about to read is not uncommon; it is, in fact, fairly typical. The story is based on actual service calls I've been on. The actions of the plumbers in the story mirror the actions of some of my past plumbers.

The story begins with a panicked homeowner calling for emergency service. She is calling from her kitchen phone as she watches a steady stream of water dripping from the dining-room chandelier. A plumber is dispatched to the location and arrives quickly. When the plumber enters the house, there is plenty of evidence that there has been a leak, but the water has stopped running out of the light fixture. The dining-room table is wet, the carpet is wet, and the homeowner is baffled by the leak disappearing before the plumber arrived.

To give you some background information, there is a full bathroom and a half-bath on the second floor of the house. Neither of the bathrooms is in the vicinity of the dining room, where the leak was noticed. Can you solve the puzzle yet?

After calming the homeowner, the plumber removes the light fixture's trim ring and shines a light into the ceiling. There is water standing on the back of the drywall ceiling. The puddle is surrounding the electrical box, giving no clear evidence of what direction it came from. With the limited viewing space available, the plumber cannot tell if there are any pipes in the joist bay. What would you do first in this situation? Before I tell you what the proper troubleshooting procedure is, let me explain how many plumbers would proceed.

A high percentage of plumbers would insist on cutting a hole in the ceiling to investigate the leak. After ruining the ceiling, these plumbers would probably not know much more than they did before opening the ceiling. It is unlikely that there would be any pipes in the joist bay, but they might find a trail of water to give some direction of where the leak originated.

While it might ultimately be necessary to cut the ceiling open, it would be a mistake to do so at such an early stage of the troubleshooting process. The plumber would be justified in cutting the ceiling, but there probably is a better way to reveal the cause of the leak.

A plumber who cut the ceiling open and found a trail of water would know what direction to go in but not how to proceed. Some plumbers would go to another point on the ceiling and make another inspection hole. This gives the homeowner two holes in her ceiling, and the plumber is probably no closer to resolving the problem.

What data do we have to assess so far? We know there are two bathrooms upstairs and that there are no pipes in the joist bay where the water ran through the light fixture. We know a significant volume of water leaked out of the ceiling, but then it stopped as quickly as it had started. What does this tell us?

We can tell now that the leak is not likely to be in the water distri-bution pipes. If it were, the leak would continue. The constant water pressure in the distribution pipes would not allow the water to cease.

Since the leak is not in the water pipes, it must be in the drainage pipes. The question is, what drainage pipes? There are no drains visible from the inspection holes in the ceiling. Many plumbers would be stuck at this point, not knowing what to do next. Some would probably hack out chunks of the ceiling until they found the leak, but they may still never find it. Leaks of this nature, especially if they haven't been occurring for an extended time, can be very difficult to pinpoint.

The water could have come from a toilet, a bathtub, a broken drain pipe, a faulty roof flashing, a water spill on an upstairs floor, or some other equally difficult circumstance. Since this is the first time such a leak has occurred in the house, there is no pattern to follow. How would you proceed?

Okay, enough of the suspense; let me tell you the right way to troubleshoot this problem. Since the water stopped on its own, it is safe to assume the leak is not from a pressurized water pipe. Before cutting the ceiling open, it would be wise to cut the power off to the light fixture and try to re-create the leak. If the leak showed up once, it will probably show up again. But, there are times when making the leak recur will seem impossible; I'll explain in a few moments.

When you assess the leak, you can see the volume of water is greater that what should have been produced by a lavatory. The amount of water present leads you to believe it must have come from a toilet, shower, or bathtub. Unless it has been raining, you can rule out the possibility of a leaking roof flashing.

The first step would be to ask the homeowner to stand watch at the light fixture. You should explain what you are going to do, and ask the homeowner to notify you if she sees water anywhere.

Ask the homeowner if anyone had used any of the upstairs plumbing within 30 minutes of the time the leak was discovered. If the plumbing had been used, and you can bet it had been, ask what fixtures were used. Did someone take a shower? Did someone take a bath? Was a toilet flushed? After getting answers to these questions, determine which bathroom was used, and start your troubleshooting in that bathroom.

We will assume that both bathrooms were used as the family was getting ready for work and school. All of the fixtures were used, and the leak was not noticed until the homeowner came downstairs.

Go upstairs and flush one of the toilets. Flush the same toilet at least twice. Wait a few minutes before moving onto the next fixture. Since the water is traveling across the ceiling before coming out, it will take a little time for it to reach the light fixture.

Once you are convinced the first toilet flushed is not the cause of the leak, flush the second toilet twice. After waiting for the water to show up at the light fixture, test the bathtub. Fill the tub full of water and then open the drain. Let all the water drain and wait to see if it shows up at the light fixture. If it doesn't, try running the shower for 5 minutes.

The riser to the shower head could have a leak in it. This leak wouldn't show up unless the shower was turned on; the riser is not under pressure unless the shower is running. If you still haven't been able to make the water reappear at the light, try the lavatories. Fill each lavatory and let them drain, one at a time.

In many cases you would have found the leak by now, but this is not an average job. You've tested all of the plumbing and nothing has caused water to come out of the light fixture. What should you do now? Well, there are still a number of things to check out.

If none of the plumbing produced the leak, it had to come from a less-obvious place. We will assume that the full bath is equipped with a bathtub that has a shower head over it. The bathroom floor is tiled and so are the walls that surround the bathing unit.

Where might the water have come from? It could have come through the floor or the tub walls. If the grouting in the tile work has gone bad, the water may have passed right through and worked its way over to the light fixture. If the tub spout or tub faucets are not properly sealed, the water may have splashed off the bather and run down into the wall cavity and onto the ceiling.

Since these types of leaks won't show up simply by running the shower, you must simulate the conditions when someone is taking a shower. No, you don't have to stand in the tub and get splashed. If the tub is equipped with a shower curtain, you're in luck.

A shower curtain is a good sign that water escaped from the bathing unit, onto the floor. The curtain can also be used to simulate the splashing that occurs when someone showers.

Start this phase of your troubleshooting by pouring water on the bathroom floor, particularly near the edge of the bathtub. Don't skimp on the water. If the water disappears, expect to hear the homeowner screaming soon. Should the water remain on the floor, move your investigation to the tub walls.

Turn on the shower and hold the shower curtain toward the front of the tub, in about the same place a bather would stand. Let the water bounce off of the shower curtain and onto the front and side tub walls. Keep this up for at least 5 minutes. When the homeowner lets you know the leak is back, you have narrowed the list of potential causes. It is either bad grouting or poor seals around the tub spout or faucet handles.

Have the homeowner come upstairs and watch through the bathtub's access door to see if the water is running through at the plumbing outlets or if it is coming through the grouting. If it is not pouring past the plumbing holes, you can assume it is running through the tile walls.

As you can see, this type of troubleshooting will be time consuming. If you have a good eye for construction, you may be able to speed up the process. For example, I can usually look at a ceiling and tell which way the floor joists are running. Since water normally doesn't cross through joist bays, I can often pinpoint precise locations for detecting the leak.

Tubs That Won't Hold Water

Tubs that won't hold water generally require an adjustment to the tub waste. It is possible for water to leak around the trim ring of the tub drain, but the problem is usually with the adjustment of the tub waste. If there is no water leaking from under the tub, you can rest assured the cause is in the tub waste adjustment.

There are several different types of tub wastes in use, and each of them can present its own type of trouble. With some types of wastes it is necessary to make adjustments in the trip mechanism. Others require the replacement of a seal or an entire stopper.

Toe-touch wastes

Toe-touch wastes don't contain any internal adjustment mechanisms. This type of waste depends on the stopper to hold water in a bathtub. If a toe-touch waste is allowing water to leak out of the tub and down the drain, there are three options available to you.

Begin your troubleshooting by turning the stopper clockwise. If the stopper is not screwed into the drain far enough, water can leak past its seal. If the stopper is screwed into the drain properly, remove it and inspect the seal. To be sure the problem is with the toe-touch stopper, install a rubber stopper in the drain and test the tub. If water doesn't leak past the rubber stopper, you have identified the cause of your problem; it is the toe-touch stopper.

Once you know the stopper is causing the trouble, you can replace the entire stopper with a new one. It may be possible to replace the seal only, but with the low cost of replacement stoppers, I would replace the whole unit to be sure of correcting the problem without a callback.

Lift-and-turn wastes

Lift-and-turn wastes (Fig. 19.1) can be treated like toe-touch wastes when troubleshooting for seeping water. Even though the two wastes don't operate in the same manner, they do seal the tub drain in the same basic way.

Mechanical strainer wastes

Mechanical strainer wastes (Fig. 19.2) are more difficult to deal with than toe-touch or lift-and-turn wastes. This type of waste depends on a plunger blocking the tee of a tub waste to maintain a static water

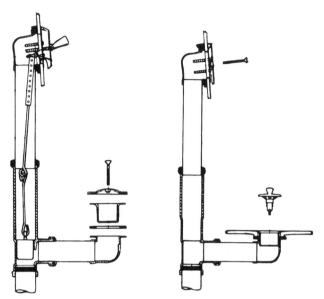

Figure 19.1 Lift-and-turn tub waste. (*Courtesy of Delta Faucet Company.*)

Figure 19.2 Trip-lever tub waste. (*Courtesy of Delta Faucet Company.*)

level in the tub. Fine-tuning the adjustment of this type of tub waste can require several attempts.

To rule out a leak around the trim ring of the tub drain, you can remove the strainer and insert a rubber stopper. This test will tell you positively if the problem is with the plunger of the tub waste.

Once you know the plunger needs to be adjusted, you must remove the screws from the overflow plate. This allows you to remove the entire trip mechanism. The plunger location can be raised or lowered by turning the threaded portion of the mechanism. Since water is leaking past the plunger, the length of the mechanism should be extended.

Reinstall the adjusted mechanism and test it. It is not uncommon to have to tinker with the adjustment several times before the perfect position is found.

Rocker-type wastes

Rocker-type wastes require similar troubleshooting to that of strainer wastes. A rocker waste, however, has a spring, rather than a plunger, that hangs in the overflow tube. The spring comes into contact with the rocker assembly and allows the tub stopper to be opened and closed.

The troubleshooting procedure for this type of waste should begin with the rocker assembly. Operate the trip lever and observe the stopper. Does it seat into the drain evenly? If not, the position of the rocker assembly should be adjusted. Does the stopper drop deeply into the tub drain? If not, the height of the stopper should be lowered.

It is not unusual for the rocker assembly to be the cause of seeping water with this type of waste; however, the problem could be with the trip mechanism. If the stopper doesn't pop up and seat with satisfactory action, the length of the trip mechanism may require adjustment. This is accomplished in the same way that was described for a strainer waste.

Tub-shoe leaks

Tub-shoe leaks can be the cause of water draining from a bathtub inadvertently. When water seeps past a tub-waste mechanism, it goes down the drain. But when the leak is at a tub shoe, the water goes into unwanted areas, such as a downstairs ceiling.

A lack of putty on the underside of the tub drain is the most common cause for tub-shoe leaks. It is possible, however, that the leak is due to a faulty or missing tub-shoe gasket. I have worked on tubs where the tub-shoe gasket was never installed.

If you are faced with a situation where there is no access to the tub shoe, you must do your work from inside the bathtub. This is really not difficult, and it saves time and trouble of cutting out a ceiling. All that is required is the removal of the tub drain.

Once the tub drain is unscrewed, the gasket can be retrieved and inspected. Replacing the gasket requires a little manual dexterity, but it can be done from inside the tub without much trouble.

I sometimes put a little pipe dope or wax on the back of the gasket to make sure it doesn't slide off the tub shoe before the drain is replaced. Losing your only gasket between the floor and ceiling can be embarrassing, and the tacky dope or wax minimizes the risk of loss.

By installing a new gasket, putting sealant compound on the drain's threads, and installing putty around the inner rim of the drain, you can be fairly certain of stopping a leaking tub shoe.

Bathtubs That Drain Slowly

Bathtubs that drain slowly are a common problem. The cause of the problem is usually either a tub waste that needs to be adjusted or a partial blockage in the drainage system. A vent that is obstructed or nonexistent can also contribute to the slow draining of a bathtub.

Hair is a frequent enemy of bathtub drains. Even tubs equipped with strainer-type drains can be slowed by accumulating hair. Mechanical tub wastes are particularly susceptible to hair attacks. Let's look at the different types of wastes and see what to do when a tub is not draining fast enough.

Toe-touch wastes

Toe-touch wastes are easy to troubleshoot. You can unscrew the stopper and inspect the crossbars in the drain for obstructions. If there aren't any, you should try installing the stopper so that it sits a little higher than it did before. The simple act of loosening the stopper may be all that is required to solve the problem of sluggish draining.

You can either identify or eliminate the stopper as the cause of your problem with ease. Put a rubber stopper in the tub drain, and fill the tub partially with water. Remove the stopper, and observe the draining action. If the water leaves the tub quickly, the stopper is your problem. When the water is slow leaving the tub through a wide-open drain, you must look further.

It is possible, but unlikely, that an obstruction has lodged in the tee of the tub waste. A small spring-type snake will eliminate this possibility quickly.

Lift-and-turn wastes

Like with a toe-touch waste, you can remove the stopper of a lift-and-turn waste to inspect the drain's crossbars for obstacles. While the stopper is out of the drain, cover the drain opening and test the tub as described above. There isn't much in either a toe-touch or lift-and-turn waste to cause trouble.

Mechanical strainer wastes

The plunger used with a mechanical strainer tub waste is frequently a cause for slow drainage. The plunger may be engulfed with hair, or it may simply be out of adjustment.

Remove the overflow plate and trip assembly. If the plunger is free of debris, there are only two likely causes for slow drainage. Either the plunger needs to be raised by shortening the lift assembly, or there is a blockage somewhere else in the drainage system.

With the plunger removed from the overflow tube, cover the drain and test the tub as described earlier. If the water drains freely, the plunger needs to be adjusted. Should the water remain sluggish without the plunger installed, look for a drainage obstruction.

By testing the tub drain with the plunger removed, you can save yourself a lot of tedious trouble. I've had plumbers spend close to an hour adjusting the height of plungers only to find out that the problem was not with the plungers at all. It pays to work smarter, not harder.

Rocker-type wastes

Rocker-type wastes can catch hair on the rocker assembly and the spring that operates it. Grasp the stopper and remove the rocker assembly. If it is not fouled with hair, remove the lift assembly from the overflow tube. With both of these parts removed, cover the drain and test the tub as described earlier.

If water runs out of the tub quickly when the waste components are removed, and the rocker and spring are clean, you can plan on having to adjust the tub waste.

Trap and Drain Obstructions

When a tub waste is not the cause of a slow-draining bathtub, look for trap and drain obstructions. In most cases this can be done by running a small spring-type snake down the overflow tube of the tub waste. A 25-foot snake should reach well into the drainage system and eliminate any obstruction that is affecting only the bathtub.

Most bathtubs are fitted with P-traps, which make snaking them easy. If you encounter a drum trap, which is illegal for most applications, you must gain access to the trap and a cleanout opening. Otherwise, your snake is not going to get past the trap, and the cable is likely to kink.

Because of their design, drum traps are impossible to snake through from an overflow opening of a tub waste. The snake can get into the trap, but it cannot get out. At first, the snake will seem to be going someplace, but all it is doing is curling up inside the trap. If you are using an electric drain cleaner, the cable will probably kink, possibly to the point where it is damaged permanently.

There are a lot of drum traps in Maine. In fact, I was working with one just a few days ago. I was called out to a tub that wouldn't drain at all. Even after I removed the interior parts of the tub waste, the fixture wouldn't drain.

Maine's plumbing systems are not vented very well, so plungers often work to clear stoppages, but not on this job. I covered the overflow hole with one plunger and plunged the drain with another. The plungers had no affect on the stopped-up drain. I was relatively certain the tub would be equipped with a drum trap, but I snaked the drain as far as I could. The cable started to kink, and I was sure there must be a drum trap on the tub drain.

There was a closet behind the head of the tub, but there was no access panel. With a mechanical tub waste and a drum trap, there should have been an access door in the wall, but there wasn't. To avoid cutting into the closet wall or the ceiling down below, I tried to clear the stoppage with water pressure from a balloon bag. Unfortunately, the results were not good. The tub started to drain ever so slightly, but it was still clogged up.

The property was a multifamily building being used as rental property. I called the property owner and got permission to cut a hole in the closet wall. When I did, I was not surprised to find a drum trap.

The trap was an old one, with a brass cleanout in it. The piping was all galvanized steel, and there were three fittings within a foot of the trap. There was really little wonder that the drain wasn't working well.

The trap was hanging between two floor joists. With the limited space between the joists, I had no room to get leverage on the cleanout with a pipe wrench. I tried to remove the cover with a basin wrench, but the cover was too tight to budge.

This type of problem is not unusual when you find an old drum trap on a bathtub. How would you get the cleanout cover off? I tried using a socket, socket extension, and rachet to remove the cover, but it still wouldn't turn. I hit the cleanout with a hammer to loosen it, but still no dice.

The next logical step was to take a cold chisel and beat the cover out of the trap. I didn't want to do this, but the limited space I had for access wasn't giving me much of an alternative. Just before I started banging away at the cleanout, I got an idea.

I dug around in my truck and found a long handle that would fit on the socket extension. With the use of the long handle, a lot of muscle, and a little luck, I was able to get the cover off the trap.

When I put the snake in the drain pipe, it hit resistance almost immediately. There were two galvanized eighth-bends on the discharge side of the trap. Only a short nipple separated the two fittings, and the clog was in the offset. A few good turns of the snake and the drain was clear.

The job turned out all right, but it could have gotten nasty if I hadn't known what to do about the drum trap. I could have fought with my snake for quite awhile, thinking it was hitting the stoppage when, in fact, it was hitting the drum trap.

If I hadn't had the socket set and long handle, I might not have been able to remove the cleanout cover. Experience also played another vital role in this particular job. The drum trap had been installed upside down, so the cleanout was on top. If I had opted to cut the ceiling from below, instead of the closet wall, I still wouldn't have had access to the cleanout.

Why did I choose to open the wall in the closet? There were two reasons. I suspected the trap might be installed upside down because I have found many like it in the past. Secondly, I knew the damage to a closet wall would not be as disruptive for the property owner as a gaping hole in the downstairs tenant's apartment. These are the types of thoughts you should consider when troubleshooting and deciding on a course of action.

Vent Blockages

Vent blockages can cause a fixture to drain slowly. If you have checked the tub waste and the drainage system without finding a cause for slow drainage, turn your attention to the vent system. We are going to go into great detail on vents and venting in Chap. 30, but it is appropriate that we take a look at how they affect bathtubs in this chapter.

Some bathtubs have individual vents, and others are either wet-vented or act as wet vents. And then, there are the tubs that are not vented at all.

If a tub has an individual vent and is draining slowly because of the vent, the problem is usually that the vent has become obstructed by leaves or other debris that has entered it. Snaking the vent pipe from the roof or attic of the property solves this type of problem.

Bathtubs that are wet-vented are usually vented through a lavatory drain. Under these conditions the vent to check out is the one near the lavatory. This type of setup is not unusual.

Tubs that serve as wet vents for other fixtures, such as toilets, can basically be treated as if they were individually vented. These tubs have individual vents, so check the vent close to the tub.

A sign that a tub vent is not working properly is a gurgling noise in the drainage system. The sound is similar to the one made when a drain is stopped up. If you have investigated all aspects of the drains and can't find a reason for them to be running slowly, take a close look at the vents.

It will be helpful to ask the property owner if the fixture has always drained slowly. If the bathtub has never drained properly, it may not be vented at all.

Overflow Leaks

Overflow leaks can be difficult for inexperienced plumbers to identify. This type of leak will not show up through normal test procedures. Overflow leaks occur when a tub is filled to capacity and someone gets into the water. The displacement of water caused by the human body can create an overflow leak.

The overflow fitting of a tub waste should be equipped with a tapered sponge gasket. This gasket prevents water that is going over the overflow plate from leaking past the drainage system. Sometimes the gasket is installed upside down, and there are occasions when the gasket is never installed. It is also possible that the gasket has lost its holding ability and is allowing water to sneak past it.

If there is an access panel at the head of the tub, an overflow leak is easy to identify. If the leak has been going on for some time, there should be evidence of water in the access opening. The leak may, however, be running through the hole that was cut in the floor to allow the installation of the tub waste.

How do you find an overflow leak? Well, you could fill the tub and get in it, but there is an easier way. Fill the tub to capacity and place a heavy object, like a 5-gallon bucket filled with water, in the bathtub. The mass will displace the water and force it to run out the overflow as it does when someone is bathing. By watching the overflow system from the access panel, you will be able to spot the leak.

If you are working alone and can't make the tub overflow without being there to hold pressure on the displacement object, you must get a little more creative. Wrap toilet tissue or a paper towel around the overflow tube. Spread out some more on the floor around the overflow. Then go back to the tub, and force the water to overflow. Any leak will show up on the paper towels and toilet tissue.

Escutcheon and Tub-Spout Leaks

Escutcheon and tub-spout leaks can be extremely hard to find unless you know to look for them. When escutcheons and tub spouts are not installed with the proper gaskets or putty, they can allow water to run right past them. The water can get into the wall and drip down to the floor. The leak can spread out and be difficult to pinpoint.

A garden hose is ideal for testing this type of situation. By spraying water on the escutcheons and tub spout, you can determine if they are leaking into the access area. If you don't have a garden hose to use, fill a pitcher or bucket with water and splash it on the escutcheons and spout. You can use paper towels in the access area to help locate any stray water.

Escutcheon and tub-spout leaks only occur when water is splashed against them. The most common cause is when someone is taking a shower and the water bounces off his or her body and onto the walls.

Valve Leaks

Valve leaks can be detected easily when there is an access panel to work with. Turn on the hot and cold water and inspect the valve. Use toilet tissue to wipe joints and discover small leaks.

Don't overlook the possibility that the valve body may be pitted or cracked. This condition occurs more often than you might think. Remember, when troubleshooting, to look for every possible problem. Inexperienced plumbers and plumbers who are in a hurry often overlook the obvious.

I've installed brand-new valves that had tiny holes in the bodies. These leaks are not always easy to find, and over the course of time, they can cause a lot of damage.

If you are troubleshooting an existing valve or replacing an old one with a new one, don't fail to inspect it closely for hairline cracks and invisible holes. Sometimes these leaks are so small that they don't even really drip. Water seeps out of the openings and spreads across the valve until it has accumulated enough to drip off the valve. If the water happens to slide down the valve and onto a supply pipe, it can easily go undetected.

Using toilet tissue to find these little leaks is the most effective way to ensure success. If you do not look for these minuscule leaks, trouble is sure to find you in the future. Once the valve is concealed in a wall, the leaks will eat away at the building materials a little at a time, for a long time to come. Then one day, the property owner will notice a damp carpet or a stained ceiling. The evidence may even show up in the lower sections of the tub wall. However the leak is found, the damages that have resulted, and they can be extensive, may be found to be your fault. Don't let this happen to you; check your work closely for leaks.

If you don't have open access to the tub valve, you can do a limited visual inspection through the holes in the wall where the valve stems protrude. Remove the handles and escutcheons, and shine a light into the cavity. Test the valve with the water both on and off. As a last resort, make an access panel to inspect the back side of the tub.

The Process of Elimination

The process of elimination is used in all forms of troubleshooting, and it should be used with bathtubs. There are very few elements of a typical bathtub that can cause problems, but putting your finger on the right cause can be difficult.

Many types of tub leaks can appear to be something they are not. For instance, water leaking past a tub spout and dripping down to the floor can look like a leak in the waste and overflow. Water that runs past the overflow gasket can saturate the floor without leaving a trace of where it came from.

Plumbers don't normally get called in on hidden tub leaks until some damage has been done. By the time a plumber arrives, water has often traveled away from the leak, making the leak harder to find.

Using the proper troubleshooting procedures and a process of elimination will reveal the leaks. It may take some time and patience, but the leaks can be found.

As you have seen, there are several types of leaks associated with bathtubs that will not show up in typical testing. I have shared my many years of experience with you to illustrate simple ways to eliminate possible causes of trouble. If you move systematically through the troubleshooting process, and remember to look for the hidden causes, you shouldn't have any difficulty troubleshooting bathtubs.

20

Troubleshooting Showers

Showers come in all shapes, sizes, and types, but they all share the same basic characteristics. Because of their design, showers don't harbor a lot of potential for baffling problems. This is not to say, however, that showers can't cause a lot of trouble. While showers don't offer many possibilities for problems, they can be difficult to troubleshoot and to work on.

Unlike bathtubs, where an access panel allows some visibility to the drainage area, the traps and drain pipes for showers are usually inaccessible. Pan-type showers can be a plumber's worst nightmare, and shower curtains can create some perplexing problems.

What Can Go Wrong with a Shower?

There aren't a lot of plumbing problems associated with showers, but the damage that can be done by a shower can be extensive. If a shower pan is leaking, it can be confused with a leaky drain. A shower arm with deteriorated threads can avoid detection from all but the best of plumbers. Showers with tile walls can present plenty of challenge for finding the cause of a leak, and shower doors and curtains can confuse the issue of troubleshooting.

Many water problems with showers are related to the users of the showers. While these problems are not pure plumbing problems, plumbers are responsible for finding the causes. Let's move on now to specific situations and see how you can improve your troubleshooting effectiveness when dealing with showers.

One-Piece Showers

One-piece showers offer the least possibly for problems of all the various types of showers. Because of their one-piece construction, leaks in the walls are nearly unheard of. Unless the unit is cracked, you can rule out wall leaks.

Valve Escutcheons

The valve escutcheons on showers are one source of mystery leaks. Just as was described in the last chapter, these escutcheons can let water penetrate the shower wall.

The escutcheons used for single-handle valves are frequently equipped with sponge-type gaskets. These gaskets are factory installed on the escutcheons and do a good job of preventing unwanted leaks.

Two-handle shower valves generally require a gasket or a putty seal to be placed behind the escutcheons. If these seals are missing or have deteriorated, water can find its way into the wall. A visual inspection can reveal the possibility of this type of leak, but a water test is the only way to be certain the escutcheons are not responsible for leaks.

Unfortunately, many showers don't have access panels that allow easy inspection of the shower valve. Bathtubs that use waste and overflows with mechanical joints are required to have access panels, but showers are not.

The lack of an access panel greatly reduces a plumber's ability to make sound judgment calls on problems related to showers. It may be necessary to cut into a wall to gain access in the case of stubborn leaks, but there are ways to make reasonable assumptions without butchering a wall. To expound on this, let's look at various circumstances on a step-by-step basis to see how cutting into walls and ceilings can be avoided.

Escutcheons

Escutcheons are notorious for creating hard-to-find leaks. When you combine the lack of an access panel with a leaky escutcheon, you have a tough situation to troubleshoot. Many plumbers would assume a wall would have to be cut to determine positively that an escutcheon is allowing water to pass through the wall. This is a logical assumption, but there is an alternative that will often work.

How can you tell if water is leaking into a wall through an escutcheon if you can't see the leak? Remove the valve handles and escutcheons. Pack the hole around each valve stem with toilet tissue, and replace the escutcheons. Spray or splash the escutcheons with water, and then remove them again and check the toilet tissue. If the paper is wet, you know the escutcheons are leaking. Dry paper, of course, indicates no leak at the escutcheons.

The toilet-paper test is not an ideal way to troubleshoot escutcheon leaks, but if you can't do anything else, it works pretty well.

Packing putty behind the escutcheons should block water, but you may also want to run a bead of caulking around the outside edges of all the escutcheons.

Shower Arms

Shower arms can rot walls, floors, and ceilings. Houses that derive their water from wells are especially susceptible to this problem.

Too many plumbers fail to inspect the inside-the-wall connection of a shower arm. I guess it's the old out of sight, out of mind thing. If you are faced with a shower leak you cannot find, check the connection between the shower arm and the shower ell.

How do you check a shower-to-ell connection without an access panel. It's really quite simple; pull the arm escutcheon forward and look through the hole. Remember to make sure water is running through the shower head when you do the inspection. It's hard to believe, but I've seen plumbers inspect shower risers and their connections without cutting the diverter on. If water is not coming out of the shower outlet, you are not going to find a leak in the shower riser.

The threads of shower arms sometimes rust out, and that is mostly what you are looking for. It's also possible that the arm was never doped up before being installed or that it just isn't tight.

Shower Ells

Shower ells can sometimes spring leaks in their castings or at their connection with the shower riser. You can check for these hidden leaks in the same way that was described for checking shower arms.

Shower Risers

Shower risers have been known to spring leaks in the pipe material, the connections, and the fitting and valve castings. For example, water with a high acidic content will often eventually eat pinholes in copper tubing. I've seen this happen a number of times; however, I've never found it to be a problem with shower risers.

If there is a problem in a shower riser, it is usually at the connection point with the shower ell or the shower valve. You can inspect the shower ell connection by following the instructions above. It's usually possible to inspect the riser-to-valve connection through one of the holes surrounding a valve stem. Again, make sure the shower head is turned on when you look for the leak.

If you work in an area subject to freezing temperatures, you might come face to face with a shower riser that has frozen and split. This is very common when dealing with seasonal cottages. Let me give you two examples of this type of problem.

The first example involves a leak that is relatively easy to find. This is the case when a shower riser has frozen and split with a large gap in the pipe. About all you have to do to find this type of problem is turn on the water to the shower riser. You will hear water rushing in the wall.

When a shower riser splits, the wall must be opened to make the necessary repair. The good thing about this type of leak is that it is easy to find. A large volume of water is released and the leak exposes itself before much damage is done. The bad thing is that the wall must be cut open.

The second type of freeze-up leak is not so easy to find, and it can do a lot of damage before it is discovered. A lot can happen to copper tubing that freezes. The tubing can blow out of fittings, swell and spilt with large gaps, or begin to split and then thaw before a large hole is made. Large splits and blowouts are easy to notice, but small holes in shower risers can go unnoticed for a long time.

I said earlier that these types of problems were common when working with seasonal cottages. People who drain down the plumbing systems in these types of properties and winterize them sometimes neglect shower risers. Even professional plumbers sometimes fail to protect shower risers when winterizing properties.

If the person draining the plumbing system fails to open the shower valves or, in the case of a tub-shower combination, the diverter valve, water remains trapped in the shower riser. This is an easy mistake for a homeowner to make, but professional plumbers shouldn't fail to drain shower risers.

When a shower riser swells under freezing conditions and begins to split, the crack can be very small. If thawing temperatures occur before a pipe is split completely, the gap does not mature, but it will still leak.

Small splits in shower risers can leak for a long while before they are found. Since the risers are not under constant pressure, the leaks only spill water when the shower is being used. If the hole in the pipe is small, the water will run down the piping and gather on the floor.

This process will continue every time the shower is used. As time goes by, the water slowly rots flooring, walls, joists, and other building materials.

These tiny leaks don't make much noise, and they are downright difficult to find without an access panel. There is a way to test for small leaks in shower risers without cutting into the wall; do you know how to do it?

If you suspect a leak in a shower riser but don't want to cut open a wall until you're sure, you can put a pressure test on the riser. Turn the shower valves off, and remove the shower head. Screw a test rig on the shower arm, and fill the riser with air pressure. Leave the test on the pipe for 10 minutes, and then check the air gauge. If the riser has a leak in it, the pressure gauge will drop, and you can feel safe in opening the wall. This little trick of the trade can save you from the embarrassment of destroying a wall unnecessarily.

Valve Leaks

Valve leaks are not uncommon, and you can inspect for most valve leaks through the holes where the valve stems come out. It is possible that the back of the valve is leaking and that you won't be able to see it.

These suggestions will help you to avoid cutting walls unnecessarily, but there will be times when opening a wall is the only option left.

If a leak is found through the frontal inspection process, open the wall to correct the leak. The only exception to this is the shower arm; it can be corrected without destroying the wall.

Sectional Showers

Sectional showers have more risk of suffering from wall leaks than one-piece showers do. The reason is simple; they are not an integral unit. Any time there is a seam in something that is meant to be watertight, there is a risk of a leak.

Some sectional showers are better than others. There is no question that many high-quality sectional units are available, but there are also a number of low-grade showers on the market.

Even the best sectional shower can leak if it is not installed properly, and many plumbers never take the time to read and heed the manufacturer's installation instructions. Too many plumbers assume their generic knowledge is all that is needed to install a sectional shower. It is beneficial to always read over the installation instructions for a given shower prior to installation. If you, as the plumber, fail to make the installation in accordance with recommended procedures, you may be held liable for damages caused by leaks at a later date.

There are, to be sure, many types of sectional showers available. Some are two-piece models, some are three-piece units, some are screwed to wall studs, and others are glued to drywall. Each type of unit offers its own version of leak possibilities. Let's take a close look at each type and discuss what should be looked for when troubleshooting a sectional shower.

Adhesive applications

Adhesive applications are common among inexpensive shower surrounds. These three- to five-piece units range in price from about $40 to over $270. I have never installed one of the cheap versions, but I have seen a lot of them leak. On the other hand, my plumbers and I have installed hundreds of the upper-end models and never experienced any problems. What does this mean? Does it mean the cheap models are no good and the expensive ones are? Not necessarily.

The problems I have found in inexpensive glue-on surrounds have always been related to either the adhesive or the finish caulking. It may be that the lower-grade material shifts and stresses more than the more expensive surrounds do, or it may just be a matter of how the walls are installed.

With the type of adhesive surrounds I've used for years, the application of the adhesive and the caulking is of paramount importance, as it is with any shower walls of this type. I have personally trained my plumbers in the proper installation methods, and for the most part, they have followed my instructions. I did have a plumber once who installed the section with the integral soap dish upside down, not once, but twice in the same home. Needless to say, that plumber is no longer with my company.

If the proper adhesive is used in accordance with the manufacturer's recommendations, there is no reason why it should pull loose. Yet, this seems to be a common problem with the less-expensive surrounds.

The overlap between sections also plays a strong role in preventing leaks. The brand I use has a substantial overlap. This forces any water that penetrates the caulking to run down the inner lap to the shower base and out into the pan. Many cheap models offer only modest overlaps.

The flexibility and installation of caulking along the seams is of prime importance. If the installer doesn't get a good seal on the joints, a leak is likely.

If you have a sectional shower that is leaking through the floor, check the seams. If you find voids in the caulking, remove the old caulking and recaulk the entire seam. It would be a good idea to recaulk all the seams.

When you inspect a glue-on sectional shower, test the shower walls to see that they are attached firmly to the walls of the building. If the walls have pulled loose, water can slip behind them and cause problems. The result can be a rotted wall, a rotted floor, or a stained ceiling.

I had a shower once that was leaking to a ceiling below it, and I couldn't find the cause. I had checked all the seams, all the interior plumbing fittings, and ultimately the drain and trap. Even after cutting out a section of the ceiling for access to the bottom of the pan, I couldn't find a leak. I was not only frustrated, I was downright angry that the shower was beating me.

My work on the shower had started in the afternoon and went into the early evening. I wasn't willing to give up, but I was stumped. It was then that the husband came home and everything clicked. As soon as I saw him, I knew what I had been missing.

The man was very tall and was balding on the front of his head. I had noticed the shower head was set higher than normal, but I wrote that off to some plumber not following standard rough-in dimensions. However, when I saw the size of the man and put that together with the height of the shower head, I knew where the leak was coming from.

I went back upstairs and checked the top edge of the sectional shower surround. Sure enough, there was no caulking around the top edge. I surmised that when the man showered, water bounced off of him and onto the top edge of the shower.

After caulking around the top edge and asking the homeowners not to use the shower overnight, I waited to see what would happen the next day.

The hole in the ceiling was left open until we could determine if the problem was corrected. After a week of routine showering there had been no leaks. The homeowners waited another week to have the hole patched, and the shower never leaked. I had found and fixed the problem. That was the first and only time I've ever encountered such a leak, but you will do well to remember the circumstances in case you ever run up against such a problem.

Vertical screw-on surrounds

Vertical, screw-on surrounds are sometimes used as waterproof shower walls. These surrounds are usually made of good, heavy material, and since they snap into place and are screwed to wall studs, they tend to be solid.

Most models use vinyl strips to seal seams and hide the heads of screws. These strips can be worth investigating if you are fighting a frustrating leak.

The track that the sectional pieces sit in can also allow water to leak past the walls. If a panel is not put into place properly, there can be a void between the panel and the track. Caulking around this area normally prevents leaks, but the caulking can go bad, especially if it is not flexible.

Shower bases sometimes have to handle heavy weight loads, and the stress can pull caulking away from seams. If the property where the shower is installed is fairly new, the settling process of new construction can stretch the caulking. Always investigate the caulking on sectional showers.

Half-and-half models

Half-and-half models look very nearly like one-piece showers. The advantage to this type of sectional unit is that it has only one seam, and it is horizontal. In theory, water should run down the wall and never enter the seam. If water accumulates in the crease of the seam, an inner lip will normally repel it. For these reasons, many plumbers don't seal the seam with caulking, and there is rarely a problem. But, I have seen problems with these types of units.

About 3 years ago, one of my plumbers was working on a job with a half-and-half sectional shower. The unit was in an upstairs bathroom and it leaked randomly. When it leaked, a substantial amount of water gathered in the ceiling below and escaped through a light fixture.

The plumber on the job was a very good service plumber, but he couldn't solve the puzzle of what was leaking and why it only leaked periodically. Having worked with me on similar jobs, my plumber went through all the right troubleshooting steps, but came up empty. He called me to the job.

I talked with the plumber and the homeowner to gain as much knowledge of the problem as possible. The fact that the shower only leaked occasionally was troubling. Everyone in the family used the shower, but it didn't always leak. After gathering all the data available, I went upstairs to inspect the shower.

I suspected the shower curtain as the problem, but my plumber had already tested the floor tile, and it didn't leak. The top edge of the shower had not been caulked, but no one in the family was extremely tall, so that didn't seem to be a factor. I went through all the normal troubleshooting procedures, and the shower never leaked.

I was about to ask the homeowner to keep track of when the shower leaked in the future, and who was using it at the time, and to call us when there was more information to work with. We had looked for all the common problems and couldn't find any. Then I noticed that the shower head was a hand-held unit mounted on a bracket.

I thought for a moment and asked the homeowner about her children. It turned out she had a young son. I speculated the height of child and came up with an idea. When the homeowner told me her son used the shower head as a hand-held unit, I was almost sure I had discovered the mystery leak.

Both my plumber and the homeowner watched intensely as I removed the shower head from its bracket. I cut the water on and aimed the shower head at the horizontal seam in the two-piece sectional shower. I moved the shower head much like a small boy might do if he were imitating the actions of a fire fighter.

After several minutes of spraying the walls we all went downstairs to see if there would be any leak. Guess what? That's right, I found the leak.

I'll admit that the seams of two-piece showers probably were never meant to stand up to the torture test I had given that one, but the uncaulked seam was the culprit. My plumber caulked the seam and the shower has never leaked since.

Leaks like this can drive a plumber crazy because they shouldn't be happening. Who would ever think of a child shooting water into the seam for fun? I got lucky.

Molded Shower Bases

Molded shower bases, in themselves, don't generally cause problems. If there is a problem with a molded base, it is usually with the drain or surround walls that are too short and caulked improperly. We've already talked about the importance of caulking and where the caulking should be, so let's concentrate on the drains for shower bases.

There are two common types of drain arrangements for molded bases and shower units. One is a metal collar that is meant to be caulked with lead or filled with a rubber gasket. The other is a screw-in drain. The two vary in their ways of leaking, so we will look at each time individually.

Metal collars

The metal collars on molded bases are factory installed, and I've never seen a collar leak. I have, however, seen plenty of collar joints leak. It is amazing what some creative homeowners and inexperienced plumbers will come up with to seal the area between the drain pipe and the collar.

Most modern plumbers use special rubber gaskets to make a legitimate seal around the drain pipe. Some old-school plumbers, myself included, still use oakum and lead. Both of these methods are fine, and usually neither leaks when installed by a professional.

The first step to take when looking for a leak in this type of base is to remove the strainer from the drain. Then you should carefully inspect the joint between the collar and the drain pipe.

Normally, water runs out of the shower base and down the drain pipe quickly enough that serious undershower leaks don't occur. However, if the joint in the collar is not a proper one, heavy water damage can be done. Let me give you an example.

I responded to a call years ago where a shower had dumped so much water into the ceiling of a rental property that a portion of the ceiling actually collapsed. Apparently the leak had gone on for quite a while before the tenant notified the landlord.

With the ceiling out of the way, I could see the underside of the shower base very well. The plywood subfloor was black with water damage, and the joists had been affected. It was obvious the leak was from the drain in the base.

I went upstairs and removed the strainer from the drain. Looking inside, I saw what appeared to be a homemade seal of glue or clear caulking. After I picked the mess out of the collar I found remnants of newspaper stuffed around the pipe. Someone had tried to seal the joint with improper materials. I suspect they used newspaper to close the void and then dumped the PVC glue or caulking in around the pipe, hoping to make a watertight seal. The idea didn't hold water.

On another occasion I removed a strainer and found lead wool sealing the drain of a shower base. The problem with this job was a lack of oakum. If oakum had been packed around the pipe and then capped with lead wool, it probably wouldn't have leaked, but without the oakum, the lead seal was ineffective.

Don't expect this type of drain to be properly sealed. Most homeowners, and some plumbers, don't realize they will need a special shower gasket or oakum and lead to seal these joints. When they attempt to make the joint, they are at a loss for what to do, and people sometimes improvise.

Screw-in drains

Screw-in drains are easy for most people to install. They look so easy to install that mistakes are often made. Putty isn't put under the rim of the drain, the fiber gasket is forgotten or lost, and the lock ring isn't always tight. Any of these conditions may result in a leak. Unfortunately, you can't check for these deficiencies without gaining access to the bottom of the drain, and this frequently means cutting out a section of ceiling.

Pan-Type Showers

Pan-type showers are the worst when it comes to pinpointing leaks. There are so many places where leaks are possible that it can be all but impossible to identify the exact location of a leak. All the normal possibilities for leaks around escutcheons and wall tile exist, plus you are faced with a base that could be leaking around the drain or through the pan itself.

Depending on the age of the shower pan, it might be made from sheet lead, copper, coated paper, or membrane material. Some pan material is not installed properly, and others give out with age and the stress of settling.

Most leakage with pan-type showers is related to the drain. Either the wrong type of drain is used or the right drain is used but installed incorrectly. Many young plumbers have limited to no experience with pan showers.

I once had a plumber install a screw-in drain in pan material. This obviously was not the right thing to do. He glued the drain on the drain pipe and left it sitting above the pan material at the height expected for the final base. Fortunately, I found the inappropriate installation before the base was poured. If I hadn't, the shower would have leaked, and my company would have probably been sued for damages. If a licensed plumber can make this type of mistake, you can imagine what do-it-yourselfers might come up with for drain options.

When the flanged drains that are meant to be used with pan showers are installed properly, they rarely leak. It is the improvised drains that you must really look out for.

There is also the possibility that a pan liner was never used for a built-up shower base. I have seen this be the case a couple of times. Inexperienced people weren't aware that a pan liner was needed, and they simply poured the base on the subfloor.

Regrouting the tile on the shower floor will take care of some bad-pan problems, but it may be necessary to remove and replace the existing floor with a new one.

If the drain is leaking, repairing it requires damaging the shower base, and trying to match tiles for the repair probably won't be feasible. Before you decide to break up a portion of the base to inspect for trouble, look below the base for a less-expensive option. There is normally much less trouble and cost involved in cutting and repairing a ceiling than there is in damaging a tiled shower base.

The key here is to know where the leak is coming from before you jump in over your head by breaking into the base. It would be a shame to destroy a base only to find that the wall tile needed new grouting.

Metal Showers

Metal showers are not very common, but they do exist, and they often develop leaks. Metal showers share many of the same leak possibilities that other showers do, but they have one unique way of leaking—they can rust out. When you are troubleshooting metal showers, be sure to look for any signs of rust or holes.

Emergency Showers

Emergency showers don't offer any particular challenge in troubleshooting. There is often no drain or base installed with an emergency station. Typically, the only troubleshooting with these units involves the valves, and that information is covered in Chap. 24.

This concludes our work with showers. We are now ready to move on to the troubleshooting methods used for spas and whirlpools.

21

Troubleshooting Spas and Whirlpools

How much do you know about troubleshooting spas and whirlpools? Do you get many calls for this type of work? Spas and whirlpools don't normally cause much trouble, but when they do, the problems can be difficult to diagnose properly. Unless you are something of a specialist on the subject, spas and whirlpools can make you scratch your head in short time.

Like faucets, there are so many different makes and models of spas and whirlpools that keeping up with all of them is nearly impossible. There are, however, many characteristics that are similar in all the various brands of both spas and whirlpools. It is that common thread that we are going to approach in this chapter.

To look at a whirlpool that is already installed, there doesn't seem to be a lot that could go wrong with it. There's the tub filler, the drain, and a few jets in the sides of the tub. Pretty simple, right? Wrong. Whirlpools have pumps, controls, suction fittings, O-rings, air lines, air-control valves, and other parts that can cause problems for the plumber in the field.

Plumbers who mostly do new work don't often have to figure out why the jets on a whirlpool are not making the water in the tub swirl. Typically, these plumbers just install the tubs and forget about them. But what happens if a new-construction plumber installs a whirlpool and the unit fails to operate properly? Who is going to troubleshoot the job and correct the problem?

If the failure occurs within the first year of installation, the plumbing company that installed the whirlpool is probably going to have to make the repairs under warranty service. Under these conditions the company has only a few options. Either a plumber from the company will have to know what to look for, where to look for it, and how to fix the problem, or the company may have to call a factory representative. The only other option will be to call in a plumber from a different company who has the ability to repair the whirlpool.

Plumbers who make their livings doing service and repair work will get calls from time to time that require them to work with spas and whirlpools. Even though plumbers are not needed to install many residential spas, they are called to correct deficiencies when the spas break down.

Do you know how to remove the apron from a spa? Does a spa have a ground-fault intercepter (GFI) breaker? Are there any strainers that may get clogged up on a spa? You may have answered all of these questions correctly, but many plumbers have never had to remove the apron from a spa. And, I'm sure a number of professional plumbers have never done any type of repair work on spas.

Whirlpools are more common in most homes than spas, but spas are becoming more popular. The fact that they can be installed without the help of a plumber, they don't have to be filled with water before every use, and the water is always heated and ready for use makes them a great alternative to large whirlpool tubs. As the trend for portable spas continues to grow, the demand for plumbers who know how to work on them will grow.

Not only are plumbers often required to know how to fix spas and whirlpools, they are frequently looked to for answers to common questions. For example, do you know what the maximum recommended temperature of water in a spa is? How long can an adult stay in the hot water of a spa safely, under normal conditions? If you can't answer these questions, your customers may doubt your ability to install or maintain their spas.

Did you know the answers to the two spa questions? Well, if you didn't, the maximum recommended temperature is 104 degrees F. It is normally accepted that healthy, normal adults should not stay in the water for more than 15 minutes at a time. You'll get more tips like this as we go along.

Spas

We will begin our look at troubleshooting spas and whirlpools with a study of spas. First we will look at the various parts and aspects of spas that are most likely to affect you on a daily basis. Then, we will delve into the problems and remedies that you may work with on spas. Let's start with a look at the major components.

Pumps

The pumps in spas are what make the water move. They do this by pulling the water in at one end and pushing it out at the other. The water coming from the pump is under great pressure as it is forced through the piping that delivers it to the outlets in the spa. The suction fitting in the spa allows the pump to pull the water back in and recirculate it.

When the pump pulls water in from the suction fitting, it directs it to a filter and then to a heater. During this process, clogs can occur in the strainer basket on the pump. Once the water has been filtered and heated, it is pumped back into the spa through the hydrojets.

Spa pumps should be sized to match the filter's capacity to trap oils and organic particles. If water is not circulated through the filter at the proper flow rate, the filter cannot do its job effectively.

Additionally, the pump should be rated in horsepower so that it is sized properly for the volume of water required by the spa. The number of hydrojets will also affect the sizing of the pump. You see, it takes a lot more effort for the pump to force water through the jets than it does to simply circulate water. As an example, a pump with a rating of 1 horsepower should be capable of handling the requirements of a spa with a capacity of up to 700 gallons of water and four hydrojets. This sizing is not carved in stone, but it is a benchmark to work from.

Many high-quality spas have two-speed pumps. These pumps are more energy efficient and are favored for their cost-saving performance. The low-speed setting on the pump is used to circulate water when the jets are not being used, and the high-speed setting is used to activate the hydrojets.

If you get into working on large spas, it will not be unusual to come across pumps rated up to 2 horsepower. It also would not be uncommon to find a two-pump system on large spas. There may be a small pump that handles the average water circulation and a larger pump that produces the force to run the hydrojets.

Filters and skimmers

Filters and skimmers are key elements in the makeup of spas. These devices not only keep the water clean, but they prolong the life of the spa. Without such devices, the spa water could become contaminated with algae, dirt, sand, and other unwanted materials. This type of contamination could be detrimental to a clean bathing experience, and it could cause problems with the pump and plumbing.

Filters that run on 2-hour cycles are common. In other words, the water gets filtered for at least 2 hours each day. Skimmers are not found on all spas, but they are a desirable feature. The skimmers pick up floating debris from the spa water and deposit it in a removable, cleanable basket.

Sometimes the skimmer is combined with the filter in a very convenient top-loading design. This allows the unit to be cleaned without draining the spa. On older, and less-expensive, models it is sometimes necessary to empty the contents of the spa to access the filters. Typically, cartridge filters can operate for up to 2 years without being cleaned.

Diatomaceous earth filters. Diatomaceous earth (DE) filters are used on large spas, such as you might find in a health club. These filters are not so easy to maintain. The makeup of a DE filter allows it to trap the much larger quantities of dirt and unwanted ingredients that may be found in the water of a spa.

Unlike cartridge filters, which can be cleaned with a garden hose, DE filters require a backwash cycle for cleaning. Once the backwash is complete, a new coating of DE should be applied.

Sand filters. Sand filters are also used on large, commercial-grade spas. Like DE filters, these high-rate sand filters require a backwash cycle when they need to be cleaned.

Heaters

Heaters are essential to the favorable operation of a spa. The heater must be sized to meet the demands of use placed on the spa. It is possible that a spa may be equipped with either a gas or electric heater. Let's talk about each type of heater for a moment.

Gas heaters. Gas heaters are desirable for their ability to heat water quickly and cost effectively. However, the need for gas piping and venting can be enough to turn customers to electric heaters.

When you are asked to work on a spa with a gas heater, you will be dealing with the following heater components: a water inlet, a water outlet, a combustion chamber, a burner, a pilot light or electronic ignition, a heat exchanger, heater controls, a vent, and gas piping.

Electric heaters. Electric heaters for spas come in both 110- and 220-volt ratings. Most residential spas can be fitted with the 110-volt versions. There are also electric heaters that are convertible to either 110 or 220 volts.

There are some drawbacks to 110-volt heaters that can affect the troubleshooter in the field. Electric heaters for small spas are small themselves. This limits their ability to heat water quickly. In fact, it is common for these heaters to be rated to produce only 2 degrees of rise in water temperature for each hour they are in use. This is bad enough, but there is more.

Not only are electric spa heaters slow, they pull a lot of electricity, usually the full amount that a 20-amp household circuit can handle safely. Therefore, the hydrojets and air blower on the spa are unable to run at the same time the heater is in use. If the spa is being used frequently, the heater may not be able to keep up with the demand for hot water. You will do well to remember this fact. This type of problem, however, is not likely to occur with electric heaters rated at 220 volts.

There is at least one brand of spa that manages to heat its water with the use of its pump. The dealer I spoke to on this particular model said that the cost for electricity to maintain a temperature of 104 degrees F in Maine, with an indoor installation, was about $6 per month. The cost doubled when I inquired about an outside installation, but it takes hardy souls to have outside spas in Maine during the winter.

Controls

Every spa you work with will be equipped with controls. Some of these controls may be operated manually, and some of them may work automatically. The controls used to start the pump or blower are normally mounted on the rim of the spa. They are sometimes mounted in the top surface of a deck that surrounds a spa. They may be touch-activated electronic switches or air switches. In the case of air switches, the act of depressing them sends air down a piece of tubing to activate the equipment switch.

Controls for the heaters used with spas are normally automatic. They are thermostatically controlled. Some spa heaters, typically those rated at 220 volts or that are gas fired, have time-clock controls. These more efficient heaters don't require as much running time to heat water, so their cycle can be set to coincide with the usage of the spa. A few sophisticated models even have a two-clock system that allows more flexibility in the time settings for heater activation.

Remote controls are also available on some types of spas. These controls are convenient for users who have spas outside of their homes. With the remote control pad, the user can program the spa without stepping outside the home.

Blowers

Many spas are equipped with air blowers. These devices produce the gushing flow of bubbles that is found in spas. While the air-induction systems don't normally require much maintenance, they can hinder the spa's ability to maintain a low heat loss. The blowers can also be a nuisance in terms of noise.

Air blowers for most spas pull in ambient air. If the air being blown into the spa is cold, such as may be the case in an outdoor setting, the air can cool the water more quickly than a small heater can compensate for the temperature changes.

Blowers run at high speeds and can produce annoying sounds. If the blower whines too loudly, it sort of ruins the atmosphere of a nice, relaxing soak in the spa. If your customer complains that a spa is too noisy, the air blower is the first place to turn your attention to.

The jets

The jets on spas are what make the warm water magical. They are also the components that most users notice problems with. Correcting the problems is often as simple as setting the air intake to a more appropriate setting.

Average hydrojets push out water at a rate of about 15 gallons per minute, but some spas are set up to allow all but one jet to be closed, producing a push from that single jet in the range of 90 gallons per minute.

Now that you are aware of the basic components that make up a spa, let's examine some of the routine maintenance responsibilities that go along with long, hot soaks in the spa.

Water Maintenance

The water maintenance in spas is not normally a plumber's responsibility; it is the job of the spa's owner. However, a lack of routine maintenance on the part of the spa owner can have an affect on a plumber called in to troubleshoot problems with the spa. Let's look at how water maintenance may affect you as a service plumber.

Potential hydrogen

Potential hydrogen, or pH as most plumbers refer to it, can cause trouble with spas. If the pH is too low, below 7 on the pH scale, parts of the spa equipment can be damaged by the acidic water. The lower the pH is, the more likely this is to occur. A low pH can also result in complaints from the users of the spa. Their eyes and skin may sting, burn, or show other ill effects from highly acidic water.

At the other end of the spectrum, a high pH count, something over 8, can cause the spa water to become cloudy. It can also produce scaling on the spa equipment, and mineral deposits can cause problems with the plumbing. The heaters of spas are particularly sus-pectable to this type of problem. For example, if the heater coils in an electric heater become encrusted with mineral deposits, the heater can require substantially more time and energy to maintain the water temperature in the spa. This is similar to what happens with the heating elements in electric water heaters for homes and businesses.

The water in a spa should be tested regularly. If problems are found, chemicals can be added, like soda ash, to correct the problems before they escalate. The most effective way to avoid mechanical and equipment problems with spas is to maintain the water at chemical levels recommended by the manufacturers.

Troubleshooting Spas

Troubleshooting spas should not be a job that intimidates you. Even if you are unfamiliar with the operation and principles of spas, don't despair. The learning curve on spas is short. To prove this point, let's examine some common complaints that come into plumber's shops on spas.

My spa is full of foam

How often have you been called by a customer who said, "My spa is full of foam"? Maybe you have never gotten such a call, but these calls are common for plumbers who are known to work with spas.

When a customer's spa is foaming at the mouth, so to speak, the customer can be more than a little panicked. They may believe that there filter has died or that their pump is malfunctioning. In all cases, they will want you, their trusted plumber, to solve their problem. The problem is, many regular plumbers have no idea what is causing the foaming action. Do you know what the cause of the problem is?

When the water in a spa begins to foam, it is an indication of poor usage of the spa. It is not that any of the equipment is failing; it is an operator error. The water being pushed out of the hydrojets stir up the water in the spa. If shampoo, body lotions, or in some cases cos-metics have gotten into the spa water, the constant agitation from the jets will whip the water and its contents into a foaming monster.

I once had a customer call me late at night who said that her spa was foaming up over its overflow rim. She was terrified that the foam would consume her entire exercise room. I knew that the problem was not serious, but she didn't.

It seems that the woman's daughter had gotten out of school early on this particular day and had spent considerable time in the spa.

During her soak she washed her hair, not a smart thing to do in a spa. The left-over shampoo was what caused the massive foaming.

If you have a customer who is frightful of foaming, suggest an antifoaming compound. There are many types available that will control the natural foaming of spas. In extreme cases, like the one I was called to, you may have to skim off the foam before the antifoam agents can take over and win the battle.

Cloudy water

Complaints about cloudy water in spas are not at all uncommon. There are a few possible causes for the cloudy water. The first logical troubleshooting step to take is an inspection of the filtration system. If the filter is set up to run at least 2 hours a day, for most residential applications, and if the filter is clean, you must look elsewhere.

Test the water for its pH rating. A high pH number, something above 7.6, can result in cloudy water. If the pH is the problem, it can be treated with chemicals, like soda ash, to control the problem and to clear up the water.

A third possible cause could be a lot of undissolved solids or an overabundance of chemicals. When this is the case, you should drain the spa and refill it with fresh water.

Hard water

Hard water can produce scaling problems with spas. The effects can come in the form of hard deposits on the sides of the spa, in the filter, or in the heater (this is one of the most common locations). The scaling is a result of an unbalanced pH rating and hard water. Appropriate chemicals introduced into the spa water will solve the problem.

Burning eyes

Some customers may complain that they have burning eyes after using their spas. This type of problem can occur when the pH rating is out of balance (a pH number that does not fall within the range of 7.4 to 7.6). It is also possible that chlorine is causing the burning sensation.

Chlorine can combine with nitrogen that is present from perspiration or cosmetics to form chloramines. The odor from this concentration can irritate human eyes. Chlorine is not the only culprit that is capable of producing irritating gases. Bromine, which is used to clean spa water, can produce bromamines. Unlike chlorine, the production of bromamines is odorless, but it can be equally irritating to the eyes.

If the pH rating for the water tests is okay, the best alternative is to drain the spa and refill it with fresh water.

Algae

Algae will build up in a spa if the unit is not covered between uses and if the water is not treated properly. When algae is a problem, chlorination or draining and refilling the spa is the logical solution.

Ugly water

Ugly water is not something that customers want in their spas. When spa water is discolored or stains the fixture, it is a sure sign that some type of metallic content is in the water. There are chemicals available to help in the removal of metals in the water, but draining and refilling the spa may be all that is necessary.

What happened to my bubbles?

If you work with spas, sooner or later you will get a call from a customer who asks, "What happened to my bubbles?" A pretty good guess on this problem is that the jets in the sides of the spa have been closed down. Kids seem to be fascinated with their abilities to turn the rings and cut off the bubbles. Parents who don't know their children have been playing with the hydrojets often don't think to check the settings before calling in professional plumbers.

It is, however, possible that the pump has lost its ability to flush water through the jets. Therefore, if the jets are open, check out the pump. If you have reason to suspect the pump, check the strainer basket. It may be that the strainer is clogged and is prohibiting the pump from picking up water from the suction outlets that can later be pumped through the jets.

The filter could, in extreme cases, be at fault for puny pressure at the jets. If the filter is clogged, the water cannot circulate properly. Under these conditions, clean the filter and see if things get back to normal.

Don't be fooled by a two-speed pump. Remember that the pump only circulates water at its lower speed, and it needs the high-speed function to kick in before it can activate the jets. In the case of dual-pump systems, you must troubleshoot both pumps, but concentrate on the larger pump first.

Tepid water

Tepid water in a spa is caused by a failing or inadequate heater. It may be that an air blower is pulling cold air into the water more quickly than the heater can compensate for the temperature change.

The heater may be too small for the spa, but this is unlikely, unless the factory-installed heater was replaced.

Constant or frequent use of a spa often causes the water temperature to drop when the heater is an electric 110-volt unit. Since these heaters don't produce a lot of hot water to begin with, and they can't run while the jets are in operation, it is easy to overpower them to the point of turning the water into a lukewarm bath rather than a hot, relaxing soak.

Another possibility is a failure of the heater's thermostat. If the thermostat is bad, the heater cannot heat the water properly. In larger units, the time-clock settings may be out of adjustment. We will address this issue in just a few minutes.

The pump won't run

When you are dealing with a spa where the pump won't run, check the circuit breakers or fuses. Most spas are equipped with GFI devices for safety purposes. Check the GFI and the panel box to see that everything is in good order. And, don't forget to see that the spa, if it is a portable model, is plugged in to a suitable electrical outlet. If all the simple electrical work is intact, check the windings on the pump and the controls.

Clogged filters

Clogged filters on spas are a common complaint. In the case of cartridge filters, you can simply replace the filter insert. For sand and DE filters, you will have to backwash them.

Time-clock problems

Time-clock problems are often associated with spas that are set up with gas-fired heaters or heaters with 220-volt ratings. If the time-clock settings get out of adjustment, the heaters on these spas will not warm the water when the customer wants it to. This results in a cool dip, instead of a hot soak.

We have covered the bases for troubleshooting spas pretty thoroughly, so now let's concentrate on the issues pertaining to whirlpools. While some of the situations are similar between whirlpools and spas, there are many differences to consider.

What Makes a Whirlpool Different?

One of the biggest differences between a whirlpool and a spa is that whirlpool tubs are not designed to hold their water for multiple uses.

Where a spa is always ready for use, whirlpools are meant to be filled with water before each use. This is a sizable difference, both in time, convenience, and cost.

The time it takes to fill a two-person whirlpool is considerable. The fact that whirlpools have to be filled before they can be used is not conducive to frequent use. It often takes more time to fill a large tub with water than people are willing to spend. In fact, it is not unusual for the filling time to exceed the soaking time.

Some large whirlpools have on-site water heaters, but many of them are filled from a large water heater that serves the rest of the plumbing fixtures. The filling of a large whirlpool not only depletes the supply of hot water quickly, but it is a costly proposition. Heating the volume of water needed for these tubs requires a lot of energy. Even the mere cost of the water used to fill the tubs can add up over the course of time. For all of these reasons, spas are gaining more attention.

I have owned large whirlpool tubs, a soaking tub, and a spa. Of all these units, the spa was always the most convenient and the most enjoyable. There is no doubt in my mind why the buying public is so pleased with the overall performance and characteristics of spas. However, as fashionable and practical as spas are, whirlpools continue to command their share of the marketplace.

Some of the other differences between spas and whirlpool tubs are not as obvious. For example, a whirlpool tub does not come equipped with a skimmer, but then, it doesn't need one. And as I mentioned earlier, many whirlpools don't have independent heaters; this is another difference between whirlpools and spas.

Since whirlpools don't retain their water for multiple uses, there is no need for the chemistry tests and treatments that are mandatory with spas.

Time clocks are not needed on whirlpools. Since the tub doesn't keep its water on board, there is no need to set a timer to turn on a heater to warm the water.

The Similarities between Whirlpools and Spas

Whirlpools and spas both rely on pumps to circulate water and to produce force at the hydrojets. Both units are fitted with hydrojets that bubble the water, and both types of tubs have suction intakes. Another similarity is the shared use of air-control devices.

Troubleshooting Whirlpools

Troubleshooting whirlpools is not much different from working with a

EMPRESS
Acrylic Whirlpool Recess Bath

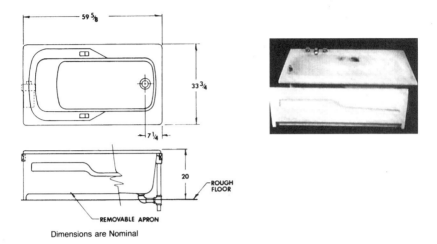

Dimensions are Nominal

Bath: 2-925W **Empress** acrylic whirlpool bath with optional 2-297 **Apron** for recess installation, four 360°
              adjustable whirlpool jets, variable air-injection control, 0-30 minute remote timer and ¾ HP whirlpool
              pump. Sloping back, with grab handle and integral arm rest on both sides.

Trim:

Waste:

Color:

Figure 21.1   Whirlpool tub with removable apron. (*Courtesy of Crane Plumbing.*)

regular bathtub. The tub filler is similar to that of a regular bathtub, and the waste and overflow is almost identical, except that it usually has a higher overflow tube.

The differences between whirlpool tubs and standard bathtubs come into play when the problems being experienced have to do with the whirlpool pump, fittings, piping, jets, and related gear (Figs. 21.1, 21.2, and 21.3).

The suction fitting on a whirlpool can become clogged and need cleaning. This doesn't happen often, but it is something you should be aware of. If the jets are not operating properly, check the strainer in the suction fitting.

If the air-volume control is not adjusted properly, a good flow of water is not going to explode out of the hydrojets. When this is the case, a few quick turns of the air-volume knob will normally solve the problem.

If the eyeball jets in the tub are giving you trouble, you can purchase a repair/replacement kit to remedy the situation.

There are O-rings at both the suction and discharge connections of

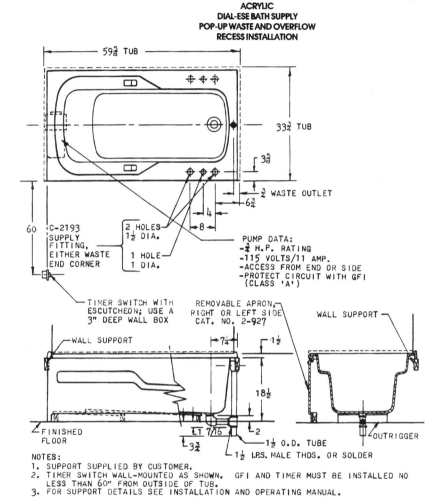

Figure 21.2 Rough-in details and description of a whirlpool tub. (*Courtesy of Crane Plumbing.*)

the whirlpool pumps. The O-rings don't normally act up, but if they do, they are easy to swap out.

The pump and control box for a whirlpool is not unlike those for other plumbing applications. You can use a standard test meter to troubleshoot the pump and controls.

Many whirlpools are equipped with personal shower attachments. These devices require a backflow preventor, but this part of the fixture shouldn't give you any trouble.

ATLANTIS WHIRLPOOL BATH
ACRYLIC
DIAL-ESE BATH SUPPLY
POP-UP WASTE AND OVERFLOW
SUNKEN INSTALLATION

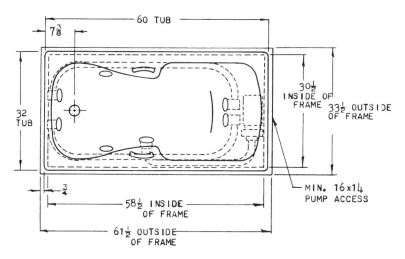

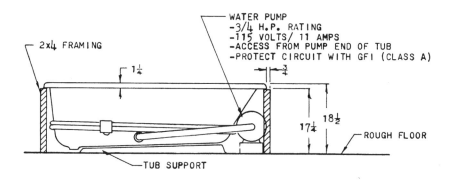

NOTES:
1. FRAME AND GROUND FAULT INTERRUPTER (G.F.I.) SUPPLIED BY CUSTOMER.
2. FOR SUPPORT DETAILS, SEE INSTALLATION INSTRUCTIONS.

Figure 21.3 Rough-in details and description of a sunken whirlpool tub. (*Courtesy of Crane Plumbing.*)

Hard water can take its toll on the mechanisms of a whirlpool, but really the consequences are no worse than they are for any normal plumbing fixture.

Usually, the most difficult and frustrating part about working on whirlpools is the limited access provided for service and repairs. The pump and controls are usually placed near the apron of the tub, but some of the fittings and tubing can be nearly impossible to see, reach, or work on.

The Basics

The basics of troubleshooting whirlpools and spas are not that different from those used on any other plumbing problems. The key is to have a good working knowledge of the brand of fixture you are working with and to move about your troubleshooting in a structured manner.

Manufacturers of the various tubs will generally supply you with detailed cut sheets on their products if you request them. Some of the companies will even throw in a few troubleshooting tips. If you order this type of information, you can do a fine job of troubleshooting both whirlpools and spas.

The next chapter is all about troubleshooting bidets. Like spas, bidets are not one of the more common fixtures for plumbers to work with. Since bidets are not an everyday household fixture, you will probably learn a lot when you turn the page.

Chapter

22

Troubleshooting Bidets

Troubleshooting bidets is not a job most plumbers do on a daily basis. In fact, some plumbers have never worked with bidets. This doesn't mean, however, that you will never be asked to service or repair one.

Bidets are not a common plumbing fixture in most American homes, but they are prevalent in many upscale housing developments. At first glance, bidets don't appear too intimidating (Figs. 22.1 and 22.2), but a number of plumbers have no idea of how they work or how to troubleshoot them.

This chapter is going to put bidets in perspective for you. We will discuss various types of bidets, their drainage systems, and their valves. Once you've finished this chapter, you will know the ins and outs of bidet plumbing. Let's start this section with a question-and-answer format to see how much you already know about bidets.

1. *Do bidets have integral traps?* Toilets have integral traps, so it would be easy to assume bidets do also, but they don't. Traps for bidets are installed below floor level as are bathtub traps.

Trap leaks under bidets can do a lot of damage, and they can be hard to find. If the slip-nut washers allow water to leak out of the trap or drainage tubing, the water can collect for a long time before it is noticed.

When a bidet is located on a second floor, the first evidence of trap leaks is usually a stained ceiling. The stain, however, will not always be found directly beneath the bidet. It is not unusual for the water from the leaks to run across the ceiling, follow electrical wires, or seek an opening, like a light fixture, to run out of.

If the water just lays on a ceiling, it can take days, or even weeks, for it to work its way through it. Small leaks can take months to show up. When these leaks are found by homeowners, severe damage has

BIARRITZ
Vitreous China Bidet

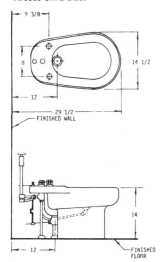

Biarritz Bidet 3-277

Bidet:	3-277 **Biarritz** vitreous china bidet with rose-spray, integral overflow and flushing rim with shelf. extended back shelf.
Trim:	C-3046 chromium plated two-valve supply fitting with backflow preventer, transfer valve and rose spray; indexed acrylic handles and pop-up waste with 1 ¼" O.D. tailpiece.
Trap:	
Supply:	
Color:	

Sectional view of Biarritz bidet, showing assembly of the built-in spray.

Figure 22.1 Bidet. (*Courtesy of Crane Plumbing.*)

usually already occurred to the ceiling. The ceiling joist may even be soaked before the leak is discovered.

Finding a hidden ceiling leak that is the result of a bidet trap is similar to tracing leaks from bathtubs and showers. The key to finding the leak in the trap is filling the bowl of the bidet to capacity and releasing all of the water at once. I've said it before, but I'll say it again. Too many plumbers fail to fill bowls to test drainage systems, and without the pressure provided from a full discharge of water, many leaks will not show themselves.

2. *Do bidets use both hot and cold water?* Yes, they do. Bidets are equipped with valves that allow the hot and cold water to be mixed to a suitable temperature.

3. *Do bidets have flush handles?* No. Unlike toilets where a water

BIARRITZ BIDET
VITREOUS CHINA
TWO VALVE SUPPLY
BACKFLOW PREVENTER
TRANSFER VALVE & POP-UP WASTE

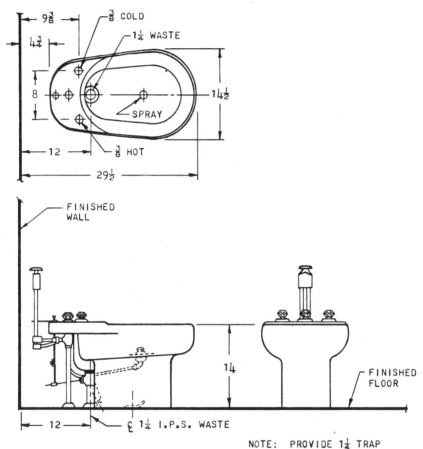

NOTE: Installation instructions and mounting bracket included with bidet.

Figure 22.2 Rough-in details and description of a bidet. (*Courtesy of Crane Plumbing.*)

reserve is flushed from a tank or flush valve to clean the bowl, bidets work on a principle similar to lavatories. They are equipped with pop-up assemblies and lift rods that allow them to hold and drain water.

4. *Do all bidets have sprayers in their bowls?* No. Some bidets use over-the-rim faucets and do not have sprayers.

5. *Are vacuum beakers required on bidets?* Bidets that use over-

the-rim faucets and that are not equipped with sprayers are not required to be fitted with vacuum breakers.

Bidets that are equipped with sprayers are required to be protected from backflow with a vacuum breaker. The vacuum breaker may be mounted on the bidet or on the wall behind the bidet.

6. *What size drain is used with a bidet?*   The tailpiece for a bidet has a diameter of $1\frac{1}{4}$ inches. The drain pipes for bidets usually have diameters of $1\frac{1}{2}$ inches.

7. *Can the valve for a bidet be rebuilt?*   Yes. Bidet valves have seats and diaphragms that can be replaced if needed. Some types of valves have cartridges and O-rings that can be replaced. The valves for bidets are similar to those used for tub/shower applications.

8. *Can the height of the spray be adjusted?*   Yes. There is a small adjustment screw located under the head of the spray. This screw allows the intensity of the spray to be adjusted.

9. *Do all bidets have integral flushing rims?*   No. Some models are not equipped with integral flushing rims.

10. *What does the transfer valve do?*   The transfer valve for a bidet diverts water to the sprayer in the bowl.

The Sprayer Doesn't Have Much Water Pressure

What would you look for if the sprayer on a bidet didn't have much water pressure? There are a number of possible causes for reduced pressure at the spray; let's look at them one at a time.

Valve turned off

The supply valve could be turned off or nearly off. Checking the supply valves is a logical first step to take. Valves get closed for lots of reasons, and sometimes people are not aware that the valves were tampered with. For example, a curious child may turn the valve toward the closed position. If this were to happen, the parents might have no idea the valve has been turned.

When you are troubleshooting, you cannot afford to overlook the obvious. I know the idea of having a valve closed inadvertently may be a long shot, but I have found this to be the case on toilets, so it could certainly happen with a bidet.

Clogged sprayer

A clogged sprayer could result in low water pressure and a lackluster fountain of water. Houses with private water supplies often develop problems with the buildup of minerals, and this could cause the flow of a bidet sprayer to be minimized.

In addition to mineral deposits, other types of debris could plug up the sprayers. Sand, gravel, loose pieces of copper shavings, and similar materials could cause the spray holes to become clogged.

Crimped hose

A crimped hose can definitely slow down the flow of water to a bidet. While it is not likely that a hose will become crimped or pinched by itself, it is worth checking. Normally, the lack of water will not be caused by a crimped hose unless the bidet was recently installed or worked on.

Obstructions

Obstructions in the water supply can slow down the delivery of water, and it doesn't take a very large object to restrict the flow of a bidet supply. You can check for obstructions in the tubing by taking sections apart and inspecting them.

Transfer valve

If the inner workings of a transfer valve goes bad, getting water to the sprayer will not be easy at all. You can think of the transfer valve as being like the diverter handle on a three-handle tub/shower valve. If it's not in good working order, the water cannot be diverted properly.

Debris in the supply valve

Debris in the supply valve is another possible cause for reduced water pressure. This situation is not common, but it can occur. If gravel, sand, or a copper shaving has become lodged in the supply valve, you will have to disassemble the valve to clear the obstruction.

Restricted supply tube

Some bidet faucets use standard supply tubes, and a restricted supply tube can affect the water pressure at the faucet. The supply tube could be crimped, or it could have an obstruction inside of it. This is a simple problem to eliminate; just remove the supply tube and inspect it. Replace the tube if necessary.

My experiences

My experiences with reduced water pressure under conditions similar to the ones described above have been numerous. To put the troubleshooting into real-life terms, let me share a few of those experiences with you.

Bread in the supply tube. I was once called to troubleshoot the reason there was no water pressure at a faucet. I went through about the same troubleshooting procedures I have explained to you. The only step I left out initially was checking the supply tube for an obstruction.

After checking the cutoff valves, the supply hoses, the diverter valve, and everything else I could think of at the time, I was stumped. Only the cold water side of the faucet was being affected, and no other fixtures in the home were suffering similar symptoms. Needless to say, I was confused and frustrated.

I was so sure I had checked everything that could be causing the problem that I went back through each phase of the troubleshooting again. The results were the same; I found nothing that would account for the low water pressure.

The house I was working in was relatively new, so I ruled out a rust closure in a galvanized supply pipe. Not wanting to give up, I sat down on the floor and stared at the fixture, trying to think of something else to try. My meditation didn't produce any new ideas. Baffled, I went downstairs to talk to the homeowner.

After questioning the homeowner on how long the problem had existed, I was surprised to find out that it had started shortly after another plumber had been to the house to repair a water line that had frozen and burst in the outside wall, below the bathroom.

With this new information, I returned to the bathroom and promptly removed the supply tube. When I looked in the tubing, I was shocked that I couldn't see any light through it. The supply tube was plugged completely with something. It was then that I used a piece of wire to rod out the supply tube. What do you think I found?

If you guessed bread, you're right. Apparently, the plumber who fixed the broken pipe had used bread to hold back water in order to make the solder joint. When the water was turned back on, the bread was forced up the pipe and lodged in the supply tube. This was an unusual occurrence, but it was not the first time I had experienced such a situation.

As a young plumber, I once used too much bread to hold water back in a pipe I was trying to solder. The water was coming with a vengeance into the joint I wanted to solder, and I packed bread into the pipe with the end of a pencil. After I repaired the broken pipe, I was unable to get water to the toilet at the end of the supply line. Ultimately, I found that the bread I had packed in so tightly was not dissolving to let the water get to the toilet. If I had not had that experience early in my career, I might not have thought to check the supply tube in my troubleshooting of the bidet. My point is this: You cannot afford to overlook any possibility when you are troubleshooting.

Copper in the supply valve.   On another occasion, I found slivers of copper tubing trapped in the supply valve to a bidet. The copper was evidently laying inside the tubing when it was installed, and the water pressure forced the copper into the supply valve, reducing the flow of water. Once I disassembled the valve and removed the copper, the problem was solved. I'll give you more of my personal experiences as we move on.

The Stains on Her Ceiling

I once had a homeowner call me to determine what was making the water stains on her ceiling. I arrived at the home and saw where the white ceiling had been discolored by water. The woman told me that she had never seen any water dripping, but that the stains just kept getting bigger. My first thought was a toilet or tub leak.

I went into the bathroom and found a shower, a whirlpool tub, two lavatories, a toilet, and a bidet. I started my troubleshooting by flushing the toilet and looking for leaks around the base; there were none. There was no condensation to be found, and all of the water supplies were dry. This led me to believe the problem was with either the tub or the shower.

This particular job turned out to be one that I am not particularly proud of. I spent a considerable amount of time trying to find the source of the water stains without success. After doing all of the routine tests on the lavatories, the whirlpool, the toilet, and the shower, I was puzzled.

I went to the bidet and turned the water on. It seemed to be working fine, and I didn't locate any leaks. At this point, I went downstairs and took some measurements to determine what fixture the stains were closest to. They were directly under the bidet. This, of course, redirected my attention to the bidet.

After wiping all of the water and drainage joints I could reach with toilet tissue, I still hadn't found the leak. Then it dawned on me. I closed the pop-up on the bidet and allowed it to fill with water. When I opened the drain, the water left slowly. I looked back under the fixture and saw the cause of the problem.

The drain for the bidet was partially blocked, causing water to drain slowly. The combination of the slow-draining fixture and a bad slip-nut washer was allowing the piping to leak where it entered the trap adapter.

I snaked the drain and replaced the worn washer. After testing the fixture again, I was confident it was fixed and that the leaking was over. To this day, I have not had any further complaints of leaks staining that woman's ceiling.

Putty under the Drain Flange

If you run across a bidet that doesn't have putty under the drain flange, you will have put your finger on a hard-to-find leak. We talked in Chap. 18 about how lavatories could empty their bowls without leaving a trace of the leak. Well, the same is true of bidets. You can apply practically all of the troubleshooting information given for drainage systems of lavatories to bidets. Let's do a rundown of all the places a bidet's drainage system may leak.

The trap

The traps for bidets can be difficult to see well. Since they are below the floor level, they can be troublesome to get to if there is a ceiling below them. It is unusual for a plastic trap to just up and start leaking, but metal traps can deteriorate with time.

If you have evidence of a leak coming from a bidet but can't find it, check the trap. Remember to fill the bowl of the bidet to capacity when checking the drainage. Just as we discussed in Chap. 18, the added pressure from a volume of water in the drainage piping can expose leaks that you might not otherwise find.

The trap adapter

You already heard about my experience with a leaking trap adapter, but don't forget what I told you. The slip-nut connection between the trap adapter and the tailpiece is not a common place to locate a leak, but it can be the source of your problem. This type of leak will generally avoid detection unless you fill the bowl to capacity and release all of the water at one time.

The threaded tailpiece

The threaded tailpiece of a bidet can leak if it is under pressure and not properly sealed. If the tailpiece is not tight, or if pipe sealant was not put on the threads, a leak can occur. It is also possible that the weakened, threaded portion of the tailpiece will corrode and create a hole. Again, this is not common, but it is usually the uncommon problems that are hardest to find.

The pivot rod

The pivot rod in the pop-up assembly provides some risk of a leak. If the retainer nut is not tight, water can easily escape around the pivot rod.

The gasket

The gasket under the drain flange can go bad and allow the fixture to leak. It is also conceivable that the nut applying pressure to the gasket may be loose, causing a leak.

The pop-up plug

The seal on the pop-up plug can be defective. When this is the case, the fixture will not hold water. You can test for this by removing the pop-up plug and inserting a rubber stopper. If the fixture still loses its water, investigate the drain flange. When the rubber stopper holds water, replace the pop-up seal or the entire pop-up plug, and your problem should be solved.

It is also possible that the pop-up plug only needs adjustment. If the lift rod is not positioned properly, the pop-up will either be held too high above the drain or not high enough. If the pop-up is too high, water will leak past it. When the pop-up is not high enough, water will drain out of the fixture slowly.

Spray assembly drain flanges

Spray assembly drain flanges that are not properly installed can generate leaks. If this part of a bidet is leaking, the cause is probably a lack of putty under the drain flange. You should loosen the mounting nut and check the installation to see that it is properly sealed.

We have just covered all the locations in the drainage system of a bidet that may leak. Now let's move into the water supply and see where leaks may exist.

Water Leaks

Water leaks can come from a number of places when working with bidets. Most of the time the leak will involve the hoses or tubing that connects the various parts of the fixture's valve and outlets. The possibilities for water leaks vary, depending upon the type of valve or faucet being used. Let's go down the list of places where you might expect to find water leaks, but keep in mind that not all bidet faucets are the same and that while some will have normal supply tubes delivering water to the faucet, others will have full-size piping feeding the valve.

Cutoff valves

As with any other type of fixture that uses standard cutoff valves, the valves for bidets can leak. The leak can occur on either side of the cutoff valve or at the packing nut. This type of leak is a pressure leak that is constant and relatively easy to find. A close visual inspection and a rubdown with toilet tissue will reveal these leaks.

Cutoffs that are screwed onto threaded nipples can leak at the threads. This can be caused by a lack of pipe dope, a loose connection, or threads that have worn through and created a hole. This situation can be fixed by removing the stop valve and making the necessary repair. The repair might be as simple as applying pipe dope and reinstalling the valve, or you might have to replace the nipple and the stop valve. Of course, the water supply to the valve must be cut off before this type of work is done.

Compression valves sometimes develop leaks from vibrations they receive during use. These leaks are normally easy to find with a visual inspection or a quick rubdown with toilet tissue. Once the leak is identified, tightening the compression nut normally stops it.

Cutoff valves that are soldered onto water distribution pipes don't generally develop leaks after having been installed for a while. It is possible, however, that water hammer could stress the joint and create a leak. Under these circumstances, the valve normally must be removed and replaced.

When stop valves are leaking around the stem of their handle, and they do frequently, all that is required to stop the leak, in most cases, is to tighten the packing nut. If the valve is extremely old, it may be necessary to remove the stem and rebuild it or replace the valve.

The compression nut that secures a supply tube in the stop valve is another common place for leaks to develop. This can happen if the supply tube is hit during routine house cleaning or if the valve is vibrated from a water hammer. These leaks can almost always be stopped simply by tightening the supply tube nut.

Supply tubes

Supply tubes can leak at either end. You've just been told what to do if the tube is leaking at the end where it meets the cutoff. When the leak is at the top of the supply tube, all that should be required is the tightening of the nut where the supply tube joins the faucet.

Faucet leaks

Faucet leaks with bidets are basically the same as they are with other common faucets. If the faucet seats are worn, water will drip from the faucet. A faulty washer or cartridge can cause a faucet leak, and bad O-rings and packing material can allow water to leak around the handles of the faucet. The procedures for correcting these problems are covered in Chap. 24.

Spray assemblies

Spray assemblies can leak where their water supply is connected to them. This is usually caused by a washer that has gone bad, but a leak can also occur if the supply nut is not tight. If you are getting a leak in this location, try tightening the nut; if that doesn't do the job, replace the washer.

Transfer valves

Transfer valves have both inlet and outlet connections that can leak. These connections are made in the same way as the connection for a spray assembly. Tighten the leaking nut first, and then replace the washer if necessary.

It is possible for a leak to develop around the stem of the transfer valve. Tightening the packing nut may stop this type of leak quickly. But if it doesn't, remove the steam and inspect the O-rings, packing, and cartridge.

Vacuum breakers

Many plumbers fail to think about the connections at vacuums breakers that may leak. In the case of bidets, the water connection to the vacuum breaker may be made with tubing, a hose, or a direct connection at the fill valve. Any of these types of connections pose potential for leaking.

In the case of tubing and hoses, tightening the connecting nuts will normally stop a leak. It may be necessary to replace a washer or a compression ferrule, but normally a few turns of the nut is all that is needed.

Vacuum breakers that connect directly to fill valves with threaded nipples can have leaks around the threads. This usually results from a lack of thread sealant, nipples that are loose, or sometimes cross-threaded connections. It is also possible that the nipple has deteriorated at the threads.

Hoses and tubing

There are usually several hoses or pieces of small tubing involved in the water connections of bidets. There is also a tee near the fill valve that can present problems. Frequently these connections are made with a nut and a washer, but they are sometimes made with compression nuts and ferrules. In either case, tightening the connecting nuts will normally stop any leaks. If leaks persist after tightening the nuts, you may have to replace a washer or compression ferrule.

As a word of caution, the small tubing used for these connections will crimp easily. When you are working with this tubing, be careful not to bend it too sharply.

Faucet Options

There are a few faucet options when dealing with bidets. One such option is the over-the-rim faucet. This type of faucet setup does not require a vacuum breaker. The faucet simply mounts on the bidet and is angled to provide water to the bowl. This type of faucet uses standard supply tubes.

Another type of faucet for bidets is wall-mounted. With these faucets the faucet handles and the vacuum breaker are mounted at the wall behind the bidet. The transfer valve is mounted on the bidet. The water supply to wall-mounted faucets is made with standard water distribution pipe.

Deck-mounted faucets are popular with bidets. The faucet handles and the transfer valve are all mounted on the deck of the bidet. The vacuum breaker rises above the fixture from behind. Many of these faucets use standard supply tubes for their water supply.

There are a few special variations available in various models of bidets. For example, there are models where the back of the fixture turns up, similar in appearance to a one-piece toilet. With this particular model, the faucet handles are deck-mounted and the transfer valve is mounted on the rise of the back section. The vacuum breaker for this style is concealed behind the back of the bidet. Standard supply tubes are the common method for supplying this type of bidet with water.

Sloppy Pop-Up Connections

Sloppy pop-up connections can be a problem when working with bidets. The piece that connects the lift rod to the pivot rod is pretty long, and it can get out of adjustment. The configuration of the pop-up assembly lends itself to problems.

If the pop-up connections are not adjusted properly, the pop-up may not open or close dependably. I've seen pop-ups on bidets work correctly three times in a row and then fail on the fourth attempt. This type of intermittent trouble can drive a plumber crazy.

When you respond to a complaint for a bidet that either won't hold or won't drain water to a customer's satisfaction, take a close look at the pop-up assembly. And don't try the pop-up just a time or two and accept it. Work the mechanism several times and don't take anything for granted.

Over-the-Rim Leaks

Over-the-rim leaks are possible with bidets that are equipped with over-the-rim faucets. The angle on these faucets is designed to prevent water from spilling over the edge of the fixture, but the problem still occurs at times.

The water from this type of leak is usually on the floor and easy to troubleshoot. There is really not any other type of bidet leak that will cause water to lie on the floor at the sides or front of the bidet.

If the base of the bidet has not been sealed with caulking, over-the-rim leaks can seep under the base of the fixture and appear to be some other type of leak.

I'll admit that over-the-rim leaks are not a common occurrence, but remember, don't rule out any possibility prematurely when troubleshooting.

In General

In general, bidets are just complicated lavatories that bolt to the floor. Many of the same principles used to service and repair lavatories, as described in Chap. 18, apply to bidets.

With bidets behind us, let's move onto the next chapter and study urinals.

23

Troubleshooting Urinals

How much do you know about troubleshooting urinals? Have you ever worked on a trough urinal? How about a stall urinal? Siphon-jet and washout urinals are more common, but many plumbers have never worked on them either. If you are accustomed to doing residential work, you don't have much call for urinal work. Residential plumbers rarely come into contact with urinals, but commercial plumbers work with them frequently.

There are not many complicated parts to deal with when repairing urinals, but this doesn't mean that troubleshooting them is simple. The fixture itself is not complex, but the flush valves for urinals do contain a number of parts that may cause trouble from time to time.

Fortunately, the parts for flush valves are typically available in kits, so entire sections can be rebuilt with relative ease. This reduces the need for identifying the exact part that is defective, and by rebuilding the section completely, the risk of callbacks is reduced.

Types of Urinals

For the benefit of those who are not at all familiar with urinals, let's start this chapter with some descriptions of various types of urinals

Stall urinals

Stall urinals have their bowls recessed beneath the floor level. A typical stall urinal will have about 4 inches of its total height below the floor. They are usually installed in a bed of sand, and they have 2-inch drains. The finished floor slopes toward the bowl of the urinal.

Stall urinals generally have a $\frac{3}{4}$-inch top spud for a flush valve to be mounted in. The water supply for an average flush valve on a urinal has a diameter of $\frac{3}{4}$ inch. Rough-in measurements vary, but the water supply is normally about $11\frac{1}{2}$ inches above the top of the urinal and about $4\frac{3}{4}$ inches to the right of center.

Trough urinals

Trough urinals are mounted on walls and have varying widths. Some are about $3\frac{1}{2}$ feet wide and others stretch to more than $5\frac{1}{2}$ feet wide. A depth of 7 inches is common, and there is usually a little more than a foot between the front lip and the back of the urinal.

The water supply for a trough urinal is typically a $\frac{1}{2}$-inch pipe. The pipe connects to a flush-pipe assembly that runs across the back of the urinal. The flush-pipe sends streams of water down the back of the urinal to clean it.

The trap arm for a trough urinal normally has a diameter of $1\frac{1}{2}$ inches. These urinals use exposed traps, usually P-traps.

Washout urinals

Washout urinals are wall-mounted and drain through a $1\frac{1}{2}$-inch tailpiece; however, some models have 2-inch drains. They have a $\frac{3}{4}$-inch top spud for a flush valve, and often have integral strainers. A standard P-trap is normally used to connect washout urinals to the drainage system.

Blowout urinals

Blowout urinals are wall-mounted, with 2-inch drains. The top spud for this type of urinal has a $1\frac{1}{2}$-inch diameter. Flush valves are required for blowout urinals.

Siphon-jet urinals

Siphon-jet urinals have 2-inch drains and $\frac{3}{4}$-inch top spuds. This type of urinal is wall-mounted.

Similarities

There are many similarities among the many types of urinals. Most urinals are equipped with flush valves. All urinals discharge through traps, and in one way or another, all urinals are set against walls. Integral traps are found on some models, and P-traps are used on others. Some have larger top spuds than others, and some have larger drains than others. While the flushing action of the various types differs, the general concept is the same. Water is introduced at the top of the urinal and leaves at the drain in the bottom of the bowl. Strainers, of one type or another, are found in most urinals.

When it comes to troubleshooting urinals, there is not a lot to look for. Most problems are either related to flush valves or drainage obstructions. Let's start our tour of troubleshooting with the flush valves.

Flush Valves

Flush valves are commonly used to flush urinals. These valves are not very complicated to work on, and many of their repair parts are sold in kits. When the kits are installed, the flush valve is rebuilt and ready for continued service. Let's do a quick rundown on the major parts of a flush valve. We will start where the water supply comes out of the wall and work our way down to the top of the urinal.

The first part of the valve is the supply flange. This part simply covers the inlet pipe for an attractive installation.

The next part we come to is the control stop. This is a key element of the flush valve. It is where the water pressure to the valve is controlled. There is a chrome cover that hides a screwdriver stop in the housing. Once the cover is removed, a screwdriver can be used to open and close the valve. When you need to cut the water off to a urinal, this is where you do it.

The horizontal chrome section between the control stop and the main valve body is called a tailpiece. The tailpiece connects to the body of the valve and carries water from the control stop to the flushing mechanism.

As we move down the body of the valve, we find a handle coupling and a handle. This is the section of the valve that gets the most use.

The long tubular section between the valve body and the urinal is called a flush connection. This is where the vacuum breaker parts are located.

The next part we come to is the spud flange. This trim cover hides the spud coupling that connects the flush valve to the urinal.

There are numerous parts inside flush valves. If you open one up, you will find springs, diaphragms, disks, nuts, washers, and assorted other parts. Luckily, these many parts are available in kits, so you don't have to know that the bad part is the handle spring. All you have to do is rebuild the handle assembly with a repair kit. This way, you are replacing all of the possible defective parts in one service call, and there is little chance of a callback.

The types of repair kits available

Many types of kits are available for rebuilding the working parts of flush valves. Since there are several, let's look at them one at a time.

Inside parts. There is a kit for replacing all the inside parts of the main body's head. These parts include a brass relief valve, a disk, a diaphragm, and a brass guide.

Washer set. The washer set that comes in kit form contains the disk, diaphragm, and washer that are installed between the brass guide and the relief valve.

Push-button assembly. A kit for the push-button assembly is also available. This kit includes numerous parts that will rebuild the entire push-button assembly. Some of the parts included are a handle coupling, socket, and spring. If you don't need to replace all of the push-button parts, you can get one of two kits that will repair portions of the assembly. One kit is available for replacing the push-button arrangement, and another kit will allow you to repair the existing push-button system.

Handle assembly. Kits are available for the complete replacement of handle assemblies. However, if you don't need the whole kit, you can buy individual parts for the handle assembly. There is also a handle-repair kit available.

Vacuum breaker repair. Vacuum breaker repair can be accomplished with a kit of parts.

Spud couplings. Spud couplings are available in kit form, too.

Outlet-coupling assembly. The kits for outlet-coupling assemblies replace existing assemblies and include all needed nuts and washers.

With the use of these kits, rebuilding a flush valve is fast and easy. Once you locate the area where the problem is likely to be, you can replace the inner workings and be on your way to the next job.

Traps

When you are working with urinals, you may be dealing with integral traps or external traps, which are usually P-traps. There is not much you have to worry about with integral traps, unless they become stopped up. There is, however, a gasket where the fixture outlet meets the drain pipe that could go bad and cause leaking.

External traps can get bumped and slip nuts can come loose. It is also possible for thin-gauge traps to deteriorate. There is not normally much call for problems with the traps of urinals, unless they are stopped up, but there are times when you may have to service the traps, so let's talk about them next.

Integral traps

Integral traps in urinals can be compared to those in toilets. Unless the trap becomes obstructed, there is nothing to be concerned with. The only area where trouble is likely to occur, other than for stoppages, is the gasket that is installed between the urinal outlet and the wall connection.

If you are faced with water leaking from behind a urinal that has a concealed trap, this is the first, and about the only item, you have to check. If the gasket is bad, you can replace it without extreme effort. You will, however, have to remove the urinal from the wall to do the job, and this involves disconnecting the flush valve.

Once the flush valve is loose, you can remove the lag bolts that hold the urinal to the wall and lift it off its bracket. This will give you access to the drain gasket.

External traps

External traps are not rare on urinals. These traps are typically standard P-traps. This creates a possibility for leaks at the trap adapter, the J-bend unions, and the tailpiece connection. While these connections rarely leak after initial installation, they can begin to leak for a number of reasons. For example, slip-nut washers can dry out and allow water to leak past them. Cleaning crews sometimes hit the traps with mop handles and knock connections loose. And, metal traps can deteriorate after exposure to untreated water. Any of these possibilities can create a leak that you may have to fix.

Troubleshooting Questions and Answers

In this section of the chapter, we are going to investigate urinals further with some troubleshooting questions and answers. I'll pose the question, and you try to answer it. You will find the answers following the questions.

1. You are called to a restaurant to repair a urinal. The restaurant owner has explained the job to you in the best way she can. Basically, the urinal is filling up nearly to the flood-level rim before the water goes down the drain.

When you go into the men's room to check out the urinal, you are expecting to find a problem in the trap or the drainage piping. You flush the urinal and the water swells up within the bowl, nearly overflowing. This particular model has an integral trap, so you decide to snake the drain before pulling the fixture off the wall.

You insert a $\frac{1}{4}$-inch bulb-head spring snake into the drain and work it through the trap. After running 25 feet of snake into the drain, you are sure the problem must be corrected. However, after retrieving the snake and flushing the urinal, the water still swills dangerously close to the flood-level rim. Confused, you put the snake down the drain again. The results are the same.

What would you do next? Do you have any idea of what the problem might be? Well, let's see.

You pull the urinal off the wall for further investigation. When you put an inspection mirror in the trap, you find the problem. A deodorizing disk that had been in the bowl of the urinal has become lodged in the trap, creating a baffle that slows down water being evacuated. Because of the shape and mobility of the deodorizing disk, your snake was able to get past the obstruction.

Once the disk is found, you are able to fish it out of the trap with a piece of wire. You reinstall the urinal and test it several times. The problem is solved.

2. This situation involves a urinal that is not flushing properly. When you depress the handle, water washes the fixture, but with very little force. What is the most likely cause of this problem?

There is a good chance that the screwdriver stop is not open far enough. Removing the stop cover and opening the valve will probably correct the problem.

3. Here, you are working with a trough urinal. The customer has complained that there is very little water pressure rinsing the fixture. The urinal is located in the bathroom at a recreational beach. The water source for the plumbing is a well.

You arrive on the job and can see that the trough is not being washed properly. What would you do first? You would probably check to see that the fixture was receiving adequate water pressure, and this would be a good decision. We will assume that you checked the water pressure at the control valve and found it to be suitable, but the pressure in the fixture is still below par. What should you do next?

Do you think that perhaps the flush tube has become clogged with mineral deposits? This would be my guess. Yes, you remove the flush tube and find that it is partially blocked with mineral deposits You clean the flush tube and the urinal works correctly.

What might you suggest to the property owner to avoid similar problems in the future? If you thought of suggesting the installation of a water softener, you are correct.

4. This question deals with a urinal that is flushing improperly.

The flush valve is not consistent in its flushing action. Sometimes it flushes pretty well, and other times it hardly flushes at all.

When you arrive on the job and flush the urinal, it works okay except that you notice the handle seems sloppy. After several flushes you experience the intermittent dependability of the flush valve. What do you think will correct this problem?

Installing a kit to rebuild the handle assembly is your best bet. By rebuilding the assembly, you can feel secure in your chances of correcting the problem.

5. This urinal is equipped with a standard urinal self-closing valve. The fixture is about 5 years old. The complaint is that water constantly dribbles into the bowl, resulting in high water bills. What can you do to fix it? This is a situation where the self-closing mechanism has become worn and must be rebuilt or replaced.

6. Here, you are called to an old school, where water is leaking onto the ceiling of the first floor. The maintenance director shows you the area where water is leaking through the ceiling and explains that the leak is not constant, but that it does occur often.

Since the leak is not constant, you rule out the water distribution system. With your attention focused on the drainage system, you move upstairs to check out the bathrooms. One of the bathrooms contains water closets, lavatories, and urinals. The other restroom contains only water closets and lavatories.

A quick inspection reveals no obvious signs of where the leak might be coming from. You flush the toilets and look for water that may seep out from around their bases. No water appears. The urinals are stall urinals. You look at their bowls to see if any cracks are present. None are. After having flushed all the toilets, you return to the first floor and look for dripping water. After a few minutes of watching and not seeing any water, you go back up to the bathrooms.

This time you flush the urinals and go back downstairs. When you enter the room where the ceiling has been damaged you see water dripping through the ceiling. You seem to have narrowed the leak down to being one of the urinals. Now the question is which urinal is it, and what is making it leak.

The ceiling is already badly damaged, so you cut a small hole in it to drain off excess water. When the dripping has stopped, you return to the bathroom and flush the first urinal. Then you go back downstairs and wait for water, but none comes. After a few minutes, you repeat the process with the second urinal. This time water does leak out of the ceiling. Now you know that a urinal is definitely the cause of the leak, and you have identified which urinal is responsible for the problem.

You return to the urinal that is causing the trouble and give the bowl a close inspection. Everything appears normal. Where do you think the leak is coming from? Did you guess the drain? If you did, you're right. Now the question is, how do you fix it?

Since the stall urinal is recessed into the floor by several inches, you don't have easy access to the drain connections. This job will require you to cut the ceiling out from below the urinal. Once you gain access to the drain, you see the leak is coming from the drain connector. You repair the leak and test the urinal. Everything remains dry, and you have solved the problem.

The Process of Elimination

The process of elimination is very effective when troubleshooting urinals. Since there are relatively few things that can go wrong with a urinal, it doesn't take long to get to the root of a problem.

The drainage system of urinals is simple, and there are only a few possibilities for problems. The drain can become blocked. A trap can cause a leak, and with some models the gasket between the urinal and the wall receptacle can be a problem.

Stoppages are easy to identify and are usually not difficult to correct. Leaks at traps are easy to see and normally simple to fix. Gasket leaks can be a bit more difficult to find and fix, but even they don't rank high on the list of hard-to-fix plumbing problems.

Urinals equipped with simple, self-closing valves don't offer many problems to plumbers. If there is trouble with the valve, it is easy to replace.

Flush valves contain more parts and present more potential for trouble than standard self-closing valves, but they are still relatively simple to troubleshoot. With repair kits available for the various sections of flush valves, rebuilding a faulty valve is not a big job.

Now that we are done with urinals, let's move on to the next chapter and learn about the troubleshooting procedures for faucets and valves.

24

Troubleshooting
Faucets and Valves

When you want to know about troubleshooting faucets and valves, you must be willing to take in a lot of information. There are so many different styles, types, brands, and models of faucets and valves on the market that it takes a lot of ambition to be able to troubleshoot all of the various situations you may run into.

Since there are so many possibilities for problems with faucets and valves, we are going to have to spend a substantial amount of time on this issue. I'm going to give you detailed explanations, case histories, and numerous drawings to aid you in developing your troubleshooting capabilities.

What types of techniques are we going to study? We will look at all the requirements involved in finding and fixing problems with faucets and valves. We will talk about lavatory faucets, laundry faucets, yard hydrants, tub and shower valves, and much more. This comprehensive chapter is going to prepare you for all the faucet and valve problems you are likely to encounter in the field.

Kitchen Faucets

Kitchen faucets come in many shapes and sizes. There are wall-mounted faucets, single-handle faucets, two-handle faucets, faucets with spray attachments, faucets without spray attachments, and specialty faucets. These faucets can have 6- or 8-inch centers. They can have one-piece bodies, or they can have individual components that must be put together to make a working faucet. Clearly, there are many possibilities for all sorts of problems with kitchen faucets.

Faucet washers

There was a time when faucet washers were found in all faucets. Times have changed and so have the inner workings of faucets. Today, faucets don't always have washers or bibb screws in them. Many modern faucets have springs, rubber seals, cartridges, and washerless stems. Gone are the days of putting an assortment of washers, screws, and seats on your truck to take care of any problem that might come along with a dripping faucet.

While the need for many more parts has evolved over time, the principles behind what makes a faucet leak are still pretty much what they were 10 or 20 years ago. We will get into these principles and practices in just a little while.

Washerless faucets

Washerless faucets are kind of like computers; they are great when they work the way they are supposed do, and they are horrendous when they don't. With washerless faucets, it is necessary to replace the entire stem unit. These stems can cost nearly as much as a replacement faucet would cost.

Cartridge-style faucets

Cartridge-style faucets are my favorite. I like these faucets because they are so simple to rebuild. It also happens that I've used one particular brand of cartridge-style faucet for about 15 years with almost no warranty callbacks, and this means a lot to me or any other plumbing contractor. There is also a side benefit to my favorite cartridge-style faucet; if for any reason the hot and cold water is piped to the wrong side of the faucet, the cartridge can be rotated to correct the problem. This is much easier than crossing supply tubes or pipes, and it doesn't show as a mistake on the plumber's part.

Ball-type faucets

Ball-type faucets are very common and popular. This style of faucet is not difficult for plumbers to repair, but there are many individual parts to be concerned with. Unlike cartridge-style faucets, where there is only the cartridge to replace, ball-type faucets have the ball, springs, and rubber parts that can all give you trouble.

Disposable faucets

Disposable faucets are becoming more and more a part of today's plumbing. Plumbers used to fix faucets, but now they often just replace them. Some old-school plumbers say this is done because the new crop of plumbers don't have the knowledge or desire to repair faucets. I disagree with this generality. Being somewhat of an old-school plumber myself, I can understand what the other veterans are talking about, but I can also see the younger plumbers' points of view.

Many modern faucets are not worth repairing. In fact, there are many faucets available that cost less than a replacement stem for more expensive faucets. For example, I can buy a complete lavatory faucet, including pop-up assembly, for less than $10. When you consider that stems can run anywhere from $6 to $17 a piece, why would you repair an old faucet when it would be cost effective to replace the entire unit? The customer gets a new faucet, and the plumber reduces the risk of a callback from a repair that doesn't last.

Troubleshooting Kitchen Faucets

Now that we have touched on the basic types of faucets in use today, let's devote some time to troubleshooting and repairing the various types. Much of what you are about to learn can be applied to more than just kitchen faucets. Many of the same troubleshooting techniques can be used for any type of faucet, and many of the repair options will be similar to those available for other types of faucets. Whether you are trying to eliminate a drip in a kitchen faucet, a lavatory faucet, a bar sink faucet, or a deck-mounted laundry faucet, most of the work will be closely related. Now with that in mind, let's jump right into the troubleshooting and repair of kitchen faucets.

Dripping from the spout

What's wrong when there is water dripping from the spout of a kitchen faucet? Usually, the problem is either with the faucet seat or the faucet stem or washer. However, the problem is not always caused by these parts being defective. Are you wondering how the faucet could be dripping if the parts are not defective? You should be, and that's good; it means you are thinking. I will explain how the problem could be caused by something other than defective faucet parts in a few moments.

When you walk in to troubleshoot a dripping faucet, you must first determine what type of faucet you are dealing with. In other words, does the faucet have washers, a ball assembly, a cartridge, or some other type of mechanism? Experienced plumbers will know the answer to this question as soon as they see the faucet in many cases. If you can't tell by looking at the exterior of the faucet, it is easy enough to determine the type by removing the handle assembly, and you're going to have to do this anyway, so even rookies are not at a great disadvantage. A visual inspection of the faucet seats, washers, and/or stem will tell you what the problem is, most of the time.

An exception. There is an exception, the one I mentioned just a few moments ago. Sometimes grit gets between the seat and the stem washer. When this happens, the stem washer cannot seat properly, and a drip results.

Plumbing systems in which iron particles are present are the ones most likely to experience a drip without having defective plumbing parts. Gravel, sand, and other impurities can also keep a faucet from closing fully. The debris in the faucet will usually be visible, but sometimes it will be small enough to avoid detection. If you suspect the leak is being caused by debris, flush the stem hole out with water. You can do this by turning on the water to the faucet slowly while the stem is removed. Water will bubble out of the hole and flush the grit out. Don't turn the water on too fast, or you will create a geyser that can shoot all the way up to a ceiling.

Faucet seats. Faucet seats sometimes become worn or pitted. When this happens, faucets drip. Small grinding wheels can be used to resurface the seats, but it is usually better to replace the damaged parts.

Corrosive elements in the water, such as iron and acid, are common causes of deteriorating faucet seats. Finding such problems in a plumbing system opens the door for the potential sale of water conditioning equipment.

Washers. Washers in faucets can wear out or be damaged by rough faucet seats. If a seat has become pitted and rough, it can cut a washer to the point where the faucet will leak. If you inspect faucet washers and find them punctured or cut, you should replace the faucet seats at the same time you replace the washers.

If the bibb screws that hold the washers to the stems are weak, you have found evidence of corrosive water conditions. This gives you yet another opportunity to recommend water treatment equipment to the customer.

Washerless stems. Washerless stems are not considered repairable. When these components go bad, they should be replaced. It may be possible to extend their life a short time by sanding the bottom of the stem with a light-grit sandpaper, but you cannot expect long-term success with this method.

Single-handle faucets. Single-handle faucets that are dripping are going to require some replacement parts unless the drip is being caused by debris lodged in the faucet. The parts could be a single cartridge, O-rings, springs, ball assemblies, or rubber seals. It is usually worthwhile to rebuild these types of faucets completely to avoid the risk of unwanted warranty callbacks.

Leaking around the base of a spout

When water is leaking around the base of a spout from a kitchen faucet, you can bet on the problem being a bad O-ring. You don't need much troubleshooting ability to solve this problem. All you have to do is remove the spout and replace the O-ring or O-rings, whichever the case may be.

Leaking around a faucet handle

When water is leaking around a faucet handle, you have one of two problems. Either the packing in the stem has gone bad, or there is a gasket or O-ring around the stem that is defective. Once you remove the handle from the faucet, you will be able to determine which problem you are faced with.

If water is leaking out around the base of the stem, you have a gasket or O-ring problem. When the water is coming out around the part of the stem that turns when the faucet is used, the problem is with the packing. Neither of these problems is difficult to correct.

With the water cut off, you can remove the stem and replace the gasket or O-ring quickly. If the problem is with the packing, you will have to remove the stem assembly and install new packing material. In either case, the job will take only a few minutes to complete.

Won't cut off

What should you look for when the water from the faucet won't cut off? Well, the first thing to do is to cut off the water supply to the faucet. With that done, you can proceed with your troubleshooting. What are you likely to find? You may find a large piece of grit between the faucet seat and the stem washer. It is also possible that the bibb screw has deteriorated and allowed the washer to float about, resulting in a steady stream of water that won't cut off. A quick look into the faucet body should tell you what is causing the problem.

Poor water pressure

Poor water pressure is a common complaint. When the pressure is low at only one fixture, the troubleshooting skills needed to solve the problem are not extensive. Before you begin a massive investigation into the inner workings of the troublesome faucet, there are a few simple things to check.

Start your troubleshooting by removing the aerator from the faucet.

Look at the screen and diverter disk to see if they are blocked by debris or mineral deposits. Turn the faucet on and see what the water pressure is like with the aerator removed. Many times this will prove to be the simple solution plumbers hope for. If the aerator is the culprit, all you have to do is clean or replace it.

When the aerator is not at fault, check the cutoff valves under the fixture. Make sure the valves are in the full-open position. You might be surprised at how often the valves are partially closed, causing low water pressure at faucets.

While you are checking the cutoffs, look at the supply tubes running up to the faucet. If a supply tube has a bad crimp in it, you have probably found the cause of the low water pressure. Also, see what type of material has been used for the water distribution piping.

If the water distribution piping is made of galvanized steel, you probably have your work cut out for you. Old galvanized water piping tends to clog itself up with rust and mineral deposits. As the pipe slowly clogs, the water pressure to the plumbing fixtures is gradually decreased. The reduction is not noticeable in the early stages, but in advanced stages the flow can be reduced to a trickle. This type of problem requires extensive work. The old pipe must be removed and new piping installed.

Once you have eliminated the external causes of low pressure, you must look inside the faucet. You may find debris blocking the faucet inlets or delivery tubes. One way to be sure the problem is in the faucet is to disconnect a supply tube and turn the water on. If you have good pressure at the tip of the supply tube, you know the problem has to be in the body of the faucet.

It may be necessary to replace some faucets that have become clogged with mineral deposits or debris. In other cases you can simply clear the blockage and put the faucet back into service.

Not a uniform flow

When a faucet is delivering water that is not in a uniform flow, the problem is with the aerator. A spraying or rough stream of water is a sure indication of a partially plugged aerator. Remove the aerator and clean or replace it.

No water at the spray head

Many plumbing customers complain that they have no water at the spray head of their kitchen faucet. There are only a few possible causes for this type of problem.

When you have a spray attachment that will not deliver water, check the spray hose. These hoses frequently become kinked under sinks. A bad kink can render the spray head useless.

If the spray hose is in good condition, take the spray head apart and inspect for mineral deposits or debris. You will often find iron deposits blocking up the works. If necessary, replace the spray head.

I can't recall ever having it happen, but it is possible that the outlet port on the bottom of the faucet could become blocked, cutting off the water supply to the spray hose. If you have replaced the old spray assembly with a new one and are still not getting water, consider the possibility of a blockage at the connection port.

Limited water at the spray head

Limited water at the spray head is more common than a complete lack of water. The same troubleshooting steps offered for solving no-water problems can be followed to solve problems with limited water at the spray head.

Water dripping from beneath a faucet

Water dripping from beneath a faucet can be frustrating. This type of problem is often misdiagnosed as being a leak at the supply tube. If water leaks from the bottom of a faucet and runs through the faucet holes in the sink, it can appear to be coming from the supply-tube connection.

Acidic water is a primary cause of leaks in the bodies of faucets. The thin copper tubing used to mix and deliver water in the bodies of faucets can be eaten up by water with a high acid content. Small pinholes can develop and allow water to escape within the faucet housing. You won't be able to put your finger on these leaks without first removing the entire faucet from the sink.

When you have a pitted or corroded faucet body, you should replace it. While it is sometimes possible to repair the leak with a spot-soldering job, you will be better off to replace the faucet.

Troubleshooting Other Faucets

Lavatory faucets

Lavatory faucets are very similar to kitchen faucets when it comes to troubleshooting and repair work. There are, however, more multipiece lavatory faucets used than there are multipiece kitchen faucets. Multipiece faucets are faucets that consist of individual handles and spouts that are connected, by plumbers, with tubing beneath the fixture. While single-body faucets are much more common, multipiece faucets are frequently found in more expensive plumbing systems. These faucets offer a few wrinkles that single-body faucets don't.

Most of the troubleshooting techniques used for multipiece faucets

Figure 24.1   Three-piece, widespread lavatory faucet. (*Courtesy of Moen, Inc.*)

are the same as those used for single-body faucets. However, the connection points under fixtures for multipiece faucets can produce leaks that don't exist in single-body faucets.

Most multipiece faucets (Fig. 24.1) are connected with small copper tubing and compression fittings. The tubing is soldered into the individual faucet components and connected to the mixing spout with compression fittings. The tubing is so small and thin that it crimps easily. This can result in low water pressure. It is also easy to break the soldered connections between the tubing and the individual components.

When you are working with multipiece faucets, be careful. Too much stress on the components can twist, crimp, or break the tubing connections. Since multipiece faucets are typically expensive, you don't want to have to replace one at your own expense.

In the interest of space and time, I will not give a blow-by-blow account of how to work with lavatory faucets. You can use the same principles and techniques described for kitchen faucets to work with lavatory faucets (Figs. 24.2 through 24.5).

Bar sinks and laundry tubs

Deck-mounted faucets for bar sinks and laundry tubs fall into the same troubleshooting and repair category as kitchen faucets. As with

Figure 24.2 Single-handle lavatory faucet. (*Courtesy of Moen, Inc.*)

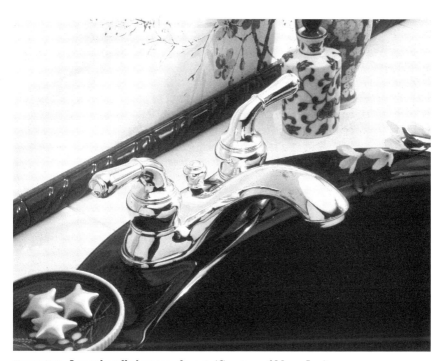

Figure 24.3 Lever-handle lavatory faucet. (*Courtesy of Moen, Inc.*)

Figure 24.4 Mini-widespread lavatory faucet. (*Courtesy of Moen, Inc.*)

Figure 24.5 Lever-handle lavatory faucet. (*Courtesy of Moen, Inc.*)

lavatory faucets, refer to the section on kitchen faucets for information on troubleshooting the faucets for bar sinks and laundry tubs.

Troubleshooting Bathtub and Shower Faucets

The faucet valves for bathtubs and showers are a little different from the faucets used for kitchen sinks (Fig. 24.6). The tools and techniques required for troubleshooting these valves are similar, but not always the same. Let's see how the differences affect your work.

Figure 24.6 Pressure balancing tub/shower valve. (*Courtesy of Symmons Industries, Inc.*)

Figure 24.7   Tub wrenches.

Dripping from the spout

When you have water dripping from a tub spout or a shower head, you can use the same basic principles described for kitchen faucets in your troubleshooting and repair work. You may, however, need a set of tub wrenches to remove the valve stems. Tub wrenches (Fig. 24.7), for those of you who aren't familiar with them, resemble deep-set sockets.

Two- and three-handle tub and shower valves are often equipped with valves that have their wrench flats concealed in the finished wall. Since you can't get a standard wrench on the flats to remove the stems, you must have tub wrenches to reach into the wall and remove the parts.

Once you have the stems for tub and shower valves removed, the rest of the troubleshooting and repair procedure is comparable to that used for kitchen faucets.

Leaking around a faucet handle

When water is leaking around a faucet handle on a tub or shower valve, the cause of the leak is usually defective packing around the stem. The stems for tub and shower valves are larger than those found in sink faucets, but they share many of the same characteristics. By removing the stem, you can replace the bad packing. It is possible that the leak is coming from a bad O-ring, so check this before reinstalling the stem.

Won't cut off

If you have a tub or shower valve that won't cut off, you can apply the same basic principles and procedures used for kitchen faucets in your troubleshooting and repair work.

Poor water pressure

Poor water pressure is rare with bathtub valves, but shower heads sometimes suffer from a lack of desirable water pressure. The cause is usually a mineral buildup in the water-saver disk that is housed in the shower head. If you remove the head and check the disk for obstructions, you will probably find some. A quick rinse under a faucet will often clear the debris. In stubborn cases you may have to poke the pieces out with a thin piece of wire or replace the disk. It is also conceivable that the holes in the shower head itself are plugged.

Uniform flow

A uniform flow is another problem you could encounter with a shower head that is partially plugged with foreign objects. While tub spouts won't give you this type of trouble, shower heads can. When you have an erratic spray from a shower head, check the head and water-saver disk for blockages.

Diverters that don't divert well

Diverters that don't divert well are a fairly common problem with three-handle tub-and-shower valves. In some cases water isn't diverted from the tub spout to the shower head. At other times water will run out of both the tub spout and the shower head simultaneously. The cause of these types of problems is the stem washer or the valve seat. Treat the condition just as you would a dripping faucet, and replace the seat or washer as needed.

Wall-mount faucets

Wall-mount faucets are not very different from their deck-mounted cousins when it comes to troubleshooting and repairing them. You can apply the same principles and procedures discussed for kitchen faucets to your work with wall-mount faucets.

Frost-free hose bibbs

Frost-free hose bibbs are similar to a sink faucet in terms of troubleshooting. The stem for a frost-free hose bibb is, however, much longer than a stem for a standard faucet. The fixture still contains a seat and a washer, but the stem is extra long. The length of the stem depends on the length of the hose bibb, but it can easily be up to a foot long.

Hose bibbs also have packing and packing nuts that can spawn leaks. You can apply the basic troubleshooting principles and practices we discussed in the section on kitchen faucets to frost-free hose bibbs.

Learn the Basics

Once you learn the basics of troubleshooting valves and faucets, you will be able to tackle just about any type of unit with a good deal of success. The general principles are about the same for all types of faucets and valves. The job only gets tricky when you fail to follow a systematic troubleshooting approach.

As routine as troubleshooting faucets is, the task can become frustrating. There will no doubt be times when the problem doesn't know it is supposed to go by the book. When this happens, you must depend on the experience and knowledge you have gained through field work and study to solve the problem. To drive this point home, let's take a little time to look at some scenarios that you might run into out in the field.

Field Conditions

Field conditions can play a major role in the level of difficulty a plumber faces in solving on-the-job problems. Repair work that should be simple can become quite complicated when there are space limitations. For example, a multipiece faucet that should be simple to repair can become a nightmare if the faucet is offset to one corner of a sink and drawers in the cabinet block your access. The same could be said for trying to replace a washer when the bibb screw has been eaten up by acidic water.

The act of replacing a faucet washer is certainly not complicated. Under normal conditions, almost any adult should be capable of handling the work. If, however, the head of the bibb screw falls apart when you touch it with a screwdriver, the simple job turns into a challenging assignment. If this were to happen to you, what would you do?

When the head of a bibb screw disintegrates, you have a few options. The least complicated option is the replacement of the faucet stem. Some customers, however, will not want to pay for a whole new stem. If customers knew how much they were going to spend for the time it takes a plumber to work with a broken screw, they would be more inclined to replace the stem, and this is something you may want to mention to cost-conscious customers.

If the customer insists on having you spend time working with the broken screw, accept the challenge and overcome it. There are two fairly simple ways of doing this. You can take a knife and cut out the old washer. This will expose the threaded portion of the old screw. Many times the shaft of the screw will not have rotted to a point where it will not turn. If you put a pair of pliers on the threaded shaft, once the washer is removed, there is a good chance you can turn the shaft out of the stem.

Should the shaft of the screw break off, don't panic. Use a small drill bit to drill out the old screw. Then you can either try to rethread the stem, or you can install snap-in washers. Snap-in washers are fitted with two ears that can be pushed into the hole you drilled. Once the ears are in the hole, their natural tension will hold the washer in place.

Every experienced plumber has a personal way to deal with a broken bibb screw. What is the right way to get the job done? Any approach that works is acceptable. You should strive for a procedure that is fast, cost-effective, dependable, and durable and that will satisfy the customer; when you meet this criteria, you have done a fine job.

Field conditions can put some plumbers into tailspins. If they have not spent a considerable amount of time doing real plumbing, they can be mystified by the actions of other plumbers, homeowners, and even inanimate plumbing. This type of situation can arise around any type of plumbing, not just faucets and valves.

I've been on countless jobs where the field conditions made doing my job all but impossible. Even now, with 20 years of experience, I still encounter circumstances in the field that test my abilities to the limits. For example, what kind of a person would install a water heater with the element access doors facing a solid wall? Obviously, someone who had no intention of ever working on the water heater again. I saw an installation like this just last week.

Why would someone thread a coupling onto the supply-tube inlets of a faucet? One of my recent jobs had such an arrangement, and there was a leak between the faucet inlet and the supply tube. If the coupling had not been installed, it would not have been possible for such a leak to exist. The coupling was not needed, and it was not standard plumbing procedure to install it. To make matters worse, the coupling had been installed so tightly that it took a basin wrench with a 14-inch pipe wrench attached to its handle to remove the couplings.

What would inexperienced plumbers think when seeing this strange coupling in a place where the books never show such a fitting? They would probably be confused, and they might try to salvage the fitting or replace it. With my experience I knew the coupling was useless, so I removed it and installed a slightly longer supply tube. The leak was fixed.

The two most important traits to develop when working in the field are experience and product knowledge. Experience only comes with time, but product knowledge can be learned as quickly as you are willing to put forth the effort. If you know how plumbing fixtures, faucets, and devices are put together, you are well on your way to being an excellent troubleshooter. On the other hand, if you do not know that a particular faucet should have a gasket in a particular place, how will you know what the problem is if the gasket is missing?

All major plumbing manufacturers are very willing to provide professionals with detailed information about their products. If you take the time to obtain cut sheets from manufacturers, you can begin your study of what makes faucets and valves work. This same approach applies to all plumbing parts, devices, and fixtures. All you have to do is ask your supplier for the information you want. If the supplier doesn't have detailed drawings, you can get the manufacturer's address from the supplier and request the information directly from the company that makes the plumbing parts. Let me give you some examples of the types of drawings I'm talking about.

Some Examples of Faucets and Valves

Here are some examples of popular faucets and valves for you to look over. As you will see, the drawings provide a lot of detail and information that can make troubleshooting and repair work much simpler. Study these illustrations and become familiar with the inner workings of the faucets and valves; this is one of the most effective ways to begin building the foundation for your troubleshooting skills. (Figs. 24.8 through 24.55)

Now that you have had an opportunity to look over the cut sheets on faucets and valves, let's move on to the next chapter and investigate plumbing appliances.

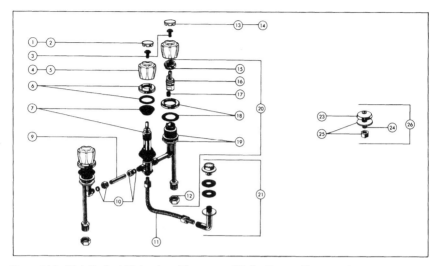

Figure 24.8   Detail of a bidet valve. (*Courtesy of Delta Faucet Company.*)

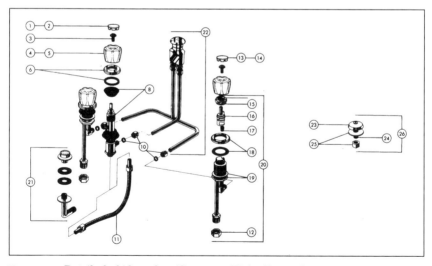

Figure 24.9   Detail of a bidet valve. (*Courtesy of Delta Faucet Company.*)

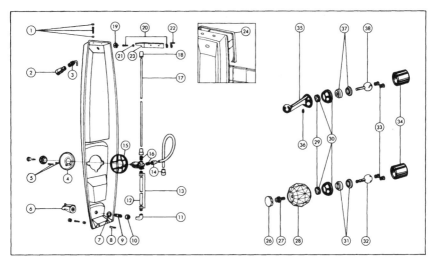

Figure 24.10 Nonpressure balance shower valve detail. (*Courtesy of Delta Faucet Company.*)

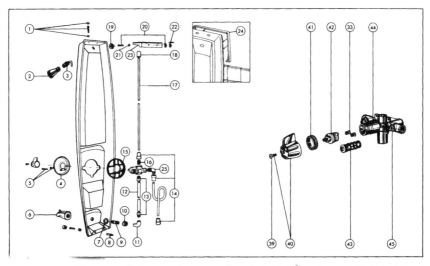

Figure 24.11 Pressure balance shower valve detail. (*Courtesy of Delta Faucet Company.*)

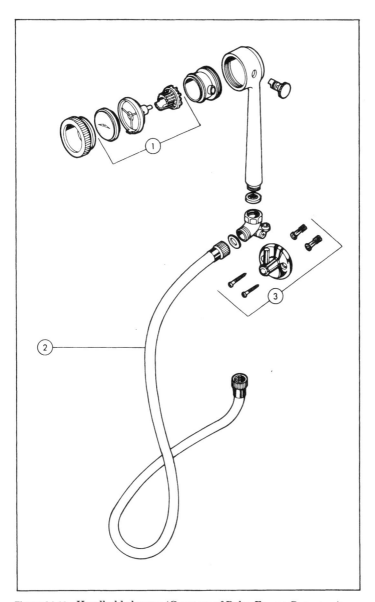

Figure 24.12 Handheld shower. (*Courtesy of Delta Faucet Company.*)

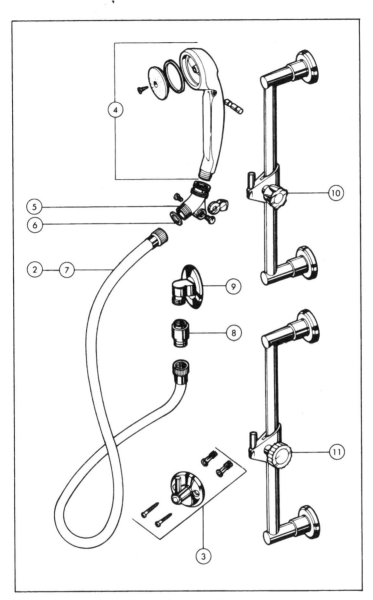

Figure 24.13 Handheld shower. (*Courtesy of Delta Faucet Company.*)

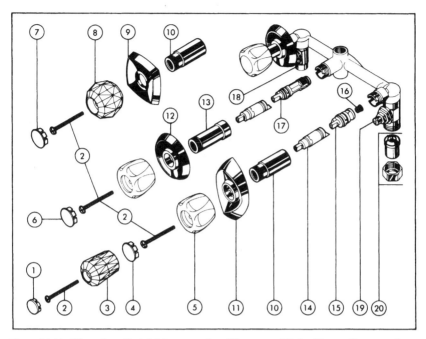

Figure 24.14 Three-handle tub/shower valve. (*Courtesy of Delta Faucet Company.*)

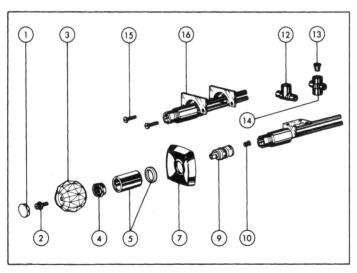

Figure 24.15 Two-handle tub/shower valve. (*Courtesy of Delta Faucet Company.*)

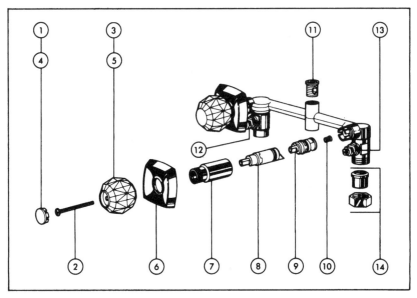

Figure 24.16   Two-handle tub/shower valve. (*Courtesy of Delta Faucet Company.*)

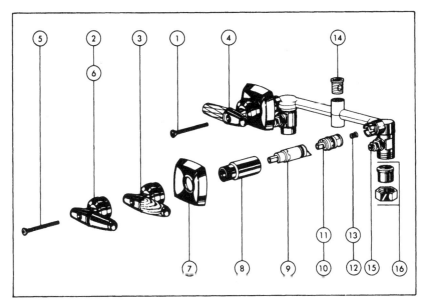

Figure 24.17   Two-handle tub/shower valve. (*Courtesy of Delta Faucet Company.*)

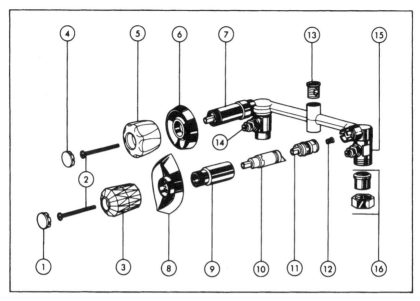

Figure 24.18   Two-handle tub/shower valve. (*Courtesy of Delta Faucet Company.*)

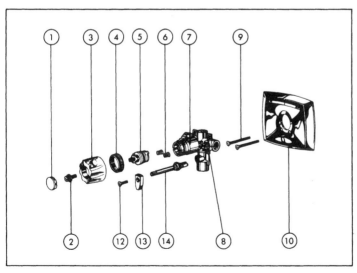

Figure 24.19   Single-handle, nonpressure balancing tub/shower valve. (*Courtesy of Delta Faucet Company.*)

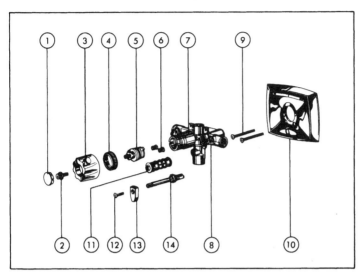

Figure 24.20 Single-handle, pressure balancing tub/shower valve. (*Courtesy of Delta Faucet Company.*)

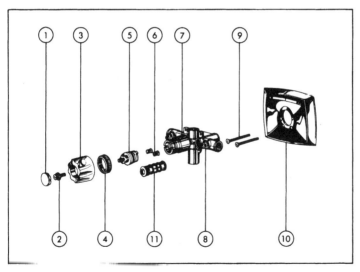

Figure 24.21 Single-handle, pressure balancing tub/shower valve. (*Courtesy of Delta Faucet Company.*)

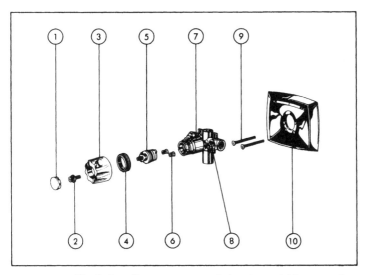

Figure 24.22 Single-handle, nonpressure balancing tub/shower valve.
(*Courtesy of Delta Faucet Company.*)

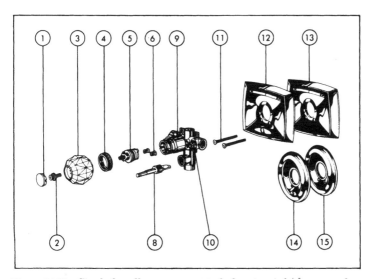

Figure 24.23 Single-handle, nonpressure balancing tub/shower valve.
(*Courtesy of Delta Faucet Company.*)

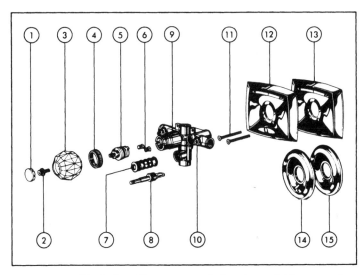

Figure 24.24   Single-handle, pressure balancing tub/shower valve. (*Courtesy of Delta Faucet Company.*)

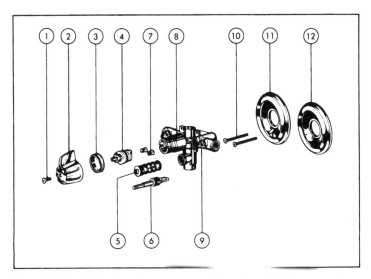

Figure 24.25   Single-handle, pressure balancing tub/shower valves. (*Courtesy of Delta Faucet Company.*)

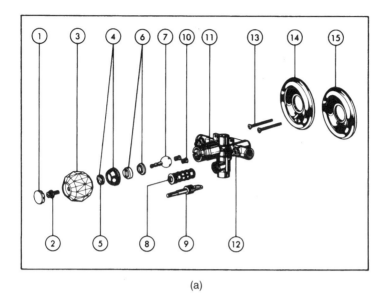

(a)

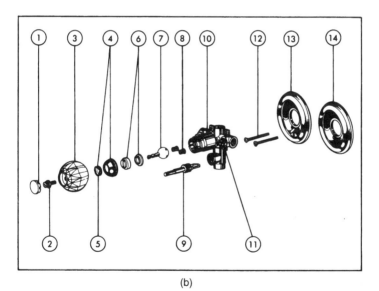

(b)

Figure 24.26 Single-handle, pressure balancing and nonpressure balancing tub/shower valves. (*Courtesy of Delta Faucet Company.*)

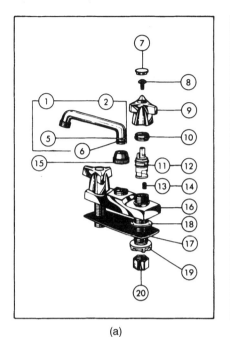

(a)

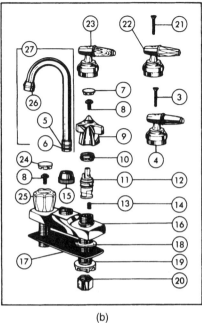

(b)

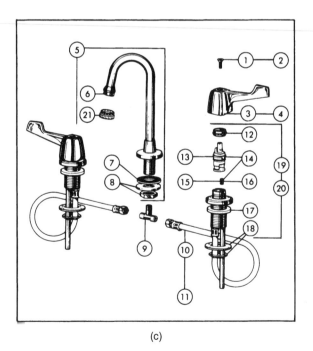

(c)

Figure 24.27 (a) Laundry faucet; (b) bar sink faucet; (c) widespread utility faucet. (*Courtesy of Delta Faucet Company.*)

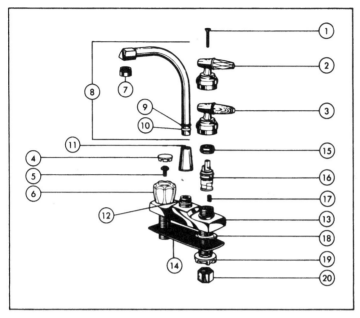

Figure 24.28a   Waterfall lavatory faucet. (*Courtesy of Delta Faucet Company.*)

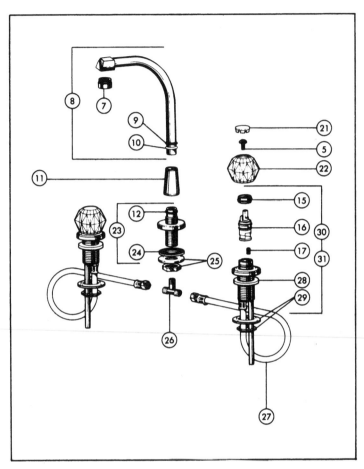

Figure 24.28b   Waterfall, widespread lavatory faucet. (*Courtesy of Delta Faucet Company.*)

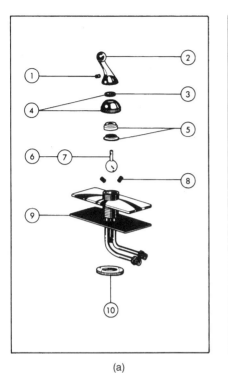

(a)

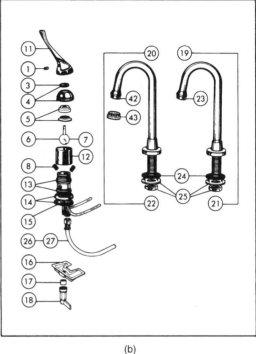

(b)

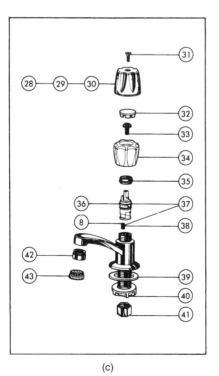

(c)

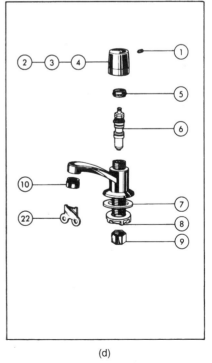

(d)

Figure 24.29 (a) Single-handle shampoo faucet; (b) single-handle utility faucet; (c) basin faucet; and (d) slow self-closing basin faucet. (*Courtesy of Delta Faucet Company.*)

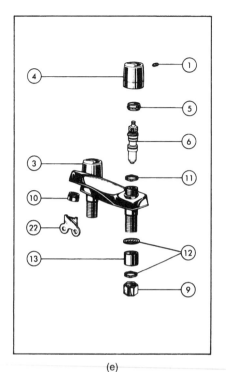
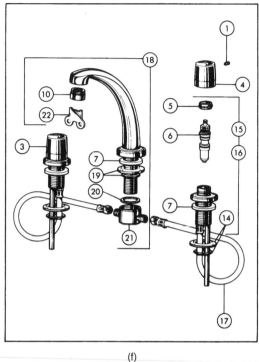

(e)   (f)

Figure 24.29 (*e*) Slow self-closing lavatory faucet; (*f*) slow self-closing, widespread lavatory faucet. (*Courtesy of Delta Faucet Company.*)

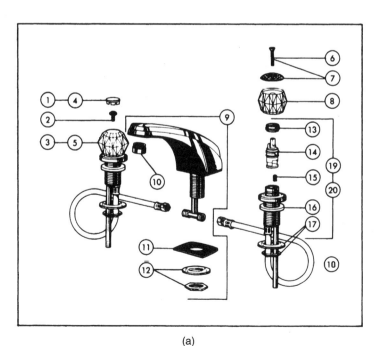

(a)

Figure 24.30 Two-handle widespread lavatory faucet. (*Courtesy of Delta Faucet Company.*)

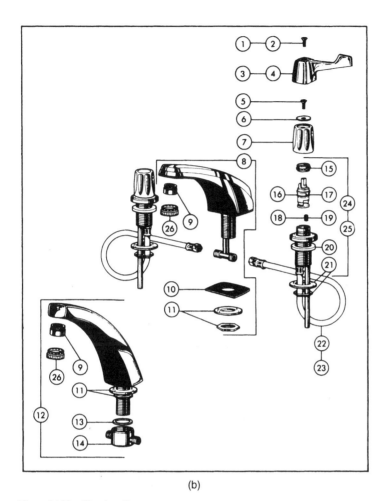

(b)

Figure 24.30   (*Continued*)

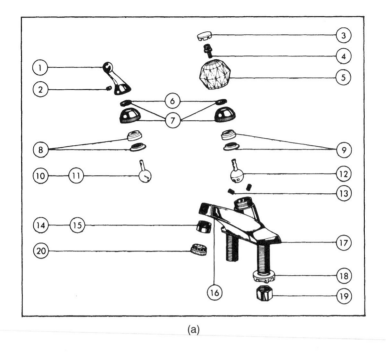

(a)

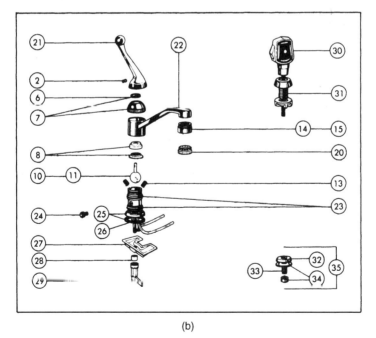

(b)

Figure 24.31   Single-handle, single-mount faucet. (*Courtesy of Delta Faucet Company.*)

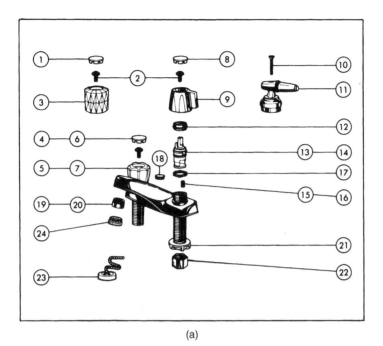

(a)

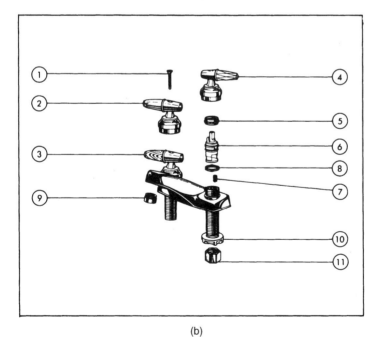

(b)

Figure 24.32   Two-handle lavatory faucet. (*Courtesy of Delta Faucet Company.*)

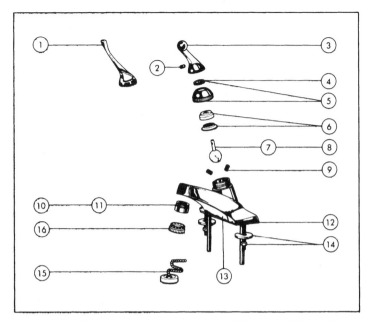

Figure 24.33 Single-handle lavatory faucet. (*Courtesy of Delta Faucet Company.*)

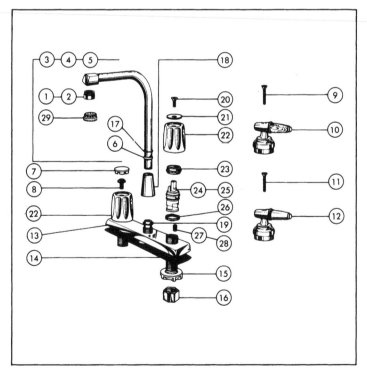

Figure 24.34 Two-handle kitchen deck-mounted waterfall faucet. (*Courtesy of Delta Faucet Company.*)

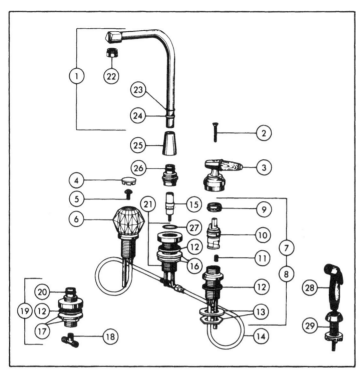

Figure 24.35 Widespread two-handle kitchen deck-mounted waterfall faucet. (*Courtesy of Delta Faucet Company.*)

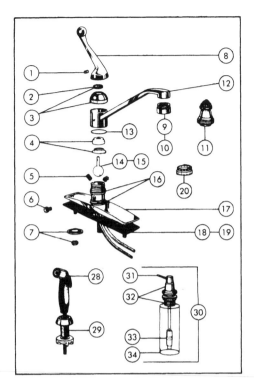

Figure 24.36 Single-handle kitchen faucet. (*Courtesy of Delta Faucet Company.*)

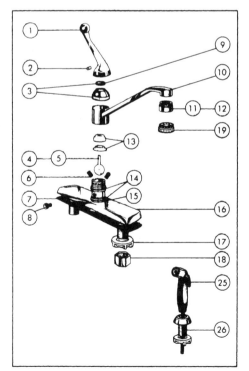

Figure 24.37 Single-handle kitchen faucet. (*Courtesy of Delta Faucet Company.*)

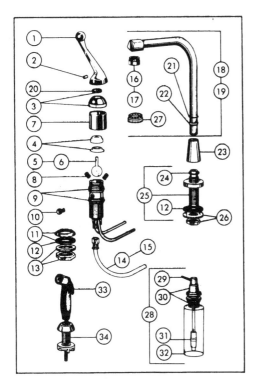

Figure 24.38 Single-handle, waterfall, single-mount utility faucet. (*Courtesy of Delta Faucet Company.*)

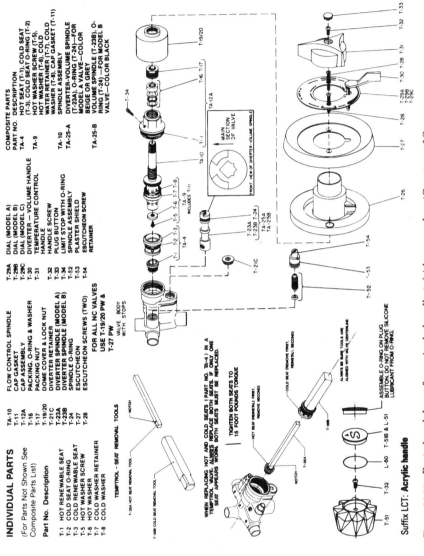

Figure 24.39 Parts breakdown for a single-handle tub/shower valve. (*Courtesy of Symmons Industries, Inc.*)

Anti-Siphon Vacuum Breaker

NIDEL® Model 34H

NIDEL® Model 34HD

FOR NON-FREEZING AREAS ONLY

NIDEL® Model 34H anti-siphon vacuum breaker is designed to protect hose connections from contamination where temperatures do not get below freezing. Uses include laundry tubs, utility sinks, boiler room hose bibbs and outside hose bibbs in non-freezing areas.

FOR FREEZING AREAS

NIDEL® Model 34HD anti-siphon vacuum breaker is designed to protect hose connections from contamination where freezing temperatures are expected. Manual draining provides easy, sure drainage during cold weather. Uses include outside hose bibbs, wash racks, dairy barns and swimming pool areas.

SPECIFICATIONS:

APPROVALS — Approved under ASSE Standard 1011, Canadian Standards Association, listed by IAPMO® and accepted by U.S. Department of Health. For use downstream from the last shut-off valve. Not for continuous line pressure.

CONSTRUCTION — Brass and stainless steel construction.

DIAPHRAM — Flexible diaphram valve suitable for hot water.

ATTACHMENT SCREWS — Break-off.

PORTS — Dual safety design closes air ports during water flow.

INLET — ¾ inch standard female hose thread.

OUTLET — ¾ inch standard male hose thread.

FINISH — Plain brass. *Optional:* Chrome plated.

Figure 24.40 Antisiphon vacuum breaker. (*Courtesy of Woodford Mfg. Co.*)

Laboratory Vacuum Breakers
NIDEL® Model 38DF

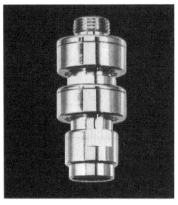

The Nidel Model 38DF vacuum breaker is designed for laboratory use to provide protection against back siphonage and back flow from aspirators or hoses left in contaminants.

DUPLEX CONSTRUCTION — Allows hose to be used above vacuum breaker.

CROSS CONNECTION PROTECTION — Guards against danger of cross connection contamination by protecting each faucet individually.

HIGH FLOW RATE — Sufficient flow rate to operate small aspirators.

EASILY ATTACHED — Gaskets provided to fit standard ⅜ inch laboratory faucets.

SPECIFICATIONS:

APPROVALS — Tested and approved by City of Los Angeles and City of Detroit Plumbing Laboratory. Accepted by U.S. Department of Health and many city and state health departments.

CONSTRUCTION — Brass and stainless steel.

INLET — ⅜ inch male pipe thread with gasket.

OUTLET — ⅜ inch straight female pipe thread.

DUPLEX VALVES — Dual-seal silicone diaphram valves for use in hot water.

PRESSURE REQUIRED — 7½ PSI minimum.

FLOW RATE — 5¾ GPM @ 25 pounds differential pressure.

FINISH — Bright chrome.
Optional — Tin plated.
Optional — Special finishes and outlets available on request.

NOTE — Install only where drippage will go into sink or drain. Not recommended for distilled water.

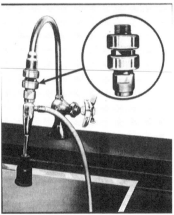

Figure 24.41 Laboratory vacuum breaker. (*Courtesy of Woodford Mfg. Co.*)

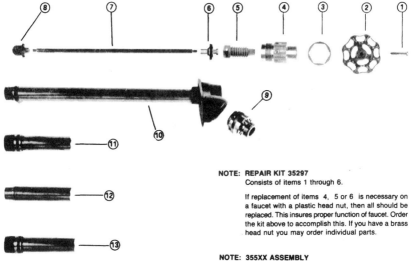

MODEL 25 PARTS

ITEM	PART NO.	DESCRIPTION
1	30234	Handle Screw
2	30239	Wheel Handle - metal
3	30236	Drain Guard
4	30241	Head Nut - brass
5	30238	Stem Screw
6	35280	Drain Valve Assembly
1, 2, 3, 4, 5 and 6		Repair Kit (see above note)
7	303XX	Operating Rod
8	30230	Plunger Assembly
9	55057	Vacuum Breaker-brass
10	3546X	"CP" Inlet Casing Assembly (specify length)
11	3545X	"P" Inlet Casing Assembly (specify length)
12	3544X	"C" Inlet Casing Assembly (specify length)
13	3547X	"CP3" Inlet Casing Assembly (specify length)

NOTE: REPAIR KIT 35297
Consists of items 1 through 6.

If replacement of items 4, 5 or 6 is necessary on a faucet with a plastic head nut, then all should be replaced. This insures proper function of faucet. Order the kit above to accomplish this. If you have a brass head nut you may order individual parts.

NOTE: 355XX ASSEMBLY
Consists of items 1 through 8 above. Order this assembly to replace all working parts of unit. **However, you must furnish wall thickness.**

OPERATING ROD
(As illustrated below)

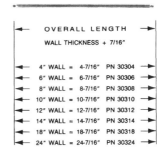

◄───	OVERALL LENGTH	───►
	WALL THICKNESS + 7/16"	

◄─── 4" WALL	= 4-7/16"	PN 30304 ───►
◄─── 6" WALL	= 6-7/16"	PN 30306 ───►
◄─── 8" WALL	= 8-7/16"	PN 30308 ───►
◄─── 10" WALL	= 10-7/16"	PN 30310 ───►
◄─── 12" WALL	= 12-7/16"	PN 30312 ───►
◄─── 14" WALL	= 14-7/16"	PN 30314 ───►
◄─── 18" WALL	= 18-7/16"	PN 30318 ───►
◄─── 24" WALL	= 24-7/16"	PN 30324 ───►

Figure 24.42 Parts breakdown for automatic-draining, freezeless wall faucets. (*Courtesy of Woodford Mfg. Co.*)

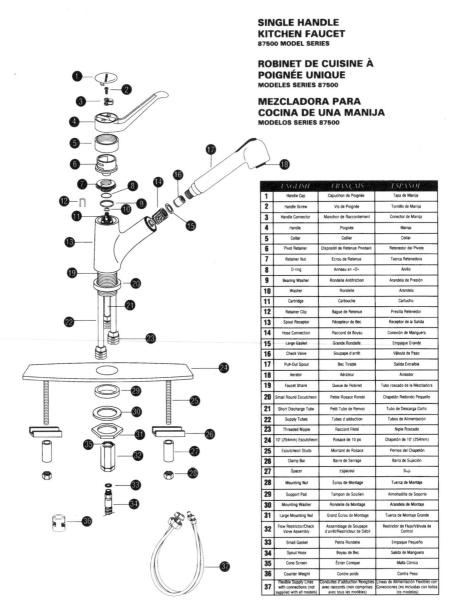

**SINGLE HANDLE
KITCHEN FAUCET**
87500 MODEL SERIES

**ROBINET DE CUISINE À
POIGNÉE UNIQUE**
MODELES SERIES 87500

**MEZCLADORA PARA
COCINA DE UNA MANIJA**
MODELOS SERIES 87500

	ENGLISH	FRANÇAIS	ESPAÑOL
1	Handle Cap	Capuchon de Poignée	Tapa de Manija
2	Handle Screw	Vis de Poignée	Tornillo de Manija
3	Handle Connector	Manchon de Raccordement	Conector de Manija
4	Handle	Poignée	Manija
5	Collar	Collier	Collar
6	Pivot Retainer	Dispositif de Retenue Pivotant	Retenedor del Pivote
7	Retainer Nut	Écrou de Retenue	Tuerca Retenedora
8	O-ring	Anneau en «O»	Anillo
9	Bearing Washer	Rondelle Antifriction	Arandela de Presión
10	Washer	Rondelle	Arandela
11	Cartridge	Cartouche	Cartucho
12	Retainer Clip	Bague de Retenue	Presilla Retenedor
13	Spout Receptor	Récepteur de Bec	Receptor de la Salida
14	Hose Connection	Raccord de Boyau	Conexión de Manguera
15	Large Gasket	Grande Rondelle	Empaque Grande
16	Check Valve	Soupape d'arrêt	Válvula de Paso
17	Pull-Out Spout	Bec Tirable	Salida Extraíble
18	Aerator	Aérateur	Aireador
19	Faucet Shank	Queue de Robinet	Tubo roscado de la Mezcladora
20	Small Round Escutcheon	Petite Rosace Ronde	Chapetón Redondo Pequeño
21	Short Discharge Tube	Petit Tube de Renvoi	Tubo de Descarga Corto
22	Supply Tubes	Tubes d'adduction	Tubos de Alimentación
23	Threaded Nipple	Raccord Fileté	Niple Roscado
24	10" (254mm) Escutcheon	Rosace de 10 po	Chapetón de 10" (254mm)
25	Escutcheon Studs	Montant de Rosace	Pernos del Chapetón
26	Clamp Bar	Barre de Serrage	Barra de Sujeción
27	Spacer	Espaceur	Buje
28	Mounting Nut	Écrou de Montage	Tuerca de Montaje
29	Support Pad	Tampon de Soutien	Almohadilla de Soporte
30	Mounting Washer	Rondelle de Montage	Arandela de Montaje
31	Large Mounting Nut	Grand Écrou de Montage	Tuerca de Montaje Grande
32	Flow Restrictor/Check Valve Assembly	Assemblage de Soupape d'arrêt/Restricteur de Débit	Restrictor de Flujo/Válvula de Control
33	Small Gasket	Petite Rondelle	Empaque Pequeño
34	Spout Hose	Boyau de Bec	Salida de Manguera
35	Cone Screen	Écran Conique	Malla Cónica
36	Counter Weight	Contre-poids	Contra Peso
37	Flexible Supply Lines with connections (not supplied with all models)	Conduites d'adduction flexibles avec raccords (non comprises avec tous les modèles)	Líneas de Alimentación Flexibles con Conecciones (no incluidas con todos los modelos)

Figure 24.43 Breakdown of single-handle kitchen faucet. (*Courtesy of Moen, Inc.*)

KITCHEN DECK FAUCETS
Single-Handle - for Eight-Inch Centers

Measurements

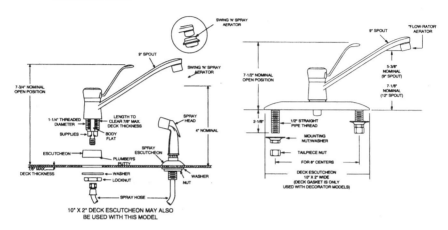

Equipped with Flow-Rator® Aerator or Swing-N-Spray Aerator

CAUTION: Always turn water off before disassembling the valve. Open valve handle to relieve water pressure and insure that complete water shut-off has been accomplished.

Before turning water on during either rough-in or trim-out, make sure that cartridge retainer clip is in place. The cartridge and retainer clip were properly installed and tested before leaving the factory. Although it is unlikely, it is nevertheless possible that through the handling of the valve by any number of persons the retainer clip may not be properly installed. This should be carefully checked at time of rough-in and trim-out. If the retainer clip is not properly installed, water pressure could force the cartridge out of the casting. Personal injury or water damage to the premises could result.

Figure 24.44   Single-handle kitchen deck faucet. (*Courtesy of Moen, Inc.*)

Disassembly

1. Turn "OFF" both hot and cold water supplies. Turn faucet on to relieve pressure and insure complete shut off.. Pull handle cap up and off (it snaps into place). Remove the handle screw (illustration 1).

2. Push the cartridge stem down and then lift and tilt handle lever and handle body off.

3. Unscrew retainer pivot nut.

4. Lift and twist spout off. The diverter can be removed at this point.

5. Pry out retainer clip with screwdriver (illustration 2).

6. Use cartridge twisting tool furnished with the new cartridge and rotate cartridge between 11 and 1 o'clock position.

7. Grasp cartridge stem with pliers. Pull cartridge out (illustration 3).

Reassembly

1. With cartridge stem UP, insert the cartridge with the ears aligned front to back (illustration 4).

2. Push the cartridge down by the ears (illustration 5) until flush with the top of the body.

3. Turn notched flat of cartridge stem toward front of sink. (Note: for cross piping installations, see instructions on back).

4. Replace the clip all the way (illustration 6). The diverter can be replaced at any time before the spout is installed.

5. Replace spout. Push down until it nearly touches the faucet escutcheon.

6. Screw on retainer pivot nut. DO NOT CROSS THREAD. Tighten firmly.

7. Press cartridge stem down. Holding handle UP, hook handle ring inside the handle body (illustration 7) into groove on the retainer pivot nut.

8. Swing handle back and forth until it drops down into place.

9. Replace handle screw. Tighten securely. Push handle cap down until it snaps into place.

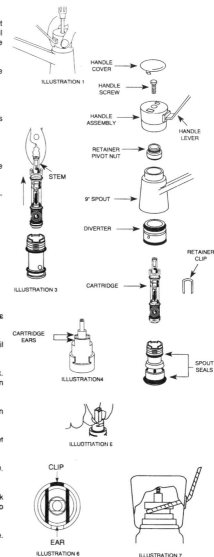

Figure 24.45 Disassembly and reassembly procedure for a cartridge-style, single-handle faucet. (*Courtesy of Moen, Inc.*)

RISER KITCHEN FAUCETS

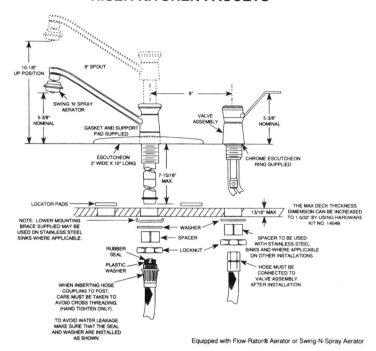

10-1/8"
UP POSITION

9" SPOUT

8"

SWING 'N' SPRAY
AERATOR

VALVE
ASSEMBLY

5-3/8"
NOMINAL

5-3/8"
NOMINAL

GASKET AND SUPPORT
PAD SUPPLIED

ESCUTCHEON
2" WIDE X 10" LONG

CHROME ESCUTCHEON
RING SUPPLIED

7-15/16"
MAX.

LOCATOR PADS

THE MAX DECK THICKNESS
DIMENSION CAN BE INCREASED
TO 1-5/32" BY USING HARDWARE
KIT NO. 14049.

13/16" MAX

NOTE: LOWER MOUNTING
BRACE SUPPLIED MAY BE
USED ON STAINLESS STEEL
SINKS WHERE APPLICABLE.

WASHER

SPACER

SPACER TO BE USED
WITH STAINLESS STEEL
SINKS AND WHERE APPLICABLE
ON OTHER INSTALLATIONS.

RUBBER
SEAL

LOCKNUT

PLASTIC
WASHER

HOSE MUST BE
CONNECTED TO
VALVE ASSEMBLY
AFTER INSTALLATION.

WHEN INSERTING HOSE
COUPLING TO POST,
CARE MUST BE TAKEN TO
AVOID CROSS THREADING.
(HAND TIGHTEN ONLY).

TO AVOID WATER LEAKAGE
MAKE SURE THAT THE SEAL
AND WASHER ARE INSTALLED
AS SHOWN.

Equipped with Flow-Rator® Aerator or Swing-N-Spray Aerator

CAUTION: **Always turn water off before disassembling the valve. Open valve handle to relieve water pressure and insure that complete water shut-off has been accomplished.**

CLIP

EAR

Before turning water on during either rough-in or trim-out, make sure that cartridge retainer clip is in place. The cartridge and retainer clip were properly installed and tested before leaving the factory. Although it is unlikely, it is nevertheless possible that through the handling of the valve by any number of persons the retainer clip may not be properly installed. This should be carefully checked at time of rough-in and trim-out. If the retainer clip is not properly installed, water pressure could force the cartridge out of the casting. Personal injury or water damage to the premises could result.

Figure 24.46   Riser-style kitchen faucet. (*Courtesy of Moen, Inc.*)

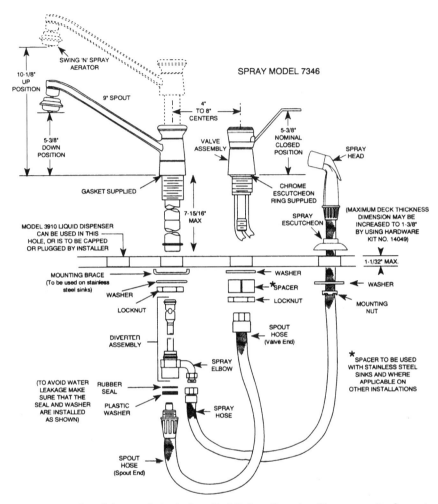

Figure 24.47   Breakdown of single-handle kitchen faucet with spray attachment. (*Courtesy of Moen, Inc.*)

**TWO-HANDLE
TUB/SHOWER FAUCET**
MODEL 82000 & 2500 SERIES.

**ROBINET À POIGNÉE DOUBLE
POUR DOUCHE/BAIGNOIRE**
MODÉLE SÉRIES 8200 ET 2500

**MEZCLADORA PARA TINA/
REGADERA DE DOS MANIJAS**
SERIES MODELOS 82000 y 2500

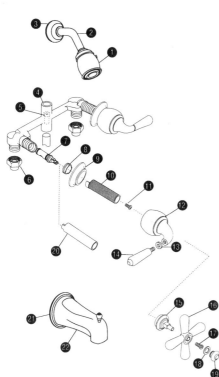

	ENGLISH	FRANÇAIS	ESPAÑOL
1	Shower Head	Pomme de Douche	Cebolieta
2	Shower Arm	Bras de Douche	Brazo de Regadera
3	Flange	Bride	Chapetón
4	Valve Body	Bâti de Soupape	Cuerpo de Válvula
5	Diverter	Dérivateur	Derivador
6	Union Adapter with Nut	Adaptateur de Raccordement avec Écrou	Adaptador de Unión con Tuerca
7	Cartridge	Cartouche	Cartucho
8	Cartridge Nut	Écrou de Cartouche	Tuerca del Cartucho
9	Escutcheon	Rosace	Chapetón
10	Stem Extension	Rallonge de Tige	Extensión de Vástago
11	Stem Extension Screw	Vis de Rallonge de Tige	Tornillo de la Extensión de Vástago
12	Handle Hub	Enjoliveur de Poignée	Centro de Manija
13	Optional Color Ring	Anneau de Couleur Facultatif	Anillo de Color Opcional
14	Lever Handle Insert (not included with all models)	Levier de Poignée Insérée (non compris avec tous les modèles)	Inserto de Manija de Palanca (no incluido con todos los modelos)
15	Handle Skirt	Jupe de Poignée	Faldón de Manija
16	Cross Handle Insert (not included with all models)	Pièce Insérée de Poignée Cruciforme (non compris avec tous les modèles)	Inserto de Manija en Cruz (no incluido con todos los modelos)
17	Cross Handle Screw	Vis de Poignée Cruciforme	Tornillo de Manija nen Cruz
18	Optional Color Ring	Anneau de Couleur Facultatif	Anillo de Color Opcional
19	Plug Button	Bouchon de Finition	Tapón
20	Cartridge Removal Tool	Outil de Démontage de Cartouche	Herramienta de Remoción
21	Tub Spout Escutcheon	Rosace de Bec de Baignoire	Chapetón de Salida de Tina
22	Tub Spout	Bec de Baignoire	Salida de Tina

Figure 24.48   Two-handle tub/shower faucet. (*Courtesy of Moen, Inc.*)

**SINGLE HANDLE
TUB/SHOWER FAUCET**
MODEL 3100 SERIES

**ROBINET À POIGNÉE UNIQUE
POUR DOUCHE/BAIGNOIRE**
SERIES MODÉLES 3100

**MEZCLADORA PARA TINA/
REGADERA DE UNA MANIJA**
SERIES MODELOS 3100

	ENGLISH	FRANCAIS	ESPAÑOL
1	Shower Head	Pomme de Douche	Cebolleta
2	Shower Arm	Bras de Douche	Brazo de Regadera
3	Flange	Bride de Douche	Chapetón
4	Retainer Clip	Bague de Retenue	Pressilla Retenedora
5	Valve Body	Bâti de Soupape	Cuerpo de Válvula
6	Balance Spool	Tambour d'équilibre	Carrete de Balance
7	Cartridge	Cartouche	Cartucho
8	Check Stop	Soupape d'arrêt	Control de Parada
9	Adjustable Temperature Limit Stop	Régulateur de Température	Parada de límite de Temperatura Ajustable
10	Plaster Ground	Plaque Murale	Base para Pared
11	Escutcheon	Rosace	Chapetón

	ENGLISH	FRANCAIS	ESPAÑOL
12	Escutcheon Screws	Vis de Rosace	Tornillos del Chapetón
13	Brake	Frein	Freno
14	Handle Stop	Arrêt de Poignée	Manija de Parada
15	Handle Stop Screw	Vis d'arrêt de Poignée	Retainer Tornillo de Manija
16	Handle Hub	Enjoliveur de Poignée	Centro de Manija
17	Lever Handle Insert	Vis de Blocage	Inserto de Manija de Palanca
18	Set Screw	Levier de Poignée Inséré	Tornillo de Sujeción
19	1/16" Hex Wrench	Clé Hexagonale de 1/16 po	Llave Hexagonal de 1/16"
20	Tub Spout Escutcheon	Rosace de Bec de Baignoire	Adjustable Temperature Limit Stop
21	Tub Spout	Rosace de Bec de Baignoire	Chapetón de Salida de la Tina
22	7/64" Hex Wrench	Clé Hexagonale de 7/64 po	Llave Hexagonal de 7/64"

Figure 24.49 Single-handle tub/shower faucet. (*Courtesy of Moen, Inc.*)

**SINGLE HANDLE
TUB/SHOWER FAUCET**
MODEL 2500 & 82000 SERIES

**ROBINET À POIGNÉE UNIQUE
POUR DOUCHE/BAIGNOIRE**
SERIES MODÉLES 2500 & 82000

**MEZCLADORA PARA TINA/
REGADERA DE UNA MANIJA**
SERIES MODELOS 2500 & 82000

3 PORT
3 OUVERTURES
3 ABERTURAS

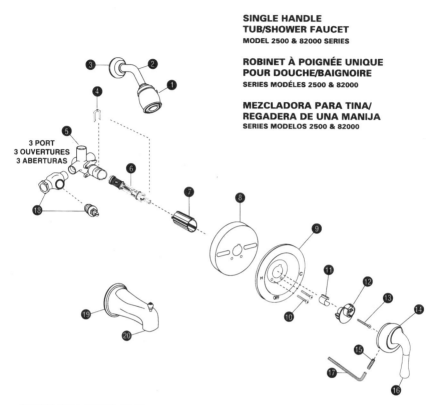

	ENGLISH	FRANÇAIS	ESPAÑOL
1	Shower Head	Pomme de Douche	Cebolleta
2	Shower Arm	Bras de Douche	Brazo de Regadera
3	Flange	Bride de Douche	Chapetón
4	Retainer Clip	Bague de Retenue	Presilla Retenedora
5	Valve Body	Bâti de Soupape	Cuerpo de Válvula
6	Cartridge	Cartouche	Cartucho
7	Stop Tube	Tube d'Arrêt	Tubo de Parada
8	Plaster Ground	Plaque Murale	Base para Pared
9	Escutcheon	Rosace	Chaptetón
10	Escutcheon Screws	Vis de Rosace	Tornillos del Chapetón

	ENGLISH	FRANÇAIS	ESPAÑOL
11	Brake	Frein	Freno
12	Handle Adapter	Adaptateur de Poignée	Adaptador de Manija
13	Handle Screw	Vis de Poignée	Tornillo de Manija
14	Handle Hub	Enjoliveur de Poignée	Centro de Manija
15	Set Screw	Vis de Blocage	Tornillo de Sujeción
16	Lever Handle Insert	Levier de Poignée	Inserto de Manija de Palanca
17	1/16" Hex Wrench	Clé Hexagonale de 1/16 po	Llave Hexagonal de 1/16"
18	Check Valve	Soupape d'Arrêt	Válvula de Control
19	Tub Spout Escutcheon	Rosace de Bec de Baignoire	Chapetón de Salida de la Tina
20	Tub Spout	Bec de Baignoire	Salida de la Tina

Figure 24.50   Single-handle tub/shower faucet. (*Courtesy of Moen, Inc.*)

**SINGLE HANDLE
TUB/SHOWER FAUCET**
MODEL 82000, 2500 SERIES

**ROBINET À POIGNÉE UNIQUE
POUR DOUCHE/BAIGNOIRE**
SERIES MODÉLES 82000, 2500

**MEZCLADORA PARA TINA/
REGADERA DE UNA MANIJA**
SERIES MODELOS 82000, 2500

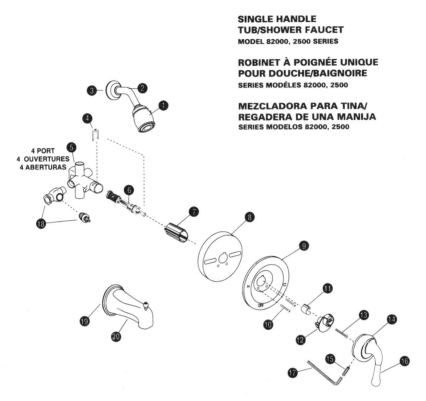

	ENGLISH	FRANÇAIS	ESPAÑOL
1	Showerhead	Pomme de Douche	Cebolleta
2	Arm	Bras de Douche	Brazo
3	Flange	Bride de Douche	Chapetón
4	Retainer Clip	Bague de Retenue	Presilla Retenedora
5	Valve Body (copper to copper)	Bâti de Soupape (cuivre/cuivre)	Cuerpo de válvula (cobre a cobre)
6	Cartridge	Cartouche	Cartucho
7	Stop Tube	Tube d'arrêt	Tubo de Parada
8	Plaster Ground	Plaque Murale	Base para Pared
9	Escutcheon	Rosace	Chapetón
10	Escutcheon Screws	Vis de Rosace	Tornillos de Chapetón

	ENGLISH	FRANÇAIS	ESPAÑOL
11	Brake	Frein	Freno
12	Handle Stop	Arrêt de Poignée	Parada de Manija
13	Handle Stop Screw	Vis d'arrêt	Tornillo de Parada
14	Handle Hub	Enjoliveur de Poignée	Centro de Manija
15	Set Screw	Vis de Blocage	Tornillo de Sujeción
16	Lever Handle Insert	Levier de Poignée	Inserto de Manija de Palanca
17	1/16" Hex Wrench	Clé Hexagonale de 1/16 po	Llave Hexagonal de 1/16"
18	Stop Valve	Soupape d'arrêt en Caoutchouc	Válvula de Parada Caucho
19	Tub Spout Escutcheon	Rosace de Bec de Baignoire	Chapetón de Salida de la Tina
20	Tub Spout	Bec de Baignoire	Salida de la Tina

Figure 24.51 Single-handle tub/shower faucet. (*Courtesy of Moen, Inc.*)

**SINGLE HANDLE
TUB/SHOWER FAUCET**
MODEL 82000, 2300

**ROBINET À POIGNÉE UNIQUE
POUR DOUCHE/BAIGNOIRE**
MODÉLES 82000, 2300

**MEZCLADORA PARA TINA/
REGADERA DE UNA MANIJA**
MODELOS SERIES 82000, 2300

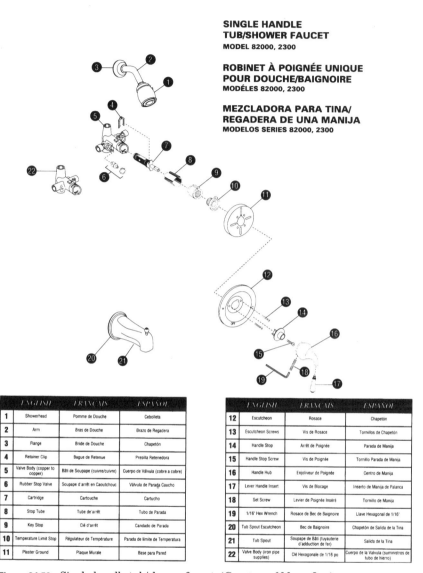

	ENGLISH	FRANÇAIS	ESPAÑOL
1	Showerhead	Pomme de Douche	Cebolleta
2	Arm	Bras de Douche	Brazo de Regadera
3	Flange	Bride de Douche	Chapetón
4	Retainer Clip	Bague de Retenue	Presilla Retenedora
5	Valve Body (copper to copper)	Bâti de Soupape (cuivre/cuivre)	Cuerpo de Válvula (cobra a cobre)
6	Rubber Stop Valve	Soupape d'arrêt en Caoutchouc	Válvula de Parada Caucho
7	Cartridge	Cartouche	Cartucho
8	Stop Tube	Tube de arrêt	Tubo de Parada
9	Key Stop	Clé d'arrêt	Candado de Parada
10	Temperature Limit Stop	Régulateur de Température	Parada de límite de Temperatura
11	Plaster Ground	Plaque Murale	Base para Pared

	ENGLISH	FRANÇAIS	ESPAÑOL
12	Escutcheon	Rosace	Chapetón
13	Escutcheon Screws	Vis de Rosace	Tornillos de Chapetón
14	Handle Stop	Arrêt de Poignée	Parada de Manija
15	Handle Stop Screw	Vis de Poignée	Tornillo Parada de Manija
16	Handle Hub	Enjoliveur de Poignée	Centro de Manija
17	Lever Handle Insert	Vis de Blocage	Inserto de Manija de Palanca
18	Set Screw	Levier de Poignée Inséré	Tornillo de Manija
19	1/16" Hex Wrench	Rosace de Bec de Baignoire	Llave Hexagonal de 1/16"
20	Tub Spout Escutcheon	Bec de Baignoire	Chapetón de Salida de la Tina
21	Tub Spout	Soupape de Bâti (tuyauterie d'adduction de fer)	Salida de la Tina
22	Valve Body (iron pipe supplies)	Clé Hexagonale de 1/16 po	Cuerpo de la Válvula (suministros de tubo de hierro)

Figure 24.52   Single-handle tub/shower faucet. (*Courtesy of Moen, Inc.*)

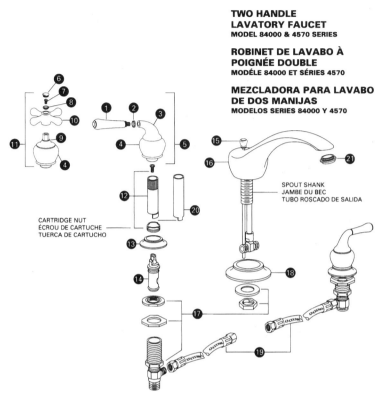

**TWO HANDLE
LAVATORY FAUCET**
MODEL 84000 & 4570 SERIES

**ROBINET DE LAVABO À
POIGNÉE DOUBLE**
MODÉLE 84000 ET SÉRIES 4570

**MEZCLADORA PARA LAVABO
DE DOS MANIJAS**
MODELOS SERIES 84000 Y 4570

SPOUT SHANK
JAMBE DU BEC
TUBO ROSCADO DE SALIDA

CARTRIDGE NUT
ÉCROU DE CARTUCHE
TUERCA DE CARTUCHO

	ENGLISH	FRANÇAIS	ESPAÑOL
1	Lever Handle Insert (not included with all models)	Levier de poignée inséré (non compris avec tous les modèls)	Inserto de Manija de Palanca (no incluido con todos los modelos)
2	Color Ring	Anneau de Couleur	Anillo de Color
3	Elbow	Coude	Codo
4	Handle Hub	Enjoliveur de Poignée	Centro de Manija
5	Lever Handle Assembly	Assemblage du Levier de Poignée	Ensamble de Manija de Palanca
6	Plug Button	Bouchon de finition	Bouchon de finition
7	Handle Screw	Vis de poignée	Vis de poignée
8	Color Ring	Anneau de Couleur	Anillo de Color
9	Handle Skirt	Jupe de Poignée	Faldón de Manija
10	Cross Handle Insert (not included on all models)	Poignée Cruciforme Insérée (non compris avec tous les modèles)	Inserto de Manija en Cruz (no incluido con todos los modelos)
11	Cross Handle Assembly	Assemblage de Poignée Cruciforme	Ensamble de Manija en Cruz

	ENGLISH	FRANÇAIS	ESPAÑOL
12	Stem Extension Kit (hot "red" and cold "blue")	Trousse de Rallonge de Tige (chaud=rouge/froid-bleu)	Juego de Extensión del Vástago (caliente 'rojo' y frío 'azul')
13	Handle Escutcheon	Rosace de Poignée	Escudo de Manija
14	Cartridge	Cartouche	Cartucho
15	Lift Rod & Knob	Bouton et Tige de Levée	Varilla Elevadora y Perilla
16	Spout Assembly	Assemblage du Bec	Ensamble de Salida
17	Hardware	Quincaillerie de Montage	Accesorios de Montaje
18	Spout Escutcheon	Rosace de Bec	Chapetón de Manija
19	Supply Hose	Boyau d'adduction	Manguera de Alimentación
20	Cartridge Removal Tool	Outil de Démontage de Cartouche	Herramienta de Remoción de Cartucho
21	Aerator	Aérateur	Aireador

Figure 24.53 Two-handle, three-piece lavatory faucet. (*Courtesy of Moen, Inc.*)

**TWO HANDLE
LAVATORY FAUCET**
MODEL 84000 & 4500 SERIES

**ROBINET DE LAVABO À
POIGNÉE DOUBLE**
MODÈLE 84000 ET SÉRIES 4500

**MEZCLADORA PARA LAVABO
DE DOS MANIJAS**
MODELO SERIES 84000 Y 4500

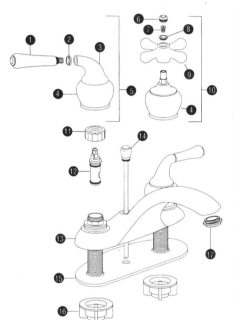

	ENGLISH	FRANÇAIS	ESPAÑOL
1	Lever Handle Insert (not included on all models)	Levier de Poignée Inséré (non compris avec tous les modèles)	Inserto de Manija de Palanca (no incluido con todos los modelos)
2	Color Ring	Anneau de Couleur	Anillo de color
3	Elbow	Coude	Codo
4	Handle Hub	Enjoliveur de Poignée	Centro de Manija
5	Lever Handle Assembly	Assemblage du Levier de Poignée	Ensamble de Manija de Palanca
6	Plug Button	Capuchon	Tapón
7	Handle Screw	Cartouche	Tornillo de Manija
8	Color Ring	Anneau de Couleur	Anillo de Color
9	Cross Handle Insert (not included on all models)	Poignée Cruciforme Insérée (non compris avec tous les modèles)	Inserto de Manija en Cruz (no incluido con todos los modelos)
10	Cross Handle Assembly	Assemblage de Poignée cruciforme	Ensamble de Manija en Cruz
11	Cartridge Nut	Écrou de Cartouche	Tuerca de Cartucho
12	Cartridge	Cartouche	Cartucho
13	Escutcheon	Rosace	Chapetón
14	Lift Rod & Knob	Tige de levée et bouton	Varilla Elevadora y Perilla
15	Deck Gasket	Support de Comptoir	Empaque de Cubierta
16	Mounting Nut	Écrou de Montage	Tuerca de Montaje
17	Aerator	Aérateur	Aireador

Figure 24.54   Two-handle lavatory faucet. (*Courtesy of Moen, Inc.*)

LAVATORY FAUCETS
FOUR-INCH CENTERSET

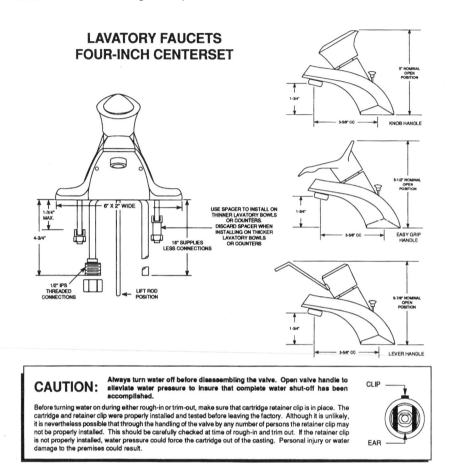

CAUTION: Always turn water off before disassembling the valve. Open valve handle to alleviate water pressure to insure that complete water shut-off has been accomplished.

Before turning water on during either rough-in or trim-out, make sure that cartridge retainer clip is in place. The cartridge and retainer clip were properly installed and tested before leaving the factory. Although it is unlikely, it is nevertheless possible that through the handling of the valve by any number of persons the retainer clip may not be properly installed. This should be carefully checked at time of rough-in and trim out. If the retainer clip is not properly installed, water pressure could force the cartridge out of the casting. Personal injury or water damage to the premises could result.

Figure 24.55 Single-handle lavatory faucet. (*Courtesy of Moen, Inc.*)

25

Troubleshooting
Plumbing Appliances

There are not many appliances that plumbers must work with, but there are a few. In some cases the appliances may only be connected to waste or water lines by plumbers. Other situations, such as those involving water heaters, require plumbers to have a good working knowledge of the appliance or device. Dishwashers, garbage disposers, and water heaters are the three major plumbing appliances most plumbers get involved with. Ice-makers are another appliance that plumbers sometimes work with.

In terms of plumbing connections, none of these appliances is very difficult to troubleshoot. However, when it comes to the inner parts of these same appliances, troubleshooting can become more difficult.

This chapter is going to go over all of the appliances mentioned above and cover the bases on effective troubleshooting procedures for each of them. We will start in the kitchen and work our way into the mechanical room.

Garbage Disposers

Garbage disposers are a common appliance in many homes. While plumbers rarely work on the motors of disposers, they are often the first service people called when a disposer is acting up. Problems with disposers can fall into three primary categories: plumbing, electrical, and appliance repair.

Our primary function in this chapter is to distinguish the causes of particular problems. While you, as a plumber, may not perform electrical work, you may be required to advise the property owner to call an electrician. Plumbers must be able to correct plumbing problems and refer other types of problems to appropriate repair people. As an example of knowing when to refer a customer to another type of service person, let's consider the following scenario.

A homeowner calls you to come out and repair a garbage disposer that is not working properly. When you arrive, you find that the disposer will cut on, but that it almost immediately trips out the reset button and goes off. After testing the disposer a few times, you are convinced the problem is most likely in the electrical wiring. Unless you are a licensed electrician, you should advise the homeowner of your opinion and suggest that perhaps an electrician should be called.

Under these conditions, you have fulfilled your role as a plumber. You responded to the call and did the initial troubleshooting. Seeing that the problem was not plumbing related, you gave your opinion and suggestions. That is about all that anyone could expect of you. I bring up the referral issue here because there are many times when working with appliances that the trouble is not in the domain of plumbers.

Now then, what can go wrong with a garbage disposer? Well, quite a bit, actually. To expand on this, let's take the probable troubleshooting phases on a one-at-a-time basis. We will do this in the form of stories about actual service calls.

In our first story, you have been called to a home to fix a garbage disposer that will not work. You arrive and turn on the garbage disposer. It makes a loud whirring sound, but doesn't dispose of its contents. What do you think is wrong with the disposer? If you said it's jammed, you're right. Now, how will you fix it?

Disposers that are suffering from jammed impeller blades can be fixed in one of three ways, depending on the type of disposer you are working with.

Expensive disposers often have a switch that allows the rotation of the impellers to be reversed, thereby freeing them from whatever is jamming them. This type of repair, when done by a plumber, often makes homeowners feel foolish and many plumbers can't look the customers in the face when they present their bill. As embarrassing as this situation can be for both parties, it is often simply a matter of turning on the reversing switch to solve the problem.

Many models of disposers are not equipped with reversing switches. A good number of these lower-grade disposers are, however, equipped with a special socket that allows the flywheel to be rotated. These sockets are usually hex shaped and located on the bottom of the unit. Wrenches for the socket are supplied with the disposers. A regular Allen wrench will get the job done if the factory-supplied wrench is not available. Once you have a wrench in the socket, all you have to do is turn the flywheel with the wrench to free the jam.

Another common method for freeing jammed impellers in disposers involves the use of a broom handle or similar tool. The broom handle is placed into the mouth of the disposer and lodged against one of the impeller blades. Then pressure is exerted on the blade as the handle is pried back to free the jam. The disposer should be turned off for all of these procedures, except for the use of a reversing switch.

In our second service call, you are called to a house where a disposer will not work. You turn on the switch and nothing happens. What would you do?

The first logical step to take is pushing the reset button that is located on the bottom of the disposer. Turn the disposer off and depress the reset button. Then turn the switch on and see if the disposer works. If it runs, test it several times to make sure it does not continue to trip the safety control.

What will you do if the disposer still fails to operate after using the reset button? Check to see if the impellers are jammed. For obvious reasons, never put your fingers or hands into the mouth of the disposer. Try turning the flywheel with a wrench or some long tool, like a broom handle. After rotating the flywheel, turn the disposer back on and see if it works.

If it is still dead, check the electrical panel to see if a fuse has blown or a circuit breaker has tripped. You'd be surprised how often this is the cause of the problem. Many homeowners, and commercial customers for that matter, never think to check their electrical panels before calling for help.

If the power to the disposer is on and it still won't run, you have two options. You can tell the customer the problem is not a plumbing problem and suggest that they call an electrician or appliance repair person. If you want to dig deeper, you can remove the wiring and test it to see that power is coming to the appliance. If full power is at the wires, the problem is likely to be with the disposer's motor.

Our next service call puts you in the position of troubleshooting foul odors being emitted from a disposer. The homeowner has used the disposer to gobble up everything from fish to fowl, and now the odors are permeating the house.

You arrive on the job and notice the smells as soon as you walk through the door. The odor is not that of sewer gas, but it certainly is unpleasant. What are you going to do first?

Are you thinking you should check the trap first? If you are, you're on the right track, sort of. If the trap is not maintaining its water seal, odors could be rising from the drain pipe. However, since the sink and disposer are being used regularly, it is unlikely the trap has lost its seal. Unless a vent has become obstructed and is allowing a siphonic action of the trap, there should be a good water seal. To be safe, you check the trap, and it is full of water, so the smells are not coming from the drain pipe. At this point you could tell the homeowner that the odor is coming from the appliance, not the plumbing and leave. This would be an acceptable course of action, but there is a better one available to you, if you know what to say. Do you?

Garbage disposers do sometimes become smelly beyond normal comfort zones. When this happens, there is a simple way to beat the smells. Fill the disposer about two-thirds full with ice cubes and cut it on. Run cold water into the disposer to flush it out. After the ice cubes have been ground up, slice a lemon in half and put it in the disposer. Cut the disposer on and allow it to devour the lemon. This home remedy works on almost any type of tough odor coming from disposers.

If a homeowner asks you whether it is best to flush a disposer with cold water or hot water, what would you say? Many people, plumbers included, would say hot water. In fact, I used to think hot water was the best to use when flushing a disposer. I found out, however, that cold water should be used. Why cold water? Cold water is said to congeal grease and carry it down the drain better than hot water. As I understand it, there is a higher likelihood of grease being caught on the sides of drainage piping if hot water is used.

Pure plumbing problems with disposers

Let's take a few moments to look at the pure plumbing problems often encountered with disposers. One of the most common is water leaking under the drain flange in the sink bowl. This is caused by a lack of putty under the flange. If this type of leak occurs on the back side of a disposer, the water from the leak can give the appearance of coming from somewhere else, especially from the trap. If you have a call where water is collecting beneath a sink bowl that is equipped with a disposer, check the drain flange carefully.

The lock-ring that holds a disposer in place can become loose, allowing water to leak. These occasions are rare, but they do sometimes happen.

Occasionally, the gasket between the disposer and the drain ell will leak. The leak could be a result of a bad washer or of the connection not being tight. A few quick turns of the lock-screw with a screwdriver will tell you if the problem is a loose connection or a bad washer.

It is also not uncommon for the slip nut that holds the discharge end of the drain ell in a trap to leak. This usually requires nothing more than tightening the slip nut, but a new slip-nut washer may be needed.

The mystery leak of a disposer

I'd like to tell you a story about the mystery leak of a disposer that perplexed two of my plumbers before the problem was solved. This situation happened a few years ago, and it is the only time I've ever heard of such a problem arising. Since this is a rare and unusual occurrence, I feel it warrants explanation.

My company received a call from a homeowner who was complaining of having water under her kitchen sink. A plumber was dispatched to the call, but he couldn't find the source of the leak. The plumber was licensed, but young and inexperienced. When he couldn't figure the problem out, a more seasoned plumber was dispatched.

The experienced plumber went through all the typical troubleshooting steps. After his best effort, he couldn't find the leak. He gave up and called me. I was out on the road and not too far away, so I went over to the house.

When I arrived, there was solid evidence that something was leaking into the base cabinet, under the sink. I conferred with my plumber and then began to troubleshoot the job.

I could see all the water connections, and none of them were even damp. I filled the left bowl of the sink with water and emptied it. Nothing leaked. Then I filled the right bowl, the one where the disposer was attached. The top to the disposer was missing, so I stuffed a rag in the opening to allow me to fill the sink. This was one step neither of my plumbers had taken.

When the right bowl was nearly full, I removed the rag and turned on the disposer. Guess what happened? The kitchen cabinet began flooding with water. The look on my plumber's face was worth the time I had spent on the call. He was somewhere between embarrassed and amazed. And I must admit, I was surprised to see what was happening. Do you know where the water was coming from?

The water was blowing out of the fitting where a dishwasher hose can be connected to a disposer. Apparently, someone had knocked out the drain plug in the side outlet of the disposer and never connected anything to it. Nor did they cap it. When the disposer was cut on, it forced water out of the uncovered drain outlet.

We capped the hole in the side of the disposer and the problem never recurred. My plumbers made two mistakes in their troubleshooting. First of all, they never bothered to turn the disposer on. Secondly, they didn't fill the bowl with the disposer to capacity and drain it. All they had done was run water down the drain from the faucet.

Even though I had no idea the water was coming out of the drain outlet, I took all of the right troubleshooting steps. I was surprised at the outcome of my test, but I was successful in finding and fixing the problem.

What is the moral to this story? It is simple, really. Take all the proper steps, don't take anything for granted, and you will probably solve any problem successfully.

Ice-Makers

Ice-makers are another good example of a situation in which the problem may not be in the plumbing. When someone's ice-maker fails to function properly, it is not unusual for them to call a plumber. From a pure plumbing point of view, the only aspect of an ice-maker that should concern a plumber is the connection at the back of the refrigerator, the tubing, and the connection to the cold-water pipe. These aspects of ice-makers can be troublemakers, but the problems are often within the ice-maker unit. When the problem is with the unit itself, the troubleshooting and repair should fall into someone else's field of expertise. This is not to say that some plumbers won't work on ice-maker units, but the units are not specifically plumbing.

If you, as a plumber, go poking around with an ice-maker unit without the knowledge and experience required to fix it, you could be buying yourself a lot of trouble. Suppose your dabbling winds up breaking the ice-maker? Guess who is going to have to pay to have it repaired or replaced? All I'm trying to say is that you should restrict your work to include only what you are an expert at. If you are not familiar with the workings of ice-makers, limit your work to the plumbing connections and tubing. There is nothing wrong with telling a customer that the problem is not plumbing related when the trouble is within the appliance.

Ice-makers are an appliance that many homeowners believe plumbers should know how to fix. Most plumbers, however, don't work on ice-makers. They make the necessary plumbing connections for the ice-makers but may not work on the unit itself. There are a few pure plumbing problems that can be associated with ice-makers, but more times than not, the trouble will be in the appliance, not in the plumbing.

What types of problems can arise for plumbers from ice-makers? Not too many, really. The saddle valve can create problems, and the tubing can leak, but that's really about all the pure plumbing involved with an ice-maker. Let's begin our short study of ice-makers with the saddle valve.

Saddle valves don't normally give plumbers much trouble. If the valve is not working properly, it is inexpensive enough to just replace it. Probably the most common complaint about saddle valves is their leaking at the packing nut. This is a simple problem, and one that is easily corrected by tightening the packing nut.

Sometimes a saddle valve will leak where the saddle comes into contact with the water distribution pipe. This is usually caused by the mounting screws being too loose or possibly by the gasket material being bad. Either cause is easily corrected.

The connection where the ice-maker tubing is attached to the refrigerator is another place where leaks are common. If the refrigerator is moved, say for cleaning purposes, the compression joint can become loose and leak. Again, this normally requires nothing more than tightening the nut.

If the tubing becomes kinked, the water flow to the ice-maker can be reduced or even cut completely off. This can also happen if the refrigerator is moved for any reason.

Plastic tubing sometimes gets cut when a refrigerator is being moved, and I've heard stories of plastic tubing just bursting for no apparent reason, though I personally have never seen that happen. Obviously, if the tubing ruptures or is cut, a serious leak will ensue.

When you are asked to troubleshoot an ice-maker problem, your first step should be to check the saddle valve. Make sure the valve is turned on. I don't know how these valves get cut off, but I've found several over the years that were in the closed position.

If the valve is open, pull the refrigerator out so that you can check the tubing connection to the appliance. Be careful not to tear the kitchen floor when you move the refrigerator.

Loosen the nut on the connection at the back of the refrigerator. If you begin getting a good spray of water, you know the appliance is receiving the water it needs to work. When this is the case, your job, as a plumber, is complete. You have proved that the appliance is being supplied with water, and that is all plumbers should be expected to do.

If for some reason there isn't any water pressure at the back of the appliance, backtrack to the saddle valve. Loosen the connection nut on the tubing at the saddle valve. If you have water at the valve and none at the appliance, you obviously have a problem in the tubing, probably a kink or possibly some type of obstruction.

If there is no water at the connection between the tubing and the saddle valve, you've either got a closed valve or a bad one. Try opening the valve, and if that doesn't work, replace it.

There really isn't much else to say about ice-makers, so let's move on to dishwashers.

Dishwashers

Dishwashers have become almost standard equipment in modern homes and commercial kitchens. While the problems that often occur with dishwashers are frequently appliance related, many people turn to plumbers for help. There are some problems associated with dishwashers that are pure plumbing problems, and many that aren't. We are going to concentrate on the pure plumbing problems and touch on the appliance-related problems.

The dishwasher won't drain

If you respond to a call where the dishwasher won't drain, what are you going to check first? There are several possibilities for why a dishwasher will not drain. Let's look at each of them individually.

Trap obstructions.   Trap obstructions are one of the first considerations when troubleshooting a dishwasher that will not drain. This is an easy defect to check for. Fill the kitchen sink bowl that uses the same drain as the dishwasher and release the water to see if the sink drains properly. If it does, you can rule out the trap. If it doesn't, disassemble the trap and investigate.

Drain-pipe obstructions.   Any drain-pipe obstructions will be found when you test the trap serving a dishwasher. If the sink using the same drain pipe that is being used by the appliance will not drain, you can count on the problem being in the drainage system. After inspecting the trap, you will know if the problem is in the trap or the drain pipe. If the pipe is clogged, you can snake it to get everything back in working order.

Air-gap obstructions.   Air-gap obstructions are rare, but they can prevent a dishwasher from draining properly. This type of condition will normally be noticed when water draining from the dishwasher spills out of the air gap. Running a piece of wire through the air gap should solve the problem.

Kinked drainage hoses. Kinked drainage hoses are another common cause of poor drainage from dishwashers. To check for this, you will have to remove the access panel on the front of the appliance. A visual inspection will tell you if the hose is crimped or not.

Plugged disposer drainage inlets. Plugged disposer drainage inlets are a surefire cause of water being unable to drain from a dishwasher. Some rookie plumbers, and a lot of do-it-yourselfers, don't know that a plug must be knocked out of a disposer drain inlet prior to connecting dishwasher drainage hosing to it.

I've been to more than a few jobs where dishwashers were flooding kitchens through the air gap because the knockout plug in a disposer hadn't been removed. To experienced plumbers, this is such a silly concept that it sometimes goes unchecked, but let me tell you, check it. Remember how I told you earlier not to take anything for granted? Well, this is one of those cases. Never assume the installer of the dishwasher did the job right.

Stopped-up strainer baskets. Stopped-up strainer baskets, on the inside of dishwashers, can also inhibit proper drainage. While this is not a pure plumbing problem, it is one you should check. If the strainer is clogged with grease, food, or other debris, a quick cleaning will get the appliance back in operation.

The water to the dishwasher won't cut off

What is likely to be the problem when you are faced with a dishwasher where the water won't cut off when filling the appliance? There are only two probable causes for this situation. Either the solenoid is bad or the inlet valve is malfunctioning. Neither of these fall into the pure plumbing category, but you should be aware of them and their symptoms.

If you choose to get involved with appliance repair, you may have to repair or replace the solenoid. If the problem is with the inlet valve, you will have to disassemble and clean it. You may even have to replace it.

No water

What's wrong when there is no water coming into a dishwasher? A common problem is a closed valve on the supply tubing. Check the supply valve and make sure it is in the open position. Assuming the valve is open, you may have a faulty solenoid, a blocked inlet screen, a crimped tubing, or on a long-shot, inadequate water pressure.

If the problem is in the solenoid, it will have to be repaired or replaced, but again, this is not pure plumbing work.

The screen in the water inlet valve may be clogged with mineral deposits. You can check this and clean the screen if necessary.

Low water pressure is so unlikely that it is hardly worth mentioning, but if you suspect the pressure in the plumbing system is too low, you can put a pressure gauge on a hose bibb to test it. As long as you have normal pressure, something in the range of 40 pounds per square inch, you should be all right.

Crimped tubing is not a likely cause when you are working with a dishwasher that has worked properly in the past, unless it was recently moved and reinstalled. Drop the front access panel and do a visual inspection of the tubing to be sure it is not crimped.

Isn't cleaning properly

What should you look for with a dishwasher that isn't cleaning dishes properly? There are five typical things to check. Water temperature is one of them. If the water being used in the dishwasher is not hot, it will not clean as well as it should. The water should be as hot as code allows. When possible, a temperature of 160 degrees F is desirable.

Hard water can make dishwashers appear to not be doing their jobs. The film and residue left by hard water often makes dishes appear to have not been cleaned well. If you suspect this to be the problem, test the water for hardness.

Water pressure has an affect on how well a dishwasher works. If the water pressure to the appliance is too low, the spray arm cannot do its job as well, and the water will not beat the dishes clean in the way it should. Low water pressure is rarely a problem, but it could be.

It is a good idea to check the spray arm to see that it is not jammed. Twirl it around by hand and make sure nothing is inhibiting its rotation.

While you are working with the spray arm, check its holes to see that they are not plugged up. If the holes are blocked, water cannot be distributed to clean the dishes properly. Holes that are blocked can be opened with a piece of wire.

Leaks

Leaks from dishwashers can come from several places. In addition to the expected plumbing leaks, water can leak past a faulty door seal. Inspect the door seal to see if it is worn, torn, or out of position.

Water leaks under dishwashers can come from the compression fitting at the dishwasher ell or from the threaded portion of the ell. It is also possible that the supply tubing may have a hole in it. These leaks can usually be found easily by removing the access panel and inspecting the areas with a flashlight.

Drainage leaks are also possible under dishwashers. If the hose clamps are loose, leaks are likely. In addition to loose clamps, the insert fitting on the discharge outlet of the dishwasher may become cracked or broken. This doesn't happen often, but it can occur.

Washing Machines

Washing machines don't present many troubleshooting challenges. There are, however, a few tips worth talking about.

Leaks around the hose connections of washing machines are usually caused by bad or missing hose washers. By turning off the water valves and removing the washing machine hoses, you can do a quick inspection of the washers. If you suspect they are causing a leak, replace them.

Have you ever been called out because a washing machine would not fill with water? If you haven't, you probably will at some time. The major cause of this problem is debris in the hose screens. The troubleshooting process is simple.

Cut off the water at the washing machine valves. Remove the hoses from the valves and place a bucket under the valves. Turn the valves on and see if you have good water pressure; you probably will. This tells you that the problem is somewhere in the hoses or in the machine itself.

Remove the hoses from the back of the washing machine. There will be cone-shaped screens in the inlet ports. Inspect the screens for debris. If the screens are blocked, clean or replace them. It is not uncommon for the screens to become clogged to the point that water pressure in the washing machine is either very low or nonexistent. This is a common problem that many inexperienced plumbers are baffled by.

While drainage leaks are not frequently a problem with washing machines, they can exist. If the drain hose is leaking, check the clamps where the hose connects to the back of the machine. Put the machine in a drain cycle and watch to see that the hose is not cracked. Cracks in rubber hose can be very difficult to see unless they are under water pressure.

Keep an eye on the washing machine drain receptor as the machine drains. If the standpipe is too short, water may be spilling out the top of the indirect waste pipe. This seems to happen most frequently when large amounts of detergent are used in the washing cycle.

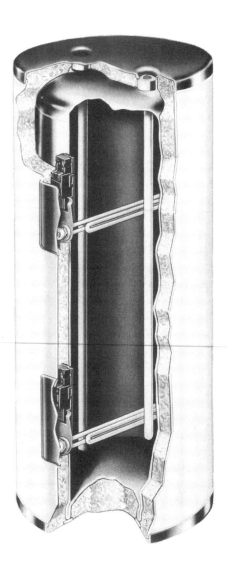

Figure 25.1 Cutaway of an electric water heater. (*Courtesy of A. O. Smith Water Products Co.*)

Electric Water Heaters

Electric water heaters (Fig. 25.1) can present a number of problems. Their thermostats can go bad (Fig. 25.2), the tanks can leak, the heating elements can burn out (Fig. 25.3), relief valves (Fig. 25.4) can pop off, and all sorts of related trouble can come up.

Gas and electric water heaters are very different in the ways they work. They both do the same job, but they don't do it the same way.

Figure 25.2 Upper thermostat for an electric water heater.

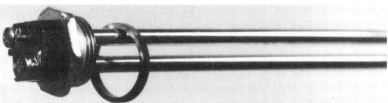

Figure 25.3 Screw-in heating element for an electric water heater.

Let's begin our troubleshooting lesson with electric water heaters and close it with gas units.

I think it goes without saying, but beware of the electrical wires and current involved when working with electric water heaters. As you probably know, there is a lot of voltage running through the wires of a water heater. Once the access cover of an electric water heater is removed, you must be extremely careful not to touch exposed wires and connections.

Relief valves that pop off. Relief valves that pop off signal one of three problems: the relief valve is bad, the water heater is building excess pressure, or the heater is building excess temperature. The problem

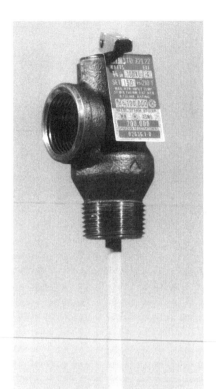

Figure 25.4 Temperature and pressure relief valve for a water heater.

is usually just a defective relief valve. In such cases, replace the relief valve and monitor it to see that the new valve works properly. If the new valve releases a discharge, investigate for extreme temperature or pressure in the tank.

The temperature of water in the heating tank can be measured with a standard thermometer. Discharge a little water from the relief valve into a container and test its temperature. If it is too high for the rating of the temperature-and-pressure relief valve, check the thermostat settings on the water heater. Turn the heat settings down and test the water again after the new temperature settings have had time to work. If the reduction on the thermostat settings does not lower the temperature of the water in the tank, replacement of the tank is usually the best course of action.

If you suspect the water heater is under too much pressure, you can test the pressure with a standard pressure gauge. The easiest way to do this is to adapt the gauge to a hose-thread adapter and attach it to the drain at the bottom of the water heater. As long as the drain is not clogged, you can get an accurate pressure reading. You could also adapt the gauge to screw into the relief valve and test the pressure by opening the relief valve.

No hot water

When no hot water is being produced by an electric water heater, there are only a few things that need to be checked. The first thing to do is to check the electrical panel to see that the fuse or circuit breaker for the water heater is not blown or tripped. For water heaters that have their own disconnect boxes, check the disconnect lever to see that it is turned on.

Assuming the water heater is receiving adequate electrical power, check the thermostats to see that they are set at a reasonable heat setting. It is highly unlikely that anyone would turn them way down, but it is possible.

The most likely cause of this problem is a bad heating element, but a bad thermostat could also be at fault. Check the continuity and voltage of these devices with a meter to determine if they should be replaced.

Limited hot water

When a water heater is producing only a limited amount of hot water, the problem is probably with the lower heating element. These elements frequently become encrusted with mineral deposits that, in time, reduce the effectiveness of the element. Check the element with your meter and replace it if necessary.

When a thermostat is set too low, a water heater will not produce adequate hot water. Check the settings on the thermostats, and check the thermostats with your meter.

Water that is too hot

Sometimes complaints come in that the hot water being produced is too hot. This is normally a simple problem to solve. Turning the settings down on the heater's thermostat will normally correct the situation. However, it is possible that the thermostat is defective. If it is, replace it.

Complaints of noise

Complaints of noise are sometimes made pertaining to water heaters. If a water heater is installed near habitable space, it is not uncommon for people to notice a rumbling in the heater. This often frightens people into calling a plumber. Do you know what causes the noise being made in the heater?

The noise is directly related to sediment that has accumulated in the water heater. Water becomes trapped between layers of the sediment, and when the water temperature reaches a certain point, the water explodes out of the layers. The miniexplosions can sound like a cracking or rumbling noise, and they should not be ignored. The steam explosions can be controlled by removing sediment from the water heater.

Sediment can be reduced in water heaters by opening the drain valve periodically. Ideally, water heaters should be drained monthly, but few are. If the buildup of sediment is extreme, a cleaning agent can be put into the water heater. The cleaning compound will reduce the scale buildup and the noise.

If you drain water from the heater and don't see any sediment, check the relief valve. It is possible the noise is coming from a steam buildup. When this is the case, you might need to replace the relief valve.

Rusty water

Rusty water can be a problem with some water heaters. If the hot water from fixtures is rusty, there is a good chance rust has accumulated in the water heater. In moderate cases the rust can be flushed out of the water heater through the drain opening. In extreme cases the entire heater should be replaced.

When the tank is leaking

When the tank is leaking (Fig. 25.5), you should plan on replacing the whole unit. It is possible to make temporary repairs for leaking tanks, but they are just that—temporary.

If you need to plug a hole temporarily, you can do it with a toggle bolt and a rubber washer. Drain the water heater to a point below the leak. Drill a hole in the tank that will allow you to insert the toggle bolt. The toggle bolt should have a metal washer against its head and a rubber washer that will come into contact with the tank. Once the toggle penetrates the tank and spreads out, tighten the bolt until the rubber washer is compressed. This will slow down or stop the leak for a while, but don't expect the repair to last indefinitely. Once a heater starts to develop pinholes, it is time to replace it.

That pretty well covers the range of problems associated with electric water heaters. As long as you know the components you are working with, water heaters are not difficult to troubleshoot (Figs. 25.6 through 25.11 and Table 25.1). Now let's turn our atten-

INSTRUCTIONS: USE THIS ILLUSTRATION AS A GUIDE WHEN CHECKING FOR SOURCES OF WATER LEAKAGE.
YOU OR YOUR DEALER MAY BE ABLE TO CORRECT WHAT APPEARS TO BE A PROBLEM.

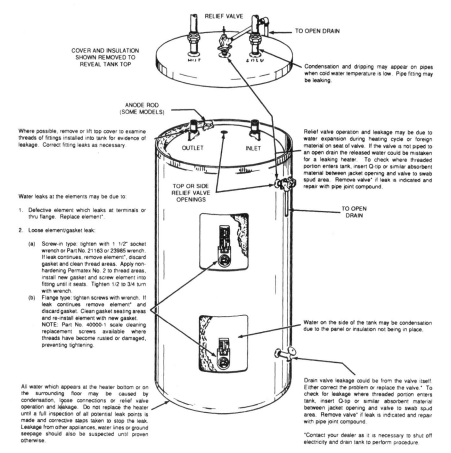

Figure 25.5 Leakage checkpoints for a water heater. (*Courtesy of A. O. Smith Water Products Co.*)

tion to gas-fired heaters and see how the troubleshooting methods differ.

Gas-Fired Water Heaters

Gas-fired water heaters (Fig. 25.12) share some of the same trouble symptoms that electric water heaters produce. However, there are many differences between the two types of heaters (Fig. 25.13). Let's look at the same problems that we studied for electric water heaters

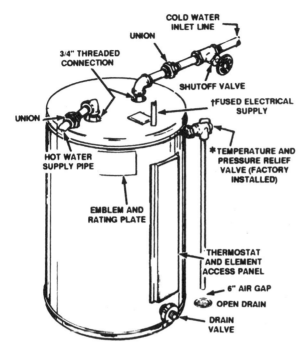

Figure 25.6   Electric water heater setup. (*Courtesy of A. O. Smith Water Products Co.*)

and see how your job will differ when troubleshooting gas units.

Relief valves

Relief valves on gas heaters can be tested and treated the same as those used on electric heaters.

No hot water

When you are getting no hot water from a gas-fired water heater, the first thing to check is the pilot light. If the pilot light is not burning, check the gas valve to see that it is turned on. If the gas valve is on and the pilot light is not burning, you must try to relight the pilot. However, before doing this, make sure that is not an accumulation of trapped gas that will explode when you light a flame.

Cut the gas valve off and ventilate the area well. When you are sure it is safe to light the pilot, turn the gas valve on and light the pilot. If the pilot will not light, make sure there is gas coming through the piping. You should be able to hear or smell it.

MODELS EDLJ/ELJF

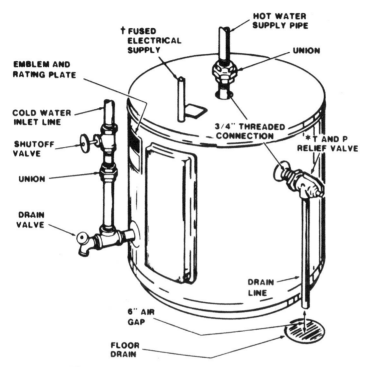

Figure 25.7 Electric water heater setup. (*Courtesy of A. O. Smith Water Products Co.*)

The thermostat may be turned off or defective. If the thermostat is set in the proper position but won't function, replace it.

The thermocoupling could also be bad. If the pilot light lights but continues to go out, replace the thermocoupling.

If none of these methods prove fruitful, check the dip tube. It should be installed in the cold-water side of the tank, and it is possible that it was put in the hot-water side by mistake.

Limited hot water

When a gas-fired water heater produces only limited hot water, the problem is usually with the gas control. Check the setting to see that it is set high enough to produce a satisfactory supply of hot water. If it is set properly but failing to work, replace it.

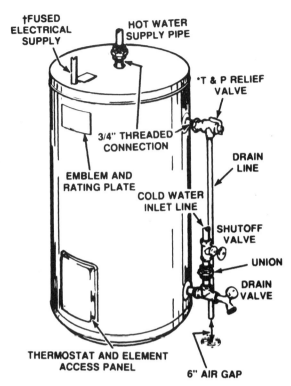

Figure 25.8 Electric water heater setup. (*Courtesy of A. O. Smith Water Products Co.*)

The dip tube could be responsible for the production of limited hot water. If it is installed in the hot-water side or is broken, you will have to either replace it or move it to the cold-water side of the tank.

Water that is too hot

When the water from a gas-fired water heater is too hot, the gas control valve is either bad or set too high. Check the setting and adjust it to a lower level. If that doesn't solve the problem, replace the valve assembly.

Noise, rusty water, and leaks in tanks

Noise, rusty water, and leaks in tanks can be treated the same for gas-fired heaters as for electric water heaters. Refer to the instructions given earlier for electric water heaters.

MODELS
ELJF-10
ELSF

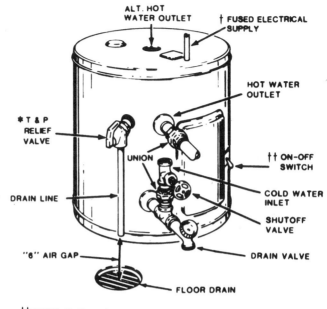

ALT. HOT WATER OUTLET

† FUSED ELECTRICAL SUPPLY

HOT WATER OUTLET

✱ T & P RELIEF VALVE

UNION

†† ON–OFF SWITCH

DRAIN LINE

COLD WATER INLET

SHUTOFF VALVE

"6" AIR GAP

DRAIN VALVE

FLOOR DRAIN

†† MODEL ELJF-10 ONLY

Figure 25.9 Electric water heater setup. (*Courtesy of A. O. Smith Water Products Co.*)

MODELS
ECTT/ETTN/ESTT

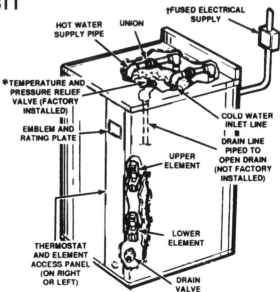

HOT WATER SUPPLY PIPE

UNION

†FUSED ELECTRICAL SUPPLY

✱TEMPERATURE AND PRESSURE RELIEF VALVE (FACTORY INSTALLED)

EMBLEM AND RATING PLATE

COLD WATER INLET LINE

DRAIN LINE PIPED TO OPEN DRAIN (NOT FACTORY INSTALLED)

UPPER ELEMENT

LOWER ELEMENT

THERMOSTAT AND ELEMENT ACCESS PANEL (ON RIGHT OR LEFT)

DRAIN VALVE

Figure 25.10 Tabletop-style water heater. (*Courtesy of A. O. Smith Water Products Co.*)

MODELS EEC/EEH/EES/EEST/PEC/PED/PEH/PEN

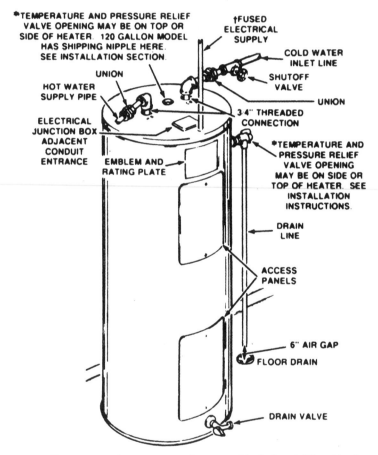

Figure 25.11 Electric water heater setup. (*Courtesy of A. O. Smith Water Products Co.*)

TABLE 25.1   Troubleshooting Guide for Electric Water Heaters

Complaint	Cause	Solution
Water leaks	Improperly sealed hot or cold supply connections, relief valve, or thermostat threads	Tighten threaded connections
	Leakage from other appliances or water lines	Inspect other appliances near water heater
Leaking t & p valve	Thermal expansion in closed water system	Install thermal expansion tank. (do not plug t & p valve)
	Improperly seated valve	Check relief valve for proper operation
Hot water odors (Caution: unauthorized removal of the anode(s) will void the warranty. For further information, contact your dealer.)	High sulfate or mineral content in water supply	Drain and flush heater thoroughly; refill
	Bacteria in water supply	Chlorinate water supply
Not enough or no hot water	Power supply to heater is not on	Turn disconnect switch on or contact electrician
	Thermostat set too low	Refer to temperature regulation
	Heater undersized	Reduce hot water use
	Incoming water is unusually cold water (winter)	Allow more time for heater to reheat
	Leaking hot water from pipes or fixtures	Check and repair leaks
	High-temperature limit switch activated	Contact dealer to determine cause; refer to temperature regulation
Water too hot	Thermostat set too high	Refer to temperature regulation
	High-temperature limit switch activated	Contact dealer to determine cause; see temperature regulation
Water heater sounds	Scale accumulation on elements	Contact dealer to clean or replace elements
	Sediment buildup in tank bottom	Drain & flush thoroughly

SOURCE: A. O. Smith Water Products Co.

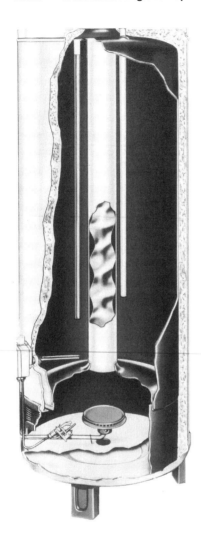

Figure 25.12 Cutaway of a gas
water heater. (*Courtesy of A. O.
Smith Water Products Co.*)

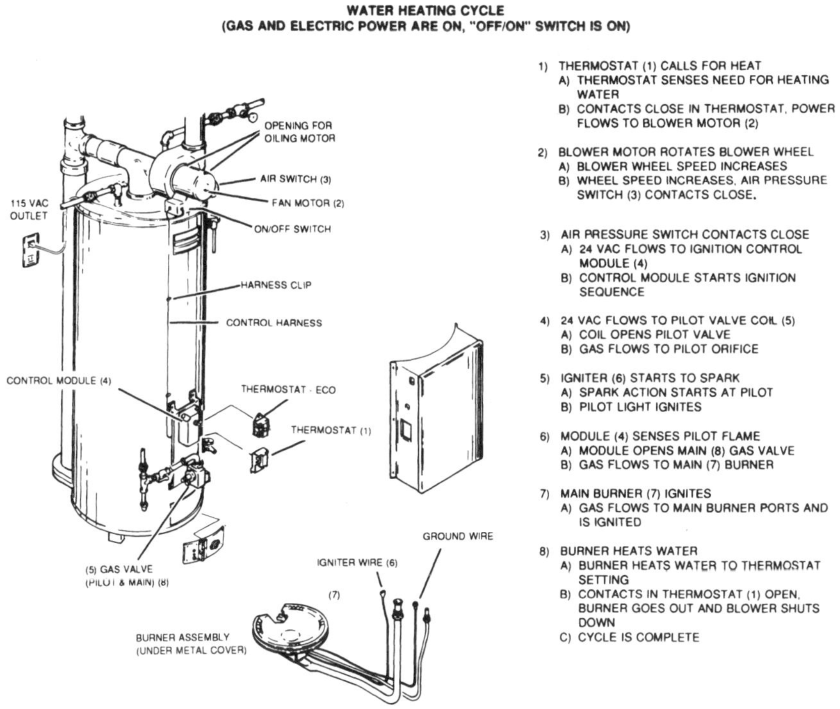

WATER HEATING CYCLE
(GAS AND ELECTRIC POWER ARE ON, "OFF/ON" SWITCH IS ON)

1) THERMOSTAT (1) CALLS FOR HEAT
   A) THERMOSTAT SENSES NEED FOR HEATING WATER
   B) CONTACTS CLOSE IN THERMOSTAT, POWER FLOWS TO BLOWER MOTOR (2)

2) BLOWER MOTOR ROTATES BLOWER WHEEL
   A) BLOWER WHEEL SPEED INCREASES
   B) WHEEL SPEED INCREASES, AIR PRESSURE SWITCH (3) CONTACTS CLOSE.

3) AIR PRESSURE SWITCH CONTACTS CLOSE
   A) 24 VAC FLOWS TO IGNITION CONTROL MODULE (4)
   B) CONTROL MODULE STARTS IGNITION SEQUENCE

4) 24 VAC FLOWS TO PILOT VALVE COIL (5)
   A) COIL OPENS PILOT VALVE
   B) GAS FLOWS TO PILOT ORIFICE

5) IGNITER (6) STARTS TO SPARK
   A) SPARK ACTION STARTS AT PILOT
   B) PILOT LIGHT IGNITES

6) MODULE (4) SENSES PILOT FLAME
   A) MODULE OPENS MAIN (8) GAS VALVE
   B) GAS FLOWS TO MAIN (7) BURNER

7) MAIN BURNER (7) IGNITES
   A) GAS FLOWS TO MAIN BURNER PORTS AND IS IGNITED

8) BURNER HEATS WATER
   A) BURNER HEATS WATER TO THERMOSTAT SETTING
   B) CONTACTS IN THERMOSTAT (1) OPEN, BURNER GOES OUT AND BLOWER SHUTS DOWN
   C) CYCLE IS COMPLETE

Figure labels: OPENING FOR OILING MOTOR, AIR SWITCH (3), 115 VAC OUTLET, FAN MOTOR (2), ON/OFF SWITCH, HARNESS CLIP, CONTROL HARNESS, CONTROL MODULE (4), THERMOSTAT - ECO, THERMOSTAT (1), GROUND WIRE, (5) GAS VALVE (PILOT & MAIN) (8), IGNITER WIRE (6), (7), BURNER ASSEMBLY (UNDER METAL COVER)

Figure 25.13 Heating cycle for a water heater. (*Courtesy of A. O. Smith Water Products Co.*)

Gas odors

Gas odors are sometimes noticed around gas-fired water heaters. This can be a dangerous situation. The problem is usually a leak in a fitting, pipe, or piece of tubing. Use soapy water on all places where a leak might occur to find the source of the smell. The soapy water will bubble when it is applied to the leaking location.

If all of the connections, pipe, and tubing check out to be okay, inspect the venting of the water heater. Inadequate or improper venting could cause gas smells to be trapped around the heater.

We have now concluded our look at troubleshooting plumbing appliances, and we are ready to move onto the troubleshooting of wastewater pumps.

26

Troubleshooting
Waste-Water Pumps

Waste-water pumps come in all shapes and sizes. The jobs they do are as varied as the configurations in which they are made. Some waste-pumping systems are equipped with high-water alarms, and some are so simple that almost anyone can install and operate them.

When you are called out to troubleshoot a waste-water pumping station, you could be faced with a seized impeller, a float ball that has stuck in one position, a check valve that refuses to close, a gate valve that has been closed, or any number of other possible problems.

The troubleshooting of waste-water pumps and the systems surrounding them is not particularly complicated. From a pure plumbing point of view, there are relatively few things that will impede the smooth operation of such systems. As with all troubleshooting, a sensible plan of attack is usually all that is necessary to make short work of finding pump problems.

When a customer's waste-water pump fails to operate properly, the customer can be quite panicked. A homeowner who is suddenly startled by a pump-station alarm going off can be difficult to calm, and business owners who are watching water levels in their sump basins rise tend to be short tempered. For a plumber, this is not a good time to be too busy to soothe the customer or to respond quickly to the call for help.

The failure of a waste-water pump can bring a business to a complete stop, and a problem pump can cripple a household. Some plumbing systems rely on their pump stations to evacuate all waste from the system; others only depend on pumps for individual fixtures or special uses.

Most pumps used to handle waste water are of the submersible type. There are, however, a large number of pedestal sump pumps in operation. Depending upon the usage of the pump, it may be small enough to mount on the drain of a laundry tub or large enough to require an extra set of hands to remove it from its basin.

Pumps that have integral floats tend to be less troublesome than those with external floats. External float frequently get stuck in either the pump-on or the pump-off position. This is a simple problem to solve, but it is an annoying one to have.

Where are waste-water pumps used? They are used in both residential and commercial applications. They can be used to facilitate the draining of washing machines, laundry tubs, rainwater piping, or entire plumbing systems.

There are five types of waste-water pumps that we are going to examine in this chapter; they are submersible sump pumps, pedestal sump pumps, sewage ejector pumps, basin waste pumps, and grinder pumps. Let's begin our journey into waste-water pumps with a discussion of submersible sump pumps.

Submersible Sump Pumps

Submersible sump pumps generally work very well, but there are times when they fail to perform properly. The most obvious difference between various types of submersible sump pumps is the location of their floats. Some styles have their floats concealed within the pump housing (Fig. 26.1). Integral floats are desirable since they are unlikely to become stuck or jammed by foreign objects.

Some pumps have their floats mounted on a metal rod that controls the path of the float. The float operation on these pumps is not quite as reliable as it is for pumps with integral floats, but the dependability is still good.

A third style of pump has a float on an arm (Fig. 26.2). This type of float sometimes gets stuck against the side of the pump basin. If the pump is installed too close to the basin, the float is sure to cause problems.

A fourth style of pump uses an independent float, one that is not put on the pump. These floats are often attached to the discharge pipe of the pump, but some units are attached to a special clip that is mounted on the pump housing. The float cable provides potential for problems, with the float becoming stuck in an undesirable position.

The normal size of the discharge outlet on a sump pump is $1\frac{1}{4}$ inches. Sometimes the discharge line is run with flexible $1\frac{1}{4}$-inch pipe, and sometimes it is piped with $1\frac{1}{2}$-inch pipe. Larger sump pumps can have discharge outlets up to 2 inches in diameter. The final discharge location for the drain of a sump pump can be hard to predict.

Storm water is not usually supposed to be piped into a sanitary

▼ Sump, Sewage, Effluent and Grinder Pumps

A.Y. McDonald Mfg. Co. offers a full line of Sump, Effluent, Ejector and Grinder Pumps for home and light industrial applications.

Sump Pumps - featuring high torque capacitor-start motors. Rustproof construction, clog proof vortex impeller, easy to service. Available in 1/4, 1/3, 1/2 and 1 1/2 H.P. models.

Sewage Ejectors and Effluent Pumps - capable of handling 3/4", 2" and 2 1/2" solids, available in 4/10 through 3 H.P. models for residential and light industrial applications.

Grinder Pumps - for disintegrating solids. For pressure sewer systems, restaurants, and industry. Replaces systems that need screening or occasionally clog. Available in a convenient 2 H.P. packaged system.

Sump Pumps

Features:

- •Rust & corrosion proof
- •Built-in check valve
- •Energy efficient
- •Water cooled
- •Capacitor start - high torque
- •Built-in thermal protection
- •Automatic & manual switch

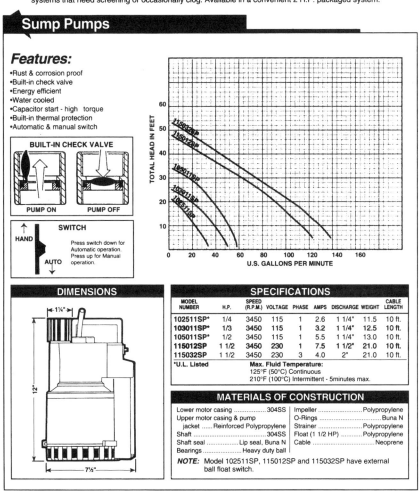

DIMENSIONS

SPECIFICATIONS

MODEL NUMBER	H.P.	SPEED (R.P.M.)	VOLTAGE	PHASE	AMPS	DISCHARGE	WEIGHT	CABLE LENGTH
102511SP*	1/4	3450	115	1	2.6	1 1/4"	11.5	10 ft.
103011SP*	1/3	3450	115	1	3.2	1 1/4"	12.5	10 ft.
105011SP*	1/2	3450	115	1	5.5	1 1/4"	13.0	10 ft.
115012SP	1 1/2	3450	230	1	7.5	1 1/2"	21.0	10 ft.
115032SP	1 1/2	3450	230	3	4.0	2"	21.0	10 ft.

*U.L. Listed   **Max. Fluid Temperature:**
125°F (50°C) Continuous
210°F (100°C) Intermittent - 5minutes max.

MATERIALS OF CONSTRUCTION

Lower motor casing 304SS	Impeller ....Polypropylene
Upper motor casing & pump	O-Rings ....Buna N
jacket ......Reinforced Polypropylene	Strainer ....Polypropylene
Shaft 304SS	Float (1 1/2 HP) ....Polypropylene
Shaft seal ....Lip seal, Buna N	Cable ....Neoprene
Bearings ....Heavy duty ball	

NOTE: Model 102511SP, 115012SP and 115032SP have external ball float switch.

Figure 26.1 Internal-float, submersible sump pump. (*Courtesy of A. Y. McDonald Mfg. Co.*)

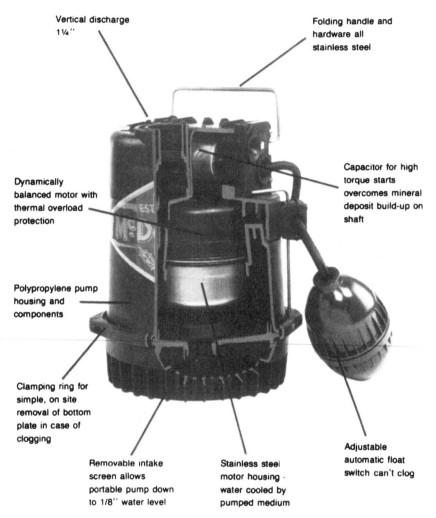

Vertical discharge 1¼"

Folding handle and hardware all stainless steel

Dynamically balanced motor with thermal overload protection

Capacitor for high torque starts overcomes mineral deposit build-up on shaft

Polypropylene pump housing and components

Clamping ring for simple, on site removal of bottom plate in case of clogging

Removable intake screen allows portable pump down to 1/8" water level

Stainless steel motor housing · water cooled by pumped medium

Adjustable automatic float switch can't clog

Figure 26.2   External-float, submersible sump pump. (*Courtesy of A. Y. McDonald Mfg. Co.*)

sewer, but it often is. The discharge from a sump pump might dump into a storm sewer or a gravel-lined hole or might even just run out on the top of the ground.

Check valves are needed when sump pumps are installed; without them, the pumps would run continuously. When the pump finishes its pumping cycle, water between the pump and the point where gravity drainage begins is trapped in the vertical pipe. Without a check valve, this water would run back into the sump and force the pump to cut back on. This on-and-off action would destroy the pump's motor. The check valves used for sump pumps may be installed in the discharge piping, or they may be built into the pump.

Some sump pumps are equipped with both manual and automatic operation switches. The automatic mode is the normal setting when the pump is installed. The voltage requirement for average sump pumps is 115 volts. Larger models require 230 volts.

The volume of water that can be pumped and the vertical distance that can be obtained from a sump pump depends on the size of the pump's motor.

A small sump pump, one with a $\frac{1}{4}$-horsepower motor, can pump approximately 33 gallons per minute with a vertical head of about 20 feet. The same type of pump with a $\frac{1}{3}$-horsepower motor will move about 50 gallons of water per minute with a head of about 25 feet. A $\frac{1}{2}$-horsepower motor on the pump will produce a volume of about 58 gallons per minute with a head of about 32 feet. A large sump pump, one with a $1\frac{1}{2}$-horsepower motor, can deliver about 140 gallons per minute at a head of about 52 feet. Now that we have a good overview of what submersible sump pumps are, let's take a look at what types of problems are often associated with them.

No check valve

A sump pump with no check valve is going to cycle on and off frequently, if not constantly. If you have a sump pump that is cutting on and off too often, investigate the check valve. You may find that one was never installed, or you might discover that the check valve is stuck in the open position. In either case, the pump will run too often.

If there is no check valve present, you should install one. If the existing check valve is stuck in the open position, you may be able to repair it, or you may have to replace it.

Closed check valve

A closed check valve will prohibit the pump from emptying its sump. If a check valve becomes stuck in the closed position, water cannot move through the drain pipe. This type of problem usually calls for the replacement of the check valve, but there are times when the flapper can be manually opened and repaired.

Frozen drain pipes

Frozen drain pipes can prevent sump pumps from doing their jobs. It is not uncommon for the drain lines from sump pumps to be run in a less-than-professional manner. This is especially true when the drain is run with flexible hose-type pipe.

If the drain from a sump pump is run on top of the ground or in a shallow trench, it is subject to freezing during cold temperatures. If the pipe does not have a good grade on it or if the pipe loops downward in spots, freezing becomes more likely.

If you are faced with a pump that runs but won't pump water, it is usually a problem with the check valve, but it could be that the drain pipe has frozen.

You can confirm if the problem is with the pump by disconnecting the drain pipe and running the pump. With the pipe disconnected, you will be able to see if the pump is producing water.

Clogged strainers

Clogged strainers routinely cause sump pumps to fail. When you have a pump that is not pumping properly, pull the pump and inspect the strainer on the bottom of the inlet opening. It is not uncommon for the strainer to be blocked with leaves, paper, or all kinds of other strange objects that shouldn't be in a sump.

If the pump you are working on does have a clogged strainer, all you should have to do is clean it. Once the strainer is clear, the pump should work correctly.

Jammed impellers

Jammed impellers can stop a pump dead in its tracks. If a pebble or some other object finds its way into the impellers, the pump can stop pumping. The motor will run, but no water will be produced. When this is the case, you must unplug the pump, gain access to the impellers, and clear them.

Floats that are stuck

Floats that are stuck can cause two different types of problems. If a float becomes stuck at its upper level, the pump it is serving will not cut off. When the float sticks at its lower level, the pump will not cut on. Both of these problems are serious and require fast attention.

Floats can be jammed by objects floating in the sump, but they most often stick without apparent cause, except for when they rub against the sump and become stuck.

The vibrations that go through a sump pump when it runs are enough to make the pump walk across the sump pit. Unless the pump is piped into place with rigid pipe, it can creep across the sump until the float begins to hit the side of the container. This results in a float that sticks and a pump that doesn't work properly.

Unplugged electrical cables

Unplugged electrical cables account for some pump problems. This situation may seem too simple to even address, but sometimes pumps become unplugged, and they don't run well without electricity. When you have a pump that won't run at all, check to see that it is plugged in and that the fuses or circuit breakers are in good condition and in the proper positions.

Pedestal Sump Pumps

Pedestal sump pumps are very common in residential properties. These pumps are inexpensive and do a pretty good job of handling modest pumping demands. The major difference in a pedestal pump and a submersible pump is that the motor of a pedestal pump is above the water level. Other than this, a pedestal sump pump can be treated like a submersible pump for troubleshooting purposes.

Sewage Ejector Pumps

Sewage ejector pumps are ones nobody wants to have trouble with. When these pumps fail to do their jobs, working conditions can get messy. Properties that depend on ejector pumps to evacuate their sewage cannot afford to have their pumps fail.

Sewage ejector pumps (Fig. 26.3) are not exactly like sump pumps, but there are similarities. For example, the impellers on a sewer pump can become jammed, just like those in a sump pump. Check valves that are used for sewer pumps can fail in the same ways as described for sump pumps. There are, however, a number of differences between the installation and operation of a sewer pump and a sump pump.

Standard sewer ejectors have 2-inch discharge outlets and are capable of handling 2-inch solids. Some plumbers install effluent pumps for use in draining sewage sumps. This is a mistake. Effluent pumps typically have 2-inch discharge outlets, but they are only rated to handle $\frac{3}{4}$-inch solids. If an effluent pump is installed for use as a sewage ejector, the size of the solids that need to be pumped can create problems. When you are troubleshooting a sewage ejector, make sure that it is, in fact, a sewage ejector and not an effluent pump.

A sewer ejector with a $\frac{1}{2}$-horsepower motor is usually all that is needed for most pump stations. These pumps can push about 175 gallons a minute up to a height of nearly 24 feet. These pumps are available in either 115 or 230 volts. The electrical cables supplied on sewage ejectors are typically about 15 feet long.

▼ *Sump, Sewage, Effluent and Grinder Pumps*

Effluent Pumps & Sewage Ejectors

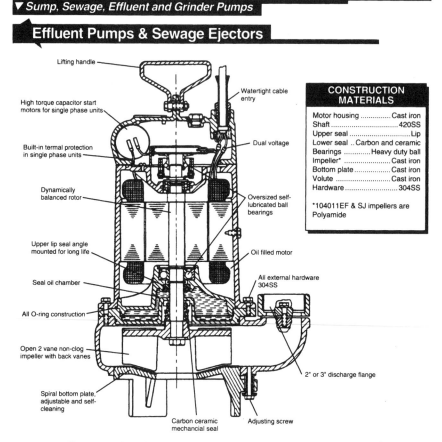

Lifting handle

High torque capacitor start
motors for single phase units

Built-in termal protection
in single phase units

Dynamically
balanced rotor

Upper lip seal angle
mounted for long life

Seal oil chamber

All O-ring construction

Open 2 vane non-clog
impeller with back vanes

Spiral bottom plate,
adjustable and self-
cleaning

Watertight cable
entry

Dual voltage

Oversized self-
lubricated ball
bearings

Oil filled motor

All external hardware
304SS

2" or 3" discharge flange

Carbon ceramic
mechanical seal

Adjusting screw

CONSTRUCTION MATERIALS	
Motor housing	Cast iron
Shaft	420SS
Upper seal	Lip
Lower seal	Carbon and ceramic
Bearings	Heavy duty ball
Impeller*	Cast iron
Bottom plate	Cast iron
Volute	Cast iron
Hardware	304SS

*104011EF & SJ impellers are Polyamide

Figure 26.3   Cutaway of a submersible effluent pump. (*Courtesy of A. Y. McDonald Mfg. Co.*)

Sewage ejector pumps are available with 3-inch discharge flanges, but a 2-inch flange is more common. Depending on the brand of pump, the float can be a part of the pump, or it can be an independent unit that plugs in with the pump through a piggyback plug. Unfortunately, the piggyback plug type of float sometimes gets tangled or stuck on the side of the sump (Fig. 26.4). Let me give you a real-world example that I faced recently.

I was at my insurance agent's office setting up a new liability policy. As we talked, my agent asked me some questions about his sewage ejector system. He was complaining that his sewer pump was cutting on too often. I made an appointment with him to check it out.

When I got to my customer's house, I flushed the toilet that was connected to the sump. The toilet was a 1½-gallon model that should

COMPLETE BASIN

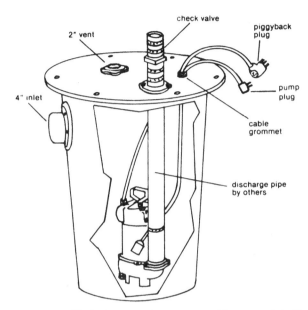

Figure 26.4 Typical sewer ejector setup. (*Courtesy of A. Y. McDonald Mfg. Co.*)

not have triggered the pump, unless the sump was already holding water.

When I flushed the toilet, the pump cut on. The pump should have reduced the level in the sump to where the next flush would not make the pump cut on. I flushed the toilet again, and the pump cut on immediately. This told me that there was a problem with the float.

I removed the cover from the sump, and sure enough, the float was out of position. The float cable was attached too low on the pipe in the sump. Because of its low position, the pump was forced to cut on sooner than it should.

I removed the float cable and reinstalled it. Once I had the cable tie holding the float cable in place, I reconnected the piping and replaced the sump cover. We flushed the toilet numerous times, and the problem was solved.

On a different job I was faced with the opposite problem. In this case the sewer pump wasn't cutting on soon enough. The customer knew the pump wasn't working right, so she was using the manual override to drain the sump periodically. Once she got tired of working the pump manually, she called me.

When I removed the cover from her sump, I found that the float was hitting the side of the sump and failing to rise to a level that would cut the pump on. I repositioned the cable on the float and the problem was solved.

These are just two of the common situations that plumbers run into when working with sewer ejectors. Gate valves can be closed, check valves can stick open or closed, impellers can be jammed, and all sorts of other problems can arise. Let's take some time now to look at various problems you may run into with sewer ejectors.

Vents

All sewer sumps should have vents that rise to open air, but not all do. I've seen sewer sumps in basements that were vented with only mechanical vents. Not only are these types of installations in violation of the plumbing code, they can cause real problems.

The contents of sewer sumps can produce some serious odors. Gas in the sump is another consideration. If the sump is not vented properly, gas can accumulate in it and present some potential dangers.

Mechanical vents do a good job of helping individual fixtures drain better, but they don't do much for sewer sumps. If you run across a sewer sump that is not properly vented to outside air, advise your customer of the danger and the need to vent it properly.

Gate valves

Gate valves are a part of every sewage ejector system. Normally, gate valves don't give people much trouble, but sometimes they get closed inadvertently. If a gate valve on a sewage ejector system is closed, the pump cannot empty the contents of its sump, and this is a serious problem. If you respond to a call where a sewer pump runs but doesn't seem to be pumping, check the gate valve to confirm that it is open.

Check valves

Check valves used with sewage ejectors are subject to the same problems as those described for sump pumps. Closed check valves don't allow sewer pumps to pump the contents from their sumps, and open check valves can make a pump cycle too frequently.

Floats

Floats on sewer ejectors are also subject to the same basic working principles as those found on sump pumps.

Impellers

The impellers in sewer pumps are subject to clogging and jamming. If improper objects are flushed down toilets that are connected to sewer sumps, impellers can have big problems. Moist towelettes, sanitary napkins, and similar items should never be flushed into a sump system; they can jam impellers very quickly.

Tank gaskets

Tank gaskets are an aspect of sump systems that many plumbers fail to give much credit to. These seals are responsible for keeping gas and odors confined in the sump and in the waste and vent system. If the seals deteriorate, sewer gas and odors can escape into living space. This is not only an unpleasant experience, but it can be a health hazard.

If you have a customer complaining about odors in their living space, check the gaskets on the sewer sump. You may be surprised to find that the plumber who installed the system never installed the tank-cover gasket. When the tank-cover gaskets and rubber grommets are not in good shape, odors and gas are able to escape into living space.

Basin Waste Pumps

Basin waste pumps (Figs. 26.5 and 26.6) work on the same basic principle as that used by sewer ejectors. By that, I mean that basin waste pumps are installed in sumps, are submersible, and pump water from the basin to a gravity-type drain.

Basin pumps typically have motors with $\frac{1}{3}$ horsepower, are installed in a 5-gallon sump, have an $1\frac{1}{2}$-inch inlet and a 2-inch vent. The discharge outlet for these pumps is usually of a 2-inch diameter. The electrical cables on these pumps are typically 10 feet long.

These small pumps are not short on pumping power. For example, a typical basin pump can pump 1050 gallons of water in 1 hour with a 20-foot lift, and that's a lot of water.

Basin-type pumps are used to handle water discharge from laundry trays, wet bars, water softeners, dehumidifiers, and washing machines, among other gray-water fixtures.

Basin pumps, or sink-tray pumps as they are often called, are normally installed above floor level. Sewer ejectors are installed below floor level, but these gray-water pumps can be installed in base cabinets or other above-floor levels.

Grinder Pumps

Grinder pumps (Fig. 26.7) are used for big sewage removal jobs. These pumps are powerful and can easily handle tough drainage jobs.

▼ *Sump, Sewage, Effluent and Grinder Pumps*

Effluent Pumps & Sewage Ejectors

DIMENSIONS

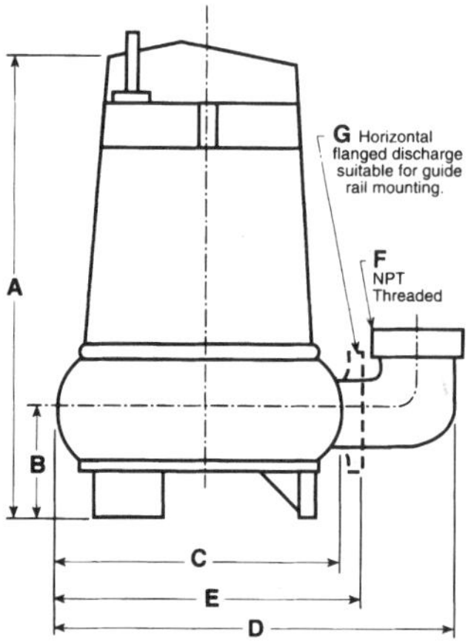

G Horizontal flanged discharge suitable for guide rail mounting.

F NPT Threaded

*U.L. Listed, C.S.A. Approved

	MODEL NUMBER	A	B	C	D	E	F	G
EFFLUENT PUMPS	104011EF*	11	2 1/4	7 3/4	9	9 1/4	2	—
	105011EF*	14 1/2	2 1/4	8	11	9 1/4	2	1 1/4
	110012EF	13 3/4	2 1/4	8	11	9 1/4	2	1 1/4
	110032EF	12	2 1/4	8	11	9 1/4	2	1 1/4
	120012EF	15 3/4	2 1/4	8	11	9 1/4	2	1 1/4
	120032EF	14 1/2	2 1/4	8	11	9 1/4	2	1 1/4
STANDARD SEWAGE EJECTORS	104011SJ*	14	2 3/4	8 1/4	10	10 1/2	2	—
	105011SJ*	17	3 3/4	9	12	10 1/2	2	2
	105032SJ	14 1/2	3 3/4	9	12	10 1/2	2	2
	110012SJ	16 1/4	3 3/4	9	12	10 1/2	2	2
	110032SJ	15	3 3/4	9	12	10 1/2	2	2
	115012SJ	18 1/2	3 3/4	9	12	10 1/2	2	2
	120032SJ	17 1/4	3 3/4	9	12	10 1/2	2	2
ENGINEERED SEWAGE EJECTORS	110012EP	16 1/4	4 3/4	9 1/4	15 1/2	10 1/2	—	3
	110032EP	16	4 3/4	9 1/4	15 1/2	10 1/2	—	3
	115012EP	19 1/4	4 3/4	9 1/4	15 1/2	10 1/2	—	3
	120032EP	18 1/4	4 3/4	9 1/4	15 1/2	10 1/2	—	3
	130032EP	18 1/4	4 3/4	9 1/4	15 1/2	10 1/2	—	3

SPECIFICATIONS

	MODEL NUMBER	RATED H.P.	SPEED R.P.M.	SOLID SIZE INCHES	OPERATING VOLTAGE	HERTZ	PHASE	DISCHARGE CONNECTION INCHES	WEIGHT	CABLE LENGTH FT.
EFFLUENT PUMPS	104011EF	4/10	3450	3/4	115 or 230	60	1	2	35	15
	105011EF	1/2	3450	3/4	115/230	60	1	2	61	15
	110012EF	1	3450	3/4	230	60	1	2	66	15
	110032EF	1	3450	3/4	230/460	60	3	2	54	15
	120012EF	2	3450	3/4	230	60	1	2	75	15
	120032EF	2	3450	3/4	230/460	60	3	2	73	15
STANDARD SEWAGE EJECTORS	104011SJ	4/10	1750	2	115 or 230	60	1	2	40	15
	105011SJ	1/2	1750	2	115/230	60	1	2	66	15
	105032SJ	1/2	1750	2	230/460	60	3	2	61	15
	110012SJ	1	1750	2	230	60	1	2	78	15
	110032SJ	1	1750	2	230/460	60	3	2	68	15
	115012SJ	1 1/2	1750	2	230	60	1	2	87	15
	120032SJ	2	1750	2	230/460	60	3	2	83	15
ENGINEERED SEWAGE EJECTORS	110012EP	1	1750	2 1/2	230	60	1	3	83	15
	110032EP	1	1750	2 1/2	230/460	60	3	3	76	15
	115012EP	1 1/2	1750	2 1/2	230	60	1	3	88	15
	120032EP	2	1750	2 1/2	230/460	60	3	3	85	15
	130032EP	3	1750	2 1/2	230/460	60	3	3	92	15

Figure 26.5 Specifications for effluent pumps and sewage ejectors. (*Courtesy of A. Y. McDonald Mfg. Co.*)

These pump systems are typically rigged with an alarm system and a total of three floats. Voltage requirements for an average grinder pump is normally 230 volts.

▼ *Sump, Sewage, Effluent and Grinder Pumps*

Sink Tray System

For maximum results where gravity drainage is not possible or practical

The "Sink Tray System" is specifically designed for residential use in pumping waste water from washing machines, laundry trays, dehumidifiers, wet bars, water softeners, etc. The system comes fully assembled and installs above the floor eliminating the need to dig a sump.

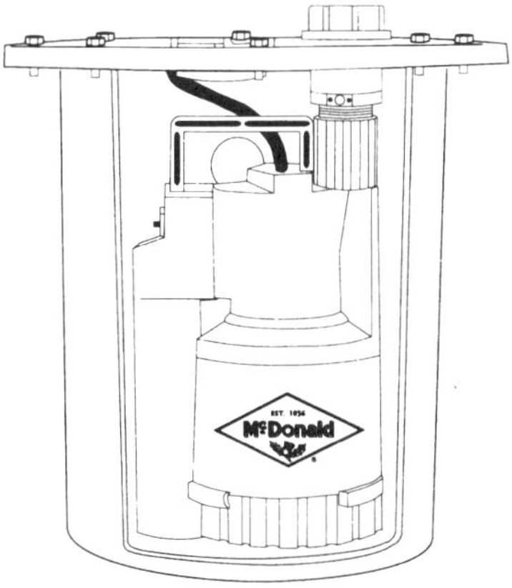

Special Features and Benefits

• Fully assembled, lightweight, self-contained and compact for easy installation.

• Installs above the floor.

• Pump includes internal float switch and check valve for trouble-free automatic operation.

• Full 1/3 HP pump for up to 1/2" solids handling.

• 10 ft. power cord with 3 prong grounded plug.

• 5 gallon corrosion resistant polyethylene tank with 1 1/2" inlet and 1 1/2" discharge compatible with all standard plumbing fittings.

• Fully vented basin for safe operation.

• Basin design ensures proper sealing for gas tight assembly.

• Quick draining action-pumps 1050 gallons of water per hour at 20 feet of lift; 2250 gallons per hour at 10 feet of lift.

Eliminates excess water from:

• Laundry trays   • Wet bars
• Dehumidifiers   • Water softeners
• Washing machines

Package Includes:

• 1/3 HP automatic totally submersible sump pump
• 5 gallon basin
• Lid
• Complete hardware package

TANK SPECIFICATIONS	
Gallons 5	Height .... 15"
Inlet .... 1 1/2"	Width .... 15"
Vent .... 2"	Weight .... 16 lbs.
Discharge .... 1 1/2"	

PUMP SPECIFICATIONS	
HP 1/3	Maximum fluid temp.
Voltage 115	...... 125° Continuous
Amp draw 2.3	...... 210° Intermittent
Cable length .... 10 ft.	

103011SPK Performance Curve

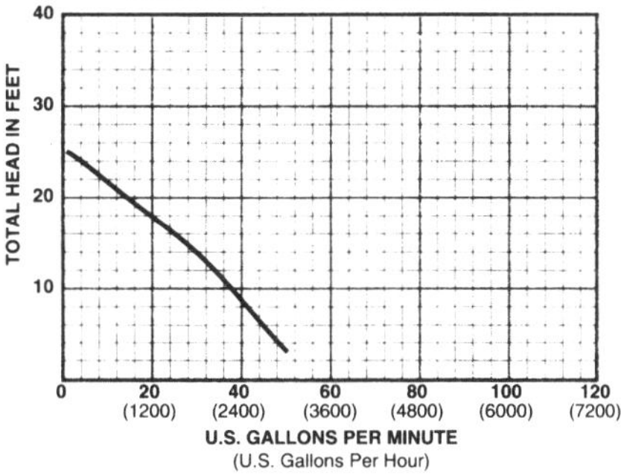

TOTAL HEAD IN FEET

U.S. GALLONS PER MINUTE
(U.S. Gallons Per Hour)

Figure 26.6 Typical basin pump setup. (*Courtesy of A. Y. McDonald Mfg. Co.*)

The discharge outlet for a typical grinder pump has a diameter of $1\frac{1}{4}$ inches, and these pumps are usually rated at 2 horsepower. In addition to the normal sump setup, grinder pumps are equipped with

Grinder Pumps

The **A.Y. McDonald Mfg. Co. Grinder Pump** comes as a completely packaged simplex system. It is designed for residential and small industrial sewage or sump applications. The pump is recommended for homes in isolated or mountainous areas, and for dewatering of dwellings located in inland protected areas where septic tanks are not permitted. System includes pump, tank, cover, check valve, discharge piping, control box and float switches.

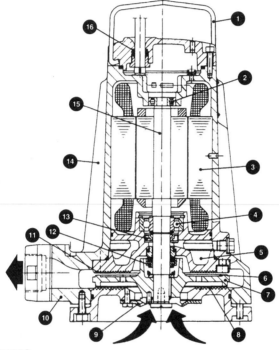

FEATURES

1. Lifting handle
2. Self-lubricating ball bearing-needs no servicing
3. Stator, insulated against heat and humidity to class F (155°C)
4. Oversized single row ball bearing
5. Oil chamber for lubrication and cooling the seal assemblies
6. Back vanes on impeller
7. Dynamically balanced impeller
8. Adjustable spiral bottom plate for handling fibrous material
9. Patented hardened rotor and stator cutter elements (Rockwell C 58-62)
10. Volute with centerline discharge suitable for mounting to guide rail bracket or discharge elbow
11. Spiral back plate
12. Mechanical lower seal enclosed in Buna N boot
13. Upper lip seal angle mounted for long life
14. Motor housing with large cooling fins
15. Rotor shaft assembly dynamically balanced
16. Watertight cable joint with strain relief

Figure 26.7 Sewage-ejector setup with alarm system. (*Courtesy of A. Y. McDonald Mfg. Co.*)

a control box. The control box is frequently equipped with a control panel, hand-off automatic switch, terminal strip, and audible alarm (Fig. 26.8).

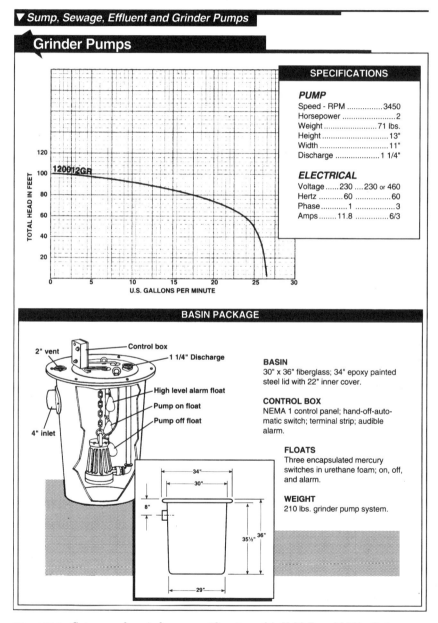

Figure 26.8 Cutaway of a grinder pump. (*Courtesy of A. Y. McDonald Mfg. Co.*)

The three floats installed with a grinder pump are all set at different levels. The lowest float is the one that cuts the pump off. Floats installed between the lowest and highest floats are the ones that cut the pumps on.

The float mounted nearest to the top of the sump is the alarm float. This float does not come into play unless the content level of the sump rises to the highest float, indicating that the pump is not working properly. The alarm float triggers an audible alarm at the control box. Any of these floats can cause a pump to fail, so don't forget to inspect the operation of the floats.

Troubleshooting Scenarios

Let's take a look at some troubleshooting scenarios that may apply to your work with waste-water pumps. Most of the stories that follow are based on actual case histories that I have dealt with. Some of them are fairly unique, so I think you will enjoy seeing what you may have to look forward to in the field.

A sump pump's sump is flooding

In our first case we are called to a house where a sump pump's sump is flooding. The customer has explained that she has never had this type of problem in the past, but her sump started flooding over the weekend. She also says the pump runs but that the water doesn't get pumped out as fast as its coming in. The result of this problem is a finished basement that is gradually being covered by ground water.

You arrive at the house and are shown to the basement. As you go down the stairs you can hear the sump pump running. The submersible pump is running, but the water in the sump is not being evacuated. What are you going to do? To help you in your evaluation, I will tell you that the weather is warm, so a frozen drain is not a consideration.

Are you thinking that the check valve is stuck in the closed position? If you are, you are making an appropriate assumption, but that is not the problem in this case. Are you about to check the impeller to see if it is jammed? Well, that's another good idea, but again, it is not the problem in this case.

What else could be causing the problem? Maybe the strainer is clogged; this is a common problem. You can pull the pump and check the strainer, and you should in most cases, but I'll save you a little time by telling you that the strainer is not clogged in this case.

As you are probably starting to see, this is not your common, run-of-the-mill pump problem. We could go on and on with potential solutions for our problem, but let's save a little room for some other case histories by getting to the point on this one.

In this particular case you could exhaust all of your troubleshooting skills within the home and never find the problem. The cause of all the trouble is outside.

This sump pump has one of the flexible drain hoses that are so popular. It runs underground to a small gravel-lined pit. It seems that the end of the drain pipe became somewhat of a lure to a 5-year-old, who thought it would be a grand idea to fill the pipe with rocks. And that's exactly what she did.

The corrugated ridges on the pipe worked to hold the larger rocks in place tightly. With the pipe filled with rocks, the pump could not discharge the contents of its sump. Finding this problem was not simple, but it only took a little logic and some process of elimination.

Not all pump problems are what you expect them to be. This example has shown you how you could look at every conceivable plumbing aspect of a pump problem and still never find the root cause of the problem. Not many plumbers would expect to find rocks stuffed into the end of the drain pipe. With this mystery out of the way, let's move onto the next story.

It won't quit running

What do you do with a pump when it won't quit running? This story could apply to a sump pump, a basin pump, an effluent pump, or a sewer pump. The symptoms and results for the following problem would be the same for any of the pumps. In the real job that this story is based on the pump was a sewer pump, so we will use that example in this story.

You are called to a house where the customer is complaining that his sewer pump seems to be running almost all the time. You go into the basement and observe the pump basin. The pump is running. What do you suppose the problem is?

Do you think it is a problem with the float? It could be, but in this case it is not. Are you thinking that the check valve is stuck in the open position? Well, you're getting warm, but you are not quite on target. What could it be?

The problem is with the check valve, but it is not that the check valve is stuck open. The check valve has been installed backward. This flaw in the installation is preventing the pump from removing the contents of the sump.

This situation may seem hard to believe, but I have rolled on many calls where this was the case. Unfortunately, some of the installations were done by plumbers in my employ. It is a mistake to install a check valve backward, but don't overlook a possibility you shouldn't overlook.

It runs, but is not pumping

What do you do with a pump when it runs but isn't pumping? There are a number of possible problems that will give this symptom, but if you use wise troubleshooting procedures, you can reduce wasted time in finding the cause of the problem.

The first thing to do is to disconnect the discharge piping and turn the pump on. If water is pumped under these conditions, you can direct your attention to potential problems with the piping. When water doesn't shoot up out of the pump, you can expect to find a clogged strainer or a jammed impeller. The problem will most likely be with the strainer.

If you have a pump that is not pushing water out of its discharge outlet, pull the pump and check the strainer. If the strainer is clear, concentrate on the impeller.

The alarm is going off

When the alarm is going off on a sewage pump system, most homeowners panic. When you arrive on such a job, what are you going to check first? The first thing to check is the alarm float; it may be stuck and malfunctioning.

Alarm floats are the highest floats in a sump basin, and they shouldn't send out false alarms. Unless the liquid level in the sump has risen to a point to push the alarm float up, the alarm shouldn't be sounding off. You can tell, however, if the problem is a faulty float or a problem pump by removing the cover on the sump pit.

If you look into the sump and see that the liquid level is well below the alarm float, you can assume the problem is with the float. However, when the liquid level is high, you must investigate the pump.

Pump stations for sewage removal in commercial properties should be equipped with a dual-pump system, but residences generally have only one pump. If the pump is not working properly, there are several types of problems you may have to troubleshoot. Let's look at them on a one-by-one basis.

A plugged-up pump is sure to set off the safety alarm on a pump station. Checking for this type of trouble is dirty work, but it must be done. You may get a good feel for the pump problem just by listening to it run. If the impellers are jammed, you may be able to hear a discrepancy in the normal running condition of the pump.

If the pump sounds like it is running properly, check the gate valve on the discharge line. Make sure it is open.

The check valve on the system could be what's causing the problem. If it is stuck in the closed position or installed backward, the pump will not be able to function properly.

One quick way to determine if the problem is in the pump or the drainage line is to disconnect the piping and observe the pump when it is running. This will tell you quickly if the pump is pumping. Be prepared to cut the pump off quickly though; otherwise you may have a large mess to clean up.

If the problem is in the pump itself, it will probably be a jammed impeller. This will require pulling the pump and doing a visual inspection. If there is a clog at the intake on the pump, you should be able to clear it with minimal trouble.

Should you establish that the problem is in the discharge piping, there are several possibilities you must consider. The gate valve is not likely to be closed, but it could be. Check valves don't usually stick in the closed position and if the system has worked properly in the past, you can rule out the possibility that the check valve is installed backward.

What does this leave you with in the way of options? If the pump is working properly and the gate and check valves are not at fault, you have some type of stoppage in the drain line or a septic system that will not accept any more discharge.

Septic systems do sometimes fill to the point where they will not accept any further discharge. If the property you are working with dumps its waste into a septic tank, remove the tank cover and inspect the liquid level. If it is above the inlet of the tank, you will have found your problem.

If the septic tank is not full, you will have to snake the drain line. The problem could be in the discharge line from the sump or in the main building drain or sewer.

Getting into the drain line to snake it will create some mess. When you remove the check valve, whatever is standing in the vertical pipe above the check valve is going to come out of the pipe, so be prepared for this. Once you have access to the pipe, you can snake it as you would any other drain.

Direct-Mount Pumps

Direct-mount pumps are the only common type of waste-water pump that we haven't discussed in this chapter. So, we will talk about them now.

Direct-mount pumps mount directly on the drains of fixtures, such as laundry tubs, bar sinks, and similar fixtures. These pumps are not as dependable or as efficient as basin-type pumps, but they are common and effective in many cases.

One of the biggest problems with direct-mount pumps is their small drain diameters. These little pumps usually have only a $\frac{3}{4}$-inch discharge outlet. If only pure water is being pumped, this is fine, but if other objects get down the drain, problems are likely to arise.

When direct-mount pumps are used on laundry tubs, they can often become jammed with foreign objects. Sand, gravel, and other small objects sometimes find their way down the drains of laundry tubs. These little particles can clog up the works in direct-mount pumps. Lint and string are also prime enemies of these small pumps. It doesn't take much to jam up the pump or the check valve used on the tiny drain lines.

If you have a direct-mount pump that is not pumping properly, inspect the check valve and the pump. You may be surprised at some of the objects you retrieve from these locations. I have found string, lint, gravel, charcoal from fish-tank filters, and other relatively common items in these locations. In the strange category, I have found plastic knives, earrings, rings, necklaces, dead goldfish, and countless other objects.

Troubleshooting Checklists

Troubleshooting checklists can help you with any kind of troubleshooting. If you can go down a checklist, there is less likelihood that you will overlook an important step in the process. You can use the following tips to help you make a troubleshooting checklist for waste-water pumps.

Pump runs but doesn't pump water

When a pump runs but doesn't pump water, there are a few things you should always look for. Let's see what they are.

- The strainer is plugged up.
- The impeller is jammed.
- The check valve is stuck in the closed position.
- The check valve is installed backward.
- The drain pipe is stopped up.
- The septic tank is full.

Pump doesn't run

When a pump isn't running at all, there are a few possibilities to consider. What are they? Well, let's see.

- A fuse is blown.
- A circuit breaker is tripped.
- The pump is not plugged in.
- The float is stuck in the off position.
- The motor is bad.
- The switch is bad.

Pump runs too often

When the pump runs too often, there are only a few things to check out; let's examine them.

- The check valve is stuck in the open position.
- The float is out of adjustment.
- There is no check valve installed.

Simple Solutions

Simple solutions are often all that are required to repair waste-water pumps. These pumps are not difficult to understand or troubleshoot, but they can send out mixed signals. If you are willing to take your time and use a pragmatic approach, finding problems with waste-water pumps will not be difficult.

27

Troubleshooting Specialty Fixtures

When we talk about troubleshooting specialty fixtures, we are talking about fixtures that are not common in every plumbing installation. For example, fluid-suction devices are common in dental offices, but they are not the type of plumbing device most plumbers service on a routine basis. When was the last time you had to work on the air or drain lines in a dental office?

Drinking stations are found in a lot of business and commercial situations, but they are not normally installed in homes. Residential plumbers making the transition into commercial work can feel at a loss when it comes to troubleshooting drinking stations or other commercial fixtures.

The problem with troubleshooting specialty fixtures is that they are special. Some of the specialty equipment is so special that you should have specifications from the manufacturer before working on it. This creates a problem for the plumber in the field who doesn't have a compilation of specifications from various manufacturers. It also creates a problem for authors who wish to help you learn to troubleshoot specialty fixtures.

Since there are so many different types of equipment available, I can't begin to give you step-by-step instructions for each type. I would like to, but I can't. What I can do, however, is go over the basics with you. As long as you develop a strong foundation of solid troubleshooting skills, you can overcome most problems. In some cases you will need specifications on the piece of equipment you are servicing, but your ingrained troubleshooting knowledge will carry you a long way in knowing where to look in the specifications for the answers you need.

What types of specialty fixtures and devices are we going to look at? We are going to start with a quick course in what you might expect to find in a dental office. From there, we will move on to drinking stations. After that, we will examine knee-kick devices, laboratory fittings, and more. But for now, let's concentrate on dental equipment.

Dental Equipment

I find that the most disturbing factors in working on dental equipment are the sounds that are always present in dental offices during working hours. There is something about hearing those drills running that makes it hard for me to concentrate on plumbing. Once the surrounding sounds are blocked out, I can get down to business.

You may not get squeamish at the sound of dental devices doing their work, but do you know what to look for when problems arise with dental equipment? I would venture a guess that only a very small percentage of plumbers have ever worked on dental equipment.

During my career, I have installed plumbing in new buildings for dental care, and I've done service and repair work on a fair amount of dental equipment. It is not the kind of work I do every day, but I find myself around it often enough to keep up with most of the changes in the industry.

The last new installation I did for a dental office involved a bottle-type fluid-suction system. At that time, bottle receptors were common practice. I'm finding now that fewer dental facilities are using the bottle systems in my area. Instead, they are equipped with suction equipment that dumps the waste into a drain connected to the sanitary sewer.

The fact that human fluids are being discharged into the sanitary sewer bothers me a little. With all the new fears of the diseases that plumbers can be subjected to, I don't like the idea of cutting open a drain pipe, thinking that is carries only normal sewage and waste, to discover the types of fluids put into the pipe from a suction system.

If I were in charge of making the plumbing rules and regulations, I would insist that there be some way for plumbers to positively identify pipes that are carrying waste other than normal sewage and graywater waste. But, I don't make the rules, so be advised. When you open a drain in a building that houses a medical office, you may find more than you bargained for in it.

When you are called in to service equipment in dental offices, you may be working with air, drain, or water lines. You can troubleshoot the primary water and drain pipes in the same way you would any other water distribution or drainage, waste and vent system. The supply piping to, and the drainage piping from, the dental equipment is a little different. And then there are the air lines. To put this into perspective, let's look at a couple of standard setups that you may be called upon to work with.

Air piping

The air piping from a compressor to the workstation at a dental chair is not complicated. The system begins with an air compressor. From the compressor, a $\frac{1}{2}$-inch air line is run. It is usually piped with type L copper tubing. In most cases, the supply pipe from the compressor is run to another piece of equipment that removes moisture from the air.

When the air piping leaves the second piece of equipment, it is run to the workstation. Typically, there will be a high-pressure cutoff switch and a quick-opening valve where the supply piping meets the work area. It is also common for the pipe size to be enlarged to $\frac{3}{4}$ inch when it arrives at its final destination.

Very few plumbers, if any, take responsibility for troubleshooting the special equipment in these setups. Normally, plumbers are only looked to for assistance in piping problems. This basically limits the knowledge needed from the plumber to the repair of leaks.

If you work with this type of equipment enough, you will find that the air compressor is likely to be equipped with its own drain. This drain enables the tank to evacuate the buildup of condensation at the end of each pumping cycle.

Evacuation systems

Evacuation systems in dental offices can come in many forms, but we will discuss one of the more current and more common ones. In this evacuation system you will run into a few things you are probably not accustomed to seeing.

The system begins at the dental chair, with the suction hose. The hose normally has a diameter of 1 inch. As the hose runs from the chair toward the suction equipment, it passes through an in-line filter. It then enters the equipment.

On the opposite side of the equipment you will find a $\frac{1}{2}$-inch cold-water supply. This supply will have a cutoff valve, and the $\frac{1}{2}$-inch tubing will be reduced to $\frac{3}{8}$-inch hose. The $\frac{3}{8}$-inch hose is what connects to the equipment.

Below the water inlet there will be a discharge outlet. A 1-inch hose is run from the outlet, through a muffler, to a $1\frac{1}{2}$-inch drain. The connection at the drain is often made with a Durham P-trap and a solid connection, not an open air-gap.

Keep in mind that the examples I have just given you are not the only configurations you are likely to find in dental offices. The age of the equipment will have a bearing on what you run into and so will the brands and models of the equipment.

Some vacuum systems require the same 1-inch piping that I described above, except that their requirements call for the piping to be done with schedule 40 PVC.

A master control with an in-line filter may also be installed. These devices allow the dentists to shut down the water supply to their equipment by throwing one switch. One of the purposes of this unit is to prevent leaks at the work area. The other is to filter water being supplied to the dental tools. These devices come in both ½- and ¾-inch sizes.

As intimidating as working in a dental office may be, the troubleshooting of the plumbing is not that different from any other type of troubleshooting you do. As long as you use a structured system to locate the cause of problems, you shouldn't have any unusual trouble when working with dental equipment. However, I strongly recommend that you obtain and refer to manufacturer's recommendations and specifications when working on any specialty equipment.

Drinking Stations

Drinking stations are, in my opinion, something of a specialty fixture; that is why we are going to discuss them in this section of the book. Since drinking stations are not standard equipment in all plumbing systems, there are many plumbers who have only a limited knowledge of how to troubleshoot them. Even though I'm calling drinking stations specialty fixtures, they are abundant enough to deserve some attention.

People often call drinking stations drinking fountains. They also refer to bubblers as drinking fountains. This can get confusing, since there are coolers, fountains, and bubblers. It can be important to know which type of drinking station you are being asked to work on. We are going to start our troubleshooting tour with drinking fountains. To aid you in your troubleshooting of water coolers and fountains, I'm including numerous drawings that illustrate the different types of equipment you may have to work with (Figs. 27.1 through Fig. 27.19)

Standard drinking fountains

The drain for an average drinking fountain has a 1¼-inch diameter. Sometimes these drains are trapped directly below the fixture, and sometimes the drain offsets into a wall before it is trapped. The drain system for a water fountain can be treated, in terms of troubleshooting, in a way similar to that used on lavatories.

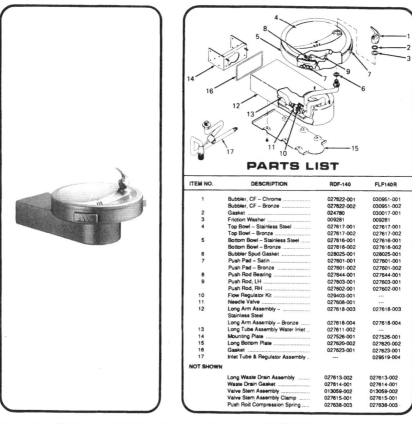

PARTS LIST

ITEM NO.	DESCRIPTION	RDF-140	FLF140R
1	Bubbler, CF – Chrome	027622-001	030951-001
	Bubbler, CF – Bronze	027622-002	030951-002
2	Gasket	024780	030017-001
3	Friction Washer	009281	009281
4	Top Bowl – Stainless Steel	027617-001	027617-001
	Top Bowl – Bronze	027617-002	027617-002
5	Bottom Bowl – Stainless Steel	027616-001	027616-001
	Bottom Bowl – Bronze	027616-002	027616-002
6	Bubbler Spud Gasket	028025-001	028025-001
7	Push Pad – Satin	027601-001	027601-001
	Push Pad – Bronze	027601-002	027601-002
8	Push Rod Bearing	027644-001	027644-001
9	Push Rod, LH	027603-001	027603-001
	Push Rod, RH	027602-001	027602-001
10	Flow Regulator Kit	029403-001	---
11	Needle Valve	027608-001	---
12	Long Arm Assembly – Stainless Steel	027618-003	027618-003
	Long Arm Assembly – Bronze	027618-004	027618-004
13	Long Tube Assembly Water Inlet	027611-002	---
14	Mounting Plate	027526-001	027526-001
15	Long Bottom Plate	027620-002	027620-002
16	Gasket	027623-001	027623-001
17	Inlet Tube & Regulator Assembly	---	029519-004
NOT SHOWN			
	Long Waste Drain Assembly	027613-002	027613-002
	Waste Drain Gasket	027614-001	027614-001
	Valve Stem Assembly	013059-002	013059-002
	Valve Stem Assembly Clamp	027615-001	027615-001
	Push Rod Compression Spring	027638-003	027638-003

Figure 27.1   Wall-mount drinking fountain. (*Courtesy of EBCO Mfg. Co.*)

The water supply for a typical drinking fountain will come out of a wall and connect to the unit. A $\frac{1}{2}$-inch supply is usually run to the cutoff valve, where the supply is reduced to $\frac{3}{8}$-inch tubing.

In terms of possible complications with drinking fountains, there are many considerations. For example, mineral deposits can build up and restrict the flow of water. The flow-regulator kit in the unit can require attention to compensate for an unsatisfactory flow of water. Then the inlet tube and regulator assembly can give you trouble. Any of these options could cause a drinking fountain to act up.

Leaks are another consideration. You might have a leak where the bubbler is attached to the interior water supply. There is a bubbler spud gasket at this location you can check. The needle valve that is inside the unit could develop a leak, and, of course, a leak could come from the connections at the inlet tube and regulator assembly.

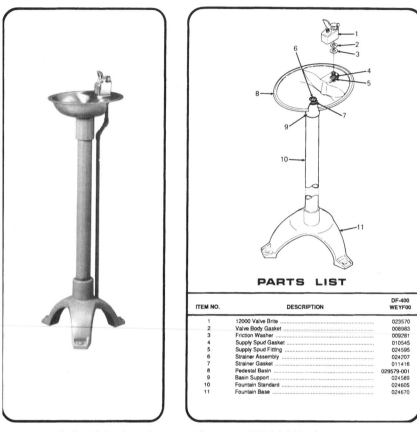

PARTS LIST

ITEM NO.	DESCRIPTION	DF-400 WEYF00
1	12000 Valve Brite	023570
2	Valve Body Gasket	008983
3	Friction Washer	009281
4	Supply Spud Gasket	010545
5	Supply Spud Fitting	024595
6	Strainer Assembly	024207
7	Strainer Gasket	011416
8	Pedestal Basin	029579-001
9	Basin Support	024589
10	Fountain Standard	024605
11	Fountain Base	024670

Figure 27.2   Pedestal drinking fountain. (*Courtesy of EBCO Mfg. Co.*)

None of the problems we have just discussed are hard to understand or to correct. As long as you know where to look, and what to look for, you can remain in control during the troubleshooting of a drinking fountain. I have provided you with some detailed drawings that show the components of a standard drinking fountain. These drawings, like others throughout this book, can be more helpful than a thousand words on the subject.

Pedestal fountains

Pedestal fountains are easy to troubleshoot in that there are not many concealed parts. The water supply runs exposed to the bubbler valve, and the only concealed parts in the water system are the gasket for the valve body and the gasket for the supply spud.

The drainage for these types of fountains is concealed in the

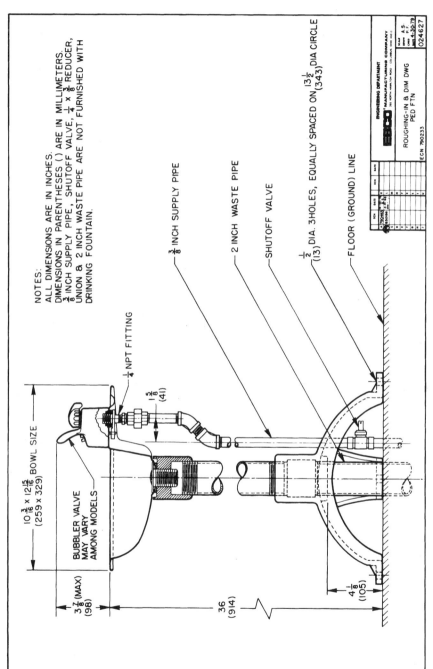

Figure 27.3 Detail of pedestal drinking fountain. (*Courtesy of EBCO Mfg. Co.*)

NOTES:
ALL DIMENSIONS ARE IN INCHES.
DIMENSIONS IN PARENTHESES ( ) ARE IN MILLIMETERS.
$\frac{3}{8}$ INCH SUPPLY PIPE, SHUTOFF VALVE, $\frac{1}{4}$ x $\frac{3}{8}$ REDUCER,
UNION & 2 INCH WASTE PIPE ARE NOT FURNISHED WITH
DRINKING FOUNTAIN.

$\frac{3}{8}$ INCH SUPPLY PIPE

2 INCH WASTE PIPE

SHUTOFF VALVE

$\frac{1}{2}$ DIA. 3 HOLES, EQUALLY SPACED ON 13$\frac{1}{2}$ DIA CIRCLE
(13)   (343)

FLOOR (GROUND) LINE

$\frac{1}{4}$ NPT FITTING

1$\frac{5}{8}$
(41)

10$\frac{3}{16}$ x 12$\frac{15}{16}$ BOWL SIZE
(259 x 329)

BUBBLER VALVE
MAY VARY
AMONG MODELS

3$\frac{7}{8}$ (MAX)
(98)

36
(914)

4$\frac{1}{8}$
(105)

ENGINEERING DEPARTMENT
EBCO MANUFACTURING COMPANY
265 NORTH HAMILTON ROAD COLUMBUS, OHIO 43213

ROUGHING-IN & DIM DWG
PED FTN

ECN 790233

024627

27.7

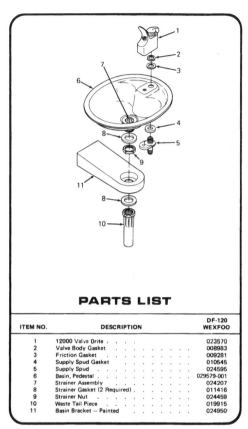

PARTS LIST

ITEM NO.	DESCRIPTION	DF-120 WEXFOO
1	12000 Valve Drite	023570
2	Valve Body Gasket	008983
3	Friction Gasket	009281
4	Supply Spud Gasket	010545
5	Supply Spud	024595
6	Basin, Pedestal	029579-001
7	Strainer Assembly	024207
8	Strainer Gasket (2 Required)	011416
9	Strainer Nut	024459
10	Waste Tail Piece	019915
11	Basin Bracket — Painted	024950

Figure 27.4 Exposed-trap drinking fountain. (*Courtesy of EBCO Mfg. Co.*)

pedestal and trapped below the floor level. The drainage for this type of unit is designed to be connected to a 2-inch drain.

Simple bubblers

Simple bubblers are easy to work with. They hang on a wall and have an exposed trap and an exposed water supply. Aside from the bubbler valve, the only other parts to be considered, other than routine plumbing fittings, are the gaskets for the valve body and the supply spud.

True water coolers

True water coolers are more complicated than simple drinking fountains. The fact that these units cool drinking water, rather than simply supply it, means that there are many more possibilities for problems. Fortunately, plumbers are rarely required to deal with the

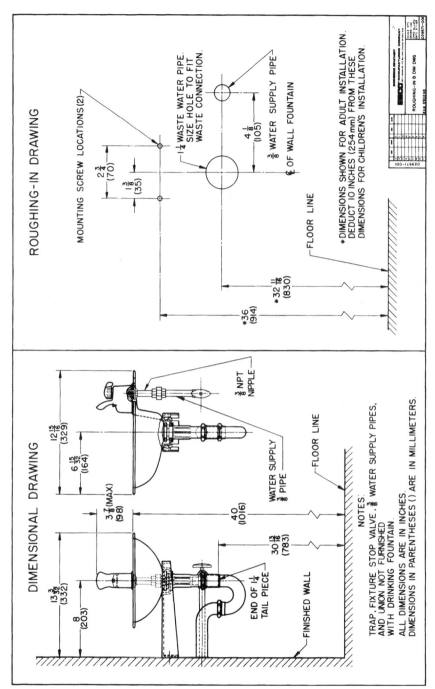

Figure 27.5 Detail of exposed-trap drinking fountain. (*Courtesy of EBCO Mfg. Co.*)

27.9

SPECIFICATIONS

Voltage	115 VAC ±10%/1PH/60 Hertz
Size (Approx)	26-1/2" H., 17-1/2" W., 18-1/2" D.
Shipping Weight (Approx)	67 LBS.
Cold Water Capacity (GPH)*	8
Compressor (HP)	1/5
Compressor (Amps)	6
(Full Load)	
Refrigerant	R-12
Refrigerant Charge	5.0 OZ.

* A.R.I. Rating: Room Temperature 90°F; Supply Water
Temperature 80°F; Drinking Water Temperature 50°F.
Based on standard test procedures.

Specifications subject to change without notice.

Figure 27.6   Wall-hung water cooler. (*Courtesy of EBCO Mfg. Co.*)

electrical or refrigeration aspects of water coolers. The basic plumbing with these units is not very different from that of standard drinking fountains.

There are several different styles of water coolers available. Some of them hang on walls, and some of them stand on floors. And, there are models available for handicap installation. While the location of various parts differs with the many models, the principles for troubleshooting them are the same. Rather than go through several paragraphs describing each type of unit, I've chosen to provide you with detailed drawings of all the common designs. The parts lists that accompany the drawings will show you where all the plumbing components are located.

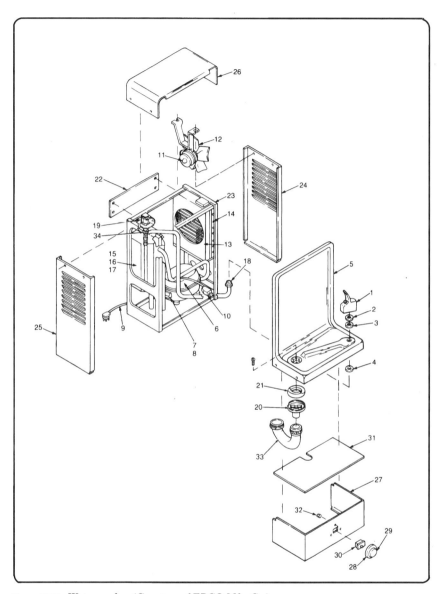

Figure 27.7   Water cooler. (*Courtesy of EBCO Mfg. Co.*)

PARTS LIST

ITEM NO.	DESCRIPTION	PLF8SKPE	PLF8SKTP
1	12000A Bubbler Valve	030774-004	030774-004
2	Valve Body Gasket	028706-006	028706-006
3	Friction Washer	028706-013	028706-013
4	Bubbler Spud Gasket	028706-001	028706-001
5	Simulated Recessed Top – Stainless Steel	028849-001	028849-001
6	Compressor Assembly (Includes Items 7 and 8)	027256-003	027256-003
7	Relay	017978	017978
8	Overload	016942-005	016942-005
9	Service Cord	026540-004	026540-004
10	Cold Control	027040-007	027040-007
11	Fan Motor	027354-009	027354-009
12	Fan Blade	022977	022977
13	Shroud	029125-001	029125-001
14	Condenser	029111-001	029111-001
15	Insulated Cooling Tank Assembly	029108-003	029108-003
16	Cooling Tank Insulation LH	028661-001	028661-001
17	Cooling Tank Insulation RH	028660-001	028660-001
18	Bubbler Supply Spud	028701-002	028701-002
19	Solenoid Valve	031011-001	031011-001
20	Waste Assembly	023046-001	023046-001
21	Waste Gasket	018050	018050
22	Hanger Bracket and Barrier Assembly	025662-003	025662-003
23	Frame Assembly	022967	022967
24	Right Side Panel – Sandstone	022979-002	022979-002
	Right Side Panel – Stainless Steel	023036	023033
	Right Side Panel – Painted	023038	023038
25	Left Side Panel – Sandstone	022980-002	022980-002
	Left Side Panel – Stainless Steel	023035	023035
	Left Side Panel – Painted	023037	023037
26	Cap – Sandstone	031068-005	031068-005
	Cap – Stainless Steel	031068-003	031068-003
	Cap – Painted	031068-002	031068-002
27	Wrapper – Sandstone	031069-204	---
	Wrapper – Stainless Steel	031069-202	---
	Wrapper – Painted	031069-201	---
	Wrapper – SAN TP	---	031069-304
	Wrapper – STN TP	---	031069-302
	Wrapper – Painted TP	---	031069-301
28	Bezel	024816	024816
29	Push Button	024817	024817
30	Switch	024764	024764
31	Support Plate	023391	023391
32	Self Threading Hex Nut	020203	020203
33	Waste Trap Assembly	027542-001	027542-001
34	Male Fitting w/Gasket – GHT X 1/4 NPT (Solenoid Valve Inlet)	031033-001	031033-001
Not Shown			
	Male Fitting Gasket	028706-024	028706-024

Figure 27.8   Parts detail of water cooler in Fig. 27.7. (*Courtesy of EBCO Mfg. Co.*)

SPECIFICATIONS

Voltage	115 VAC ±10%/1PH/60 Hertz
Size	25″ H., 18″ W., 19″ D.
Shipping Weight (Approx)	60 LBS.
Cold Water Capacity	7.8 GPH
Compressor	1/5 HP
Refrigerant	R-500
Refrigerant Charge	4 OZ.
Compressor (Full Load)	6 Amps

Specifications subject to change without notice.

Figure 27.9   Wall-hung water cooler. (*Courtesy of EBCO Mfg. Co.*)

Specialty Valves and Fittings

Next we are going to talk about specialty valves and fittings. These devices are common in some plumbing systems, such as hospitals, but rare in most. While you may never have a need to work with the equipment we are about to discuss, the background knowledge you gain from the following paragraphs may prove helpful.

Knee-action valves

Knee-action valves (Figs. 27.20 through 27.23) are most commonly found in hospitals. These specialty fittings allow the operation of fixtures without the use of hands. This allows medical personnel to avoid contamination from faucet handles, and it also leaves both hands free. Knee-action valves can also be used to accommodate handicapped people.

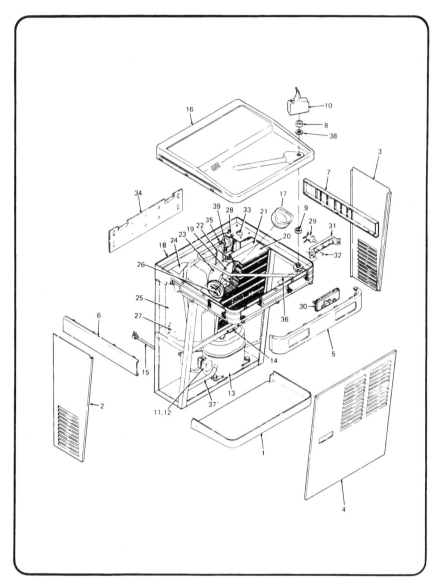

Figure 27.10   Water cooler. (*Courtesy of EBCO Mfg. Co.*)

PARTS LIST

ITEM NO.	DESCRIPTION	PLF8WMD-002	PLF8WMQ-002 PLF8WMQA-003
1	Front Housing – Sandstone	026780-002	026780-002
	Front Housing – Stainless Steel	026780-004	026780-004
	Front Housing – Tan	026780-005	026780-005
2	Left Side Panel – Sandstone	026798-003	026798-003
	Left Side Panel – Stainless Steel	026798-007	026798-007
	Left Side Panel – Tan	026798-009	026798-009
3	Right Side Panel – Sandstone	026798-004	026798-004
	Right Side Panel – Stainless Steel	026798-008	026798-008
	Right Side Panel – Tan	026798-010	026798-010
4	Front Panel – Sandstone	026802-002	026802-002
	Front Panel – Stainless Steel	026802-004	026802-004
	Front Panel – Tan	026802-005	026802-005
5	Front Bezel	026869-001 OPEN	026869-001 OPEN
6	Left Side Bezel	026886-003 CLOSED	026886-001 OPEN
7	Right Side Bezel	026886-004 CLOSED	026886-002 OPEN
8	Valve Body Gasket	028706-006	028706-006
9	Washer (Under Top)	028706-001	028706-001
10	Bubbler Valve Assembly	030774-004	030774-004
11	Overload	017977	017977
12	Relay	017978	017978
13	Compressor	027256-002	027256-002
14	Cold Control	027040-007	027040-007
15	Service Cord	026540-004	026540-004
16	Top	028835-001	028835-001
17	Waste Gasket	027205	027205
18	Frame Assembly	026976-003	026976-003
19	Fan Blade	024033	024033
20	Condenser	026967	026967
21	Fan Shroud	026968	026968
22	Fan Motor Bracket	026969	026969
23	Fan Motor Assembly	027354-009	027354-009
24	Cooling Tank Insulation (RH)	028660-001	028660-001
25	Cooling Tank Insulation (LH)	028661-001	028661-001
26	Waste Assembly	026981-001	026981-001
27	Cooling Tank Assembly (Insulated)	031045-001	031045-001
28	Solenoid Valve	031011-001	031011-001
29	Switch	028591-001	028591-001
30	Switch Pad	028587-001	028587-001
31	Switch Bracket	028588-001	028588-001
32	Compression Spring	027638-005	027638-005
33	Timer (QA models)	---	030973-001
34	Hanger Bracket	016698-002	016698-002
35	Connector Fitting	023834-002	023834-002
36	Tube Assembly Sol to Blb	031035-001	031035-001
37	Heat Exchanger	028785-001	028785-001
38	Friction Washer	028706-013	028706-013
39	Male Fitting w/Gasket –GHT X 1/4 NPT	031033-001	031033-001

NOT SHOWN

	Male Fitting Gasket	028706-024	028706-024

Figure 27.11   Parts detail of water cooler in Fig. 27.10. (*Courtesy of EBCO Mfg. Co.*)

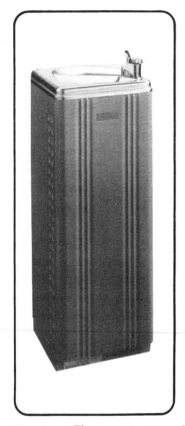

SPECIFICATIONS

7P MODELS

Voltage	115 VAC ±10%/1PH/60 Hertz
Size (Approx)	40" H., 15" W., 15" D.
Shipping Weight (Approx)	73 LBS.
Cold Water Capacity	7 GPH
Compressor	1/8 HP
Compressor Amps (Full Load)	2.5
Refrigerant	R-500
Refrigerant Charge	4-3/8 OZ.

13P AND 13PL MODELS

Voltage	115 VAC ±10%/1PH/60 Hertz
Size (Approx)	40" H., 15" W., 15" D.
Shipping Weight (Approx)	80 LBS.
Cold Water Capacity	13 GPH
Compressor	1/5 HP
Compressor Amps (Full Load)	5.0
Refrigerant	R-12
Refrigerant Charge	5 OZ.

*A.R.I. Rating: Room Temperature 90°F; Supply Water Temperature 80°F; Drinking Water Temperature 50°F.

Specifications subject to change without notice.

Figure 27.12   Floor-mount water cooler. (*Courtesy of EBCO Mfg. Co.*)

The activator on these units can be made to mount on either a wall, the crossbar of a chair carrier, or the door of a laboratory cabinet.

The cutoff valves for knee-action valves are built into the unit. Nipples are used to connect the valves to water supplies in walls. When pressure is applied on the knee bracket, it causes the valve to open the supply channels, bringing water to the device in use. Once pressure is removed from the knee bracket, the valve closes and stops the flow of water. The action is similar to push-button faucets.

Repair kits and parts are available for the various makes and models. If you have a fixture that is not producing water in the desired manner, you can cut off the stop valves on the knee-action valve and troubleshoot the mechanisms. Once the defective part is identified, it can be replaced easily.

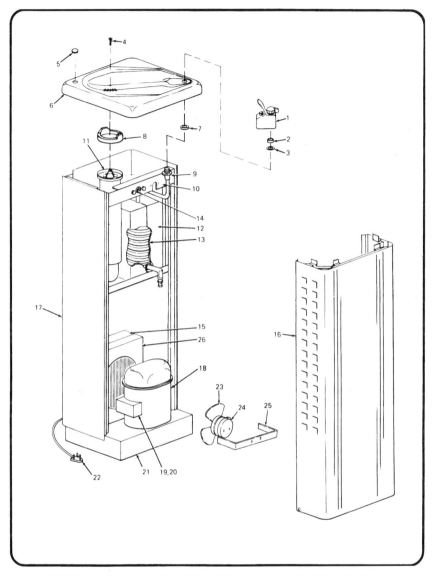

Figure 27.13 Detail of water cooler in Fig. 27.12. (*Courtesy of EBCO Mfg. Co.*)

PARTS LIST

ITEM NO.	DESCRIPTION	7P-003 PLF7P PLF7P-001	7P-001D PLF7P-D PLF7P-D001	13P-003 PLF13P PLF13P-001	13PL-003 PLF13PL PLF13PL-001
1	Bubbler Valve 12000 – Used on all models except PLF-001 Series	026551-002	026551-002	026551-002	026551-002
1	Bubbler Valve 12000A – Used on PLF-001 Series only	030774-002	030774-002	030774-002	030774-002
2	Valve Body Gasket	008983	008983	008983	008983
3	Friction Washer	009281	009281	009281	009281
4	Drain Screw	026655-002	026655-002	026655-002	026655-002
5	Hole Cover	016994	016994	016994	016994
6	Top	028845-001	028845-001	028845-001	028845-001
7	Supply Spud Gasket	028706-001	028706-001	028706-001	028706-001
8	Waste Gasket	027389-001	027389-001	027389-001	027389-001
9	Bubbler Supply Spud	028701-002	028701-002	028701-002	028701-002
10	Upper Cabinet Bracket	027407-001	027407-001	027407-001	027407-001
11	Precooler Assembly	027412-001	027412-001	027412-001	027412-001
12	Cooling Tank Insulation – 2 Rqd	028230-002	028230-002	---	---
	Cooling Tank Insulation – R.H.	---	---	028660-001	028660-001
	Cooling Tank Insulation – L.H.	---	---	028661-001	028661-001
13	Insulated Cooling Tank	029673-001	029673-001	029096-001	029096-002
14	Cold Control	027040-007	027040-007	027040-007	027040-007
15	Condenser	018252	018252	018252	018252
16	Front Panel – Tan	024442-005	024442-005	024442-005	024442-009
	Front Panel – Stainless Steel	024442-003	024442-003	024442-003	024442-011
17	Cabinet – Tan	015570-002	015570-002	015570-002	015569-002
	Cabinet – Stainless Steel	015570-005	015570-005	015570-005	015569-005
18	Compressor (Includes Items 19 and 20)	027245-001	026868-003	027256-002	027256-002
19	Overload	027268-002	---	017977	017977
20	Relay	027267-003	026867	017978	017978
21	Base – Tan	028199-001	028199-001	028199-001	028199-001
	Base – Stainless Steel	028199-002	028199-002	028199-002	028199-002
22	Service Cord	026540-003	026540-003	026540-005	026540-005
23	Fan Blade	019199	019199	022977	022977
24	Fan Motor	027354-020	027354-020	027354-008	027354-008
25	Fan Bracket	017530	017530	017530	017530
26	Fan Shroud	017821	017821	029656-001	029656-001
ACCESSORY PARTS					
	Glass Filler Installation Kit	029098-001	029098-001	029098-001	029098-001
	Foot Pedal Kit	026645-014	026645-014	026645-014	026645-013

Figure 27.14 Parts detail of water cooler in Fig. 27.12. (*Courtesy of EBCO Mfg. Co.*)

SPECIFICATIONS

Models CP3M-001D, CP3M-P, CP3M-Z,
PLF3CM, PLF3CM-D, & PLF3CM-P

Voltage	115 VAC ±10%/1PH/60 Hertz
Size (Approx)	16" H., 17" W., 13" D.
Shipping Weight (Approx)	49 LBS.
Cold Water Capacity	3 GPH
Compressor - All Models	1/8 HP
Compressor (Full Load) 115 Volt	2.5 Amps
Refrigerant	R-12
Refrigerant Charge - All Models	3-1/8 OZ.

Models CP3M-D5/6 & PLF3CMY-D

Voltage	220/240 Volt 50/60 Hertz
Size (Approx)	16" H., 17" W., 13" D.
Shipping Weight (Approx)	49 LBS.
Cold Water Capacity	3 GPH
Compressor - All Models	1/8 HP
Compressor (Full Load) 220 Volt	1.2 Amps
Refrigerant	R-12
Refrigerant Charge - All Models	3-1/8 OZ.

* A.R.I. Rating: Room Temperature 90°F; Supply Water Temperature 80°F; Drinking Water Temperature 50°F.

Specifications subject to change without notice.

Figure 27.15   Wall-hung water cooler. (*Courtesy of EBCO Mfg. Co.*)

Pedal valves

Pedal valves (Figs. 27.24 through 27.27) are used for purposes similar to those of knee-action valves. The installation of pedal valves is usually made at floor level, but wall-hung models are available. These valves can be equipped to work with a single water supply or with a dual water supply.

Pedal valves that are designed to mix hot and cold water have two inlet openings and one outlet opening. When the pedals are depressed, they cause a valve in the body to open, providing water to a fixture. These specialty fittings are easy to disassemble and troubleshoot.

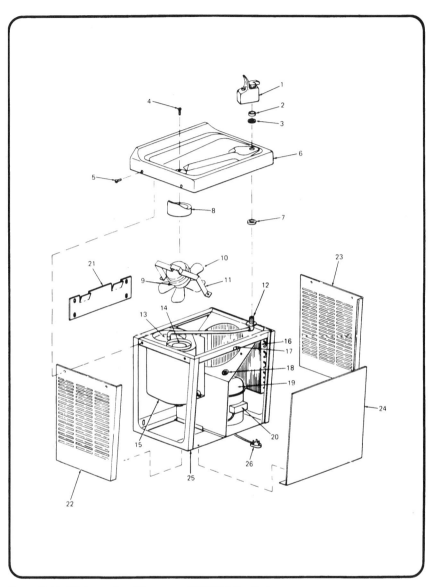

Figure 27.16   Water cooler. (*Courtesy of EBCO Mfg. Co.*)

PARTS LIST

ITEM NO.	DESCRIPTION	CP3M PLF3CM SERIES	CP3M-D5/6 PLF3CMY-D SERIES
1	12000 Bubbler Valve	026551-002	026551-002
	12000A Bubbler Valve (PLF-001 Series)	030774-002	030774-002
2	Valve Body Gasket	008983	008983
3	Friction Washer	009281	009281
4	Drain Screw	026655-002	026655-002
5	Top Screw	026630-005	026630-005
6	Top	028846-003	028846-003
7	Gasket (Under Top)	028706-001	028706-001
8	Waste Gasket	018050	018050
9	Fan Motor Assembly	027354-021	027354-006
10	Fan Blade	020243	020243
11	Fan Motor Bracket	026969	026969
12	Bubbler Supply Spud	029177-001	029177-001
13	Cooling Tank Assembly	029437-002	029437-002
14	Waste Assembly	019988	019988
15	Cooling Tank Insulation L/H	019990L	019990L
	Cooling Tank Insulation R/H	019990R	019990R
16	Condenser	028072-001	028072-001
17	Panel Clip	018156-001	018156-001
18	Cold Control	027040-007	027040-005
19	Compressor (Includes Item 20)	See Below	026883-003
20	Relay	See Below	026309
21	Hanger Bracket & Barrier Assembly	025662-002	025662-002
22	Left Side Panel – Tan	026935-003	026935-003
	Left Side Panel – Stainless Steel	026935-027	026935-027
23	Right Side Panel – Tan	026935-004	026935-004
	Right Side Panel – Stainless Steel	026935-028	026935-028
24	Front Panel – Tan	026925-001	026925-001
	Front Panel – Stainless Steel	026925-014	026925-014
25	Frame Assembly	029436-001	029436-001
26	Service Cord	026540-013	028625-002

COMPRESSORS

DANFOSS CP3M-D AND PLF3CM-D

19	*Compressor	026868-003
20	Relay	026867
	Overload	Internal

PANASONIC CP3M-P AND PLF3CM-P

19	*Compressor	028570-003
20	Relay	027709-002
	Overload	Internal

TECUMSEH CP3M-Z AND PLF3CM

19	*Compressor	027245-001
20	Relay	027267-003
	Overload	027268-002

*NOTE: Compressors include relay and overload.

Figure 27.17   Parts detail of water cooler in Fig. 27.16. (*Courtesy of EBCO Mfg. Co.*)

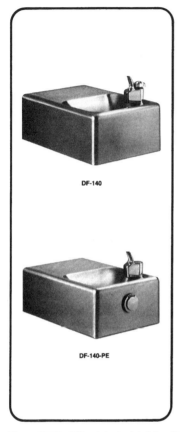

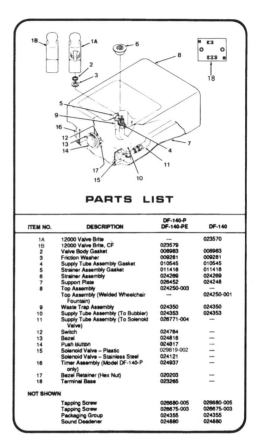

DF-140

DF-140-PE

PARTS LIST

ITEM NO.	DESCRIPTION	DF-140-P DF-140-PE	DF-140
1A	12000 Valve Brite	---	023570
1B	12000 Valve Brite, CF	023579	
2	Valve Body Gasket	008983	008983
3	Friction Washer	009281	009281
4	Supply Tube Assembly Gasket	010545	010545
5	Strainer Assembly Gasket	011416	011416
6	Strainer Assembly	024269	024269
7	Support Plate	026452	024248
8	Top Assembly	024250-003	---
	Top Assembly (Welded Wheelchair Fountain)	---	024250-001
9	Waste Trap Assembly	024350	024350
10	Supply Tube Assembly (To Bubbler)	024353	024353
11	Supply Tube Assembly (To Solenoid Valve)	026771-004	---
12	Switch	024764	---
13	Bezel	024816	---
14	Push Button	024817	---
15	Solenoid Valve – Plastic	029819-002	---
	Solenoid Valve – Stainless Steel	024121	---
16	Timer Assembly (Model DF-140-P only)	024937	---
17	Bezel Retainer (Hex Nut)	020203	---
18	Terminal Base	023265	---
NOT SHOWN			
	Tapping Screw	026680-005	026680-005
	Tapping Screw	026675-003	026675-003
	Packaging Group	024355	024355
	Sound Deadener	024880	024880

Figure 27.18   Wheelchair drinking fountain. (*Courtesy of EBCO Mfg. Co.*)

Bedpan flushers

Bedpan flushers (Figs. 27.28 and 27.29) are available in numerous configurations. For example, one type of flusher is a diverter flushing fitting. It has a vertical swing spout and a removable spray outlet. This type of fitting is available with or without an antiseptic injector unit.

Another type of bedpan flusher resembles a wall-mounted laundry-tub faucet. It has a vacuum breaker on top of its spout and a spray assembly.

As with most specialty fixtures, these faucets can be treated with the same troubleshooting skills given for other faucets and fixtures throughout this book.

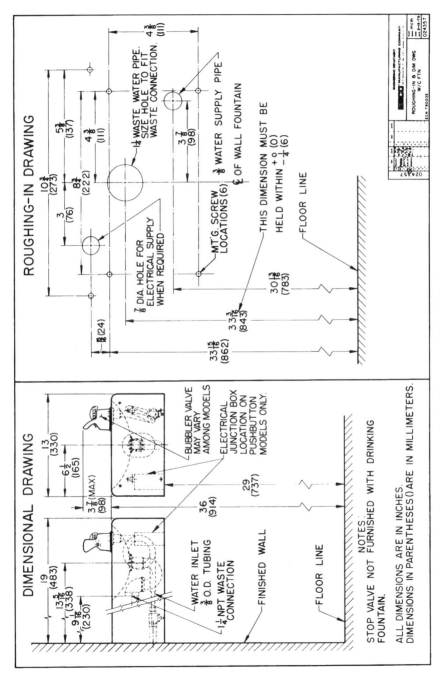

Figure 27.19 Detail of drinking fountain in Fig. 27.18. (*Courtesy of EBCO Mfg. Co.*)

NORWICH
Vitreous China Lavatory

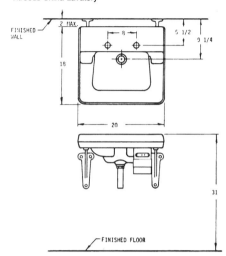

Norwich 1H-163 on Enameled Brackets

Lavatory: 1H-163 **Norwich** vitreous china lavatory with rectangular basin, splash lip, front overflow and two soap depressions. Punched for goose-neck spout, soap valve and fixture-mounted knee-action mixing valve.

Support:

Trim:

Waste:

Trap:

Supplies:

Color:

Soap Valve:

Exposed metal trim is polished chromium plated, unless otherwise described.

Figure 27.20   Lavatory with knee-action valve. (*Courtesy of Crane Plumbing.*)

NORWICH LAVATORY
VITREOUS CHINA
EXPOSED ENAMELED IRON BRACKETS
KNEE ACTION VALVE AND GOOSENECK SPOUT
PUSH BUTTON SOAP VALVE

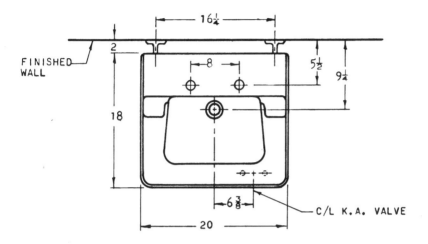

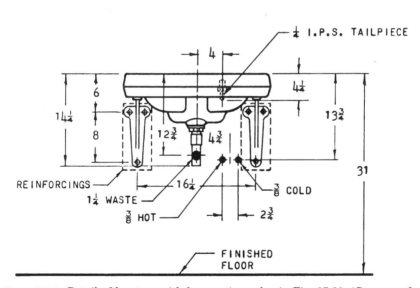

Figure 27.21 Detail of lavatory with knee-action valve in Fig. 27.20. (*Courtesy of Crane Plumbing.*)

NEU-MEDIC™
Vitreous China Surgeon's Scrub-Up Sink

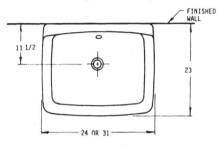

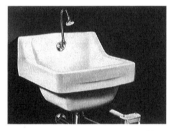

Neu-Medic 5H-256 Surgeon's Scrub-up Sink
Vitreous China

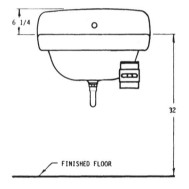

Sink: 5H-256 **Neu-Medic** vitreous china surgeon's sink with 6¼" high back and return ends.

Supports:

Trim:

Supplies:

Waste:

Trap:

Exposed metal trim is polished chromium plated, unless otherwise described.

Figure 27.22   Surgeon's scrub sink with knee-action valve. (*Courtesy of Crane Plumbing.*)

NEU-MEDIC™ WASH-UP SINK
VITREOUS CHINA
KNEE ACTION VALVE AND GOOSENECK SPOUT
OPEN STRAINER
CAST OR BENT TUBE "P" TRAP

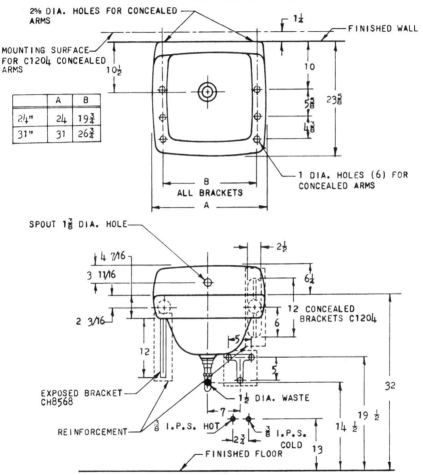

	A	B
24"	24	19¾
31"	31	26¾

NOTE: ROUGHING-IN DIMENSIONS FOR CHAIR CARRIER SHOULD BE OBTAINED DIRECT FROM MANUFAC-
TURER OF CARRIER.

Figure 27.23 Detail of sink with knee-action valve in Fig. 27.22. (*Courtesy of Crane Plumbing.*)

NORWICH
Vitreous China Lavatory
On Chair Carrier

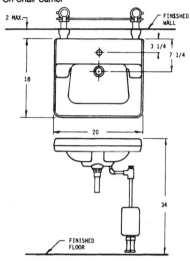

Norwich 1H-164 with Concealed Arms

Lavatory: 1H-164 **Norwich** vitreous china lavatory with rectangular basin, splash lip, front overflow and two soap depressions. Punched for goose-neck spout.

Support:

Trim:

Waste:

Trap:

Supplies:

Color:

Exposed metal trim is polished chromium plated, unless otherwise described.

Figure 27.24 Lavatory with a pedal control. (*Courtesy of Crane Plumbing.*)

NEWCOR
Vitreous China Flushing Rim Service Sink
Siphon Jet Action

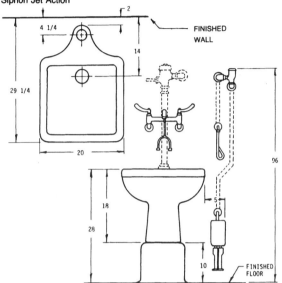

Newcor 7H-538 Flushing Rim
Service Sink Vitreous China

Sink: 7H-538 **Newcor** vitreous china, flushing rim siphon jet service sink; bolt caps.

Trim:

Pedestal:

Valve:

Supply:

Rim Guard:

*Bedpan
Cleanser:*

Note: For efficient operation of the sink, a minimum flowing water pressure of 20 P.S.I. is required at the valve.

Exposed metal trim is polished chromium plated, unless otherwise described.

Figure 27.25 Flushing rim service sink. (*Courtesy of Crane Plumbing.*)

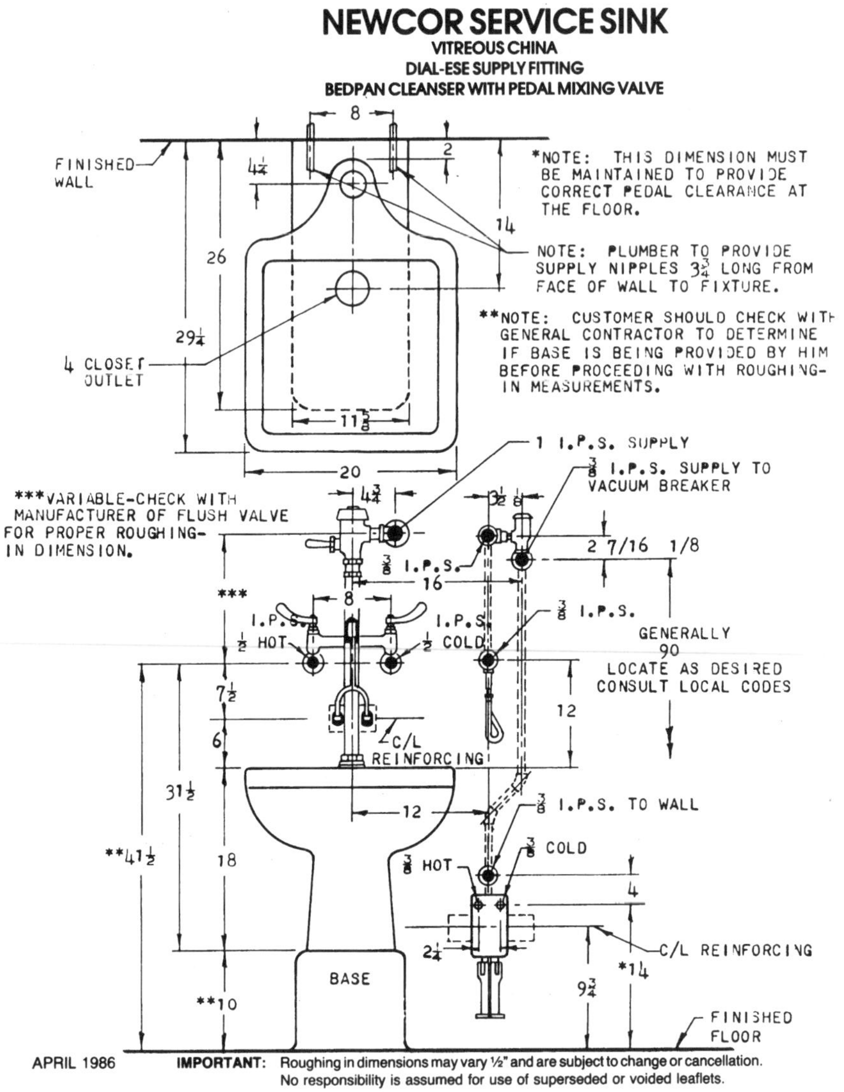

NEWCOR SERVICE SINK
VITREOUS CHINA
DIAL-ESE SUPPLY FITTING
BEDPAN CLEANSER WITH PEDAL MIXING VALVE

APRIL 1986   **IMPORTANT:** Roughing in dimensions may vary ½" and are subject to change or cancellation.
No responsibility is assumed for use of superseded or voided leaflets.

Figure 27.26 Detail of flushing rim service sink in Fig. 27.25. (*Courtesy of Crane Plumbing.*)

WHIRLTON
Siphon Jet Quiet
Action Water Closet

3H-704
Water Economy

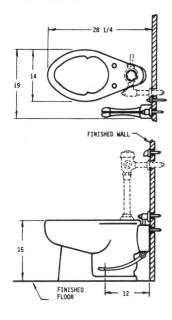

3H-704 Elongated Rim
With Bedpan Lugs

Bowl: 3H-704 **Whirlton** vitreous china water economy, siphon jet, whirlpool quiet action, elongated rim, 1½" top spud bowl with integral bedpan lugs; bolt caps

Valve:

*Bedpan
Cleanser:*

Seat:

Color:

Note: Bedpan is not included.

Exposed metal trim is polished chromium plated, unless otherwise described.

Figure 27.27 Water closet with bedpan lugs and pedal control. (*Courtesy of Crane Plumbing.*)

SANI-SINK
Vitreous China Flushing Rim Service Sink
Blowout Action

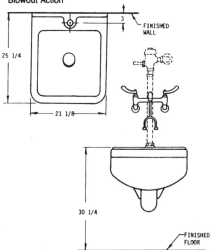

Sani-Sink 7H-544 Flushing Rim
Service Sink Vitreous China

Sink: 7H-544 **Sani-Sink** vitreous China wall-hung flushing rim blowout service sink with 1 ½" top inlet.

Supports:

Trim:

Valve:

Exposed metal trim is polished chromium plated, unless otherwise described.

Figure 27.28   Flushing rim service sink. (*Courtesy of Crane Plumbing.*)

SANI-SINK
FLUSHING RIM BLOWOUT SINK
VITREOUS CHINA - TOP SUPPLY
DIAL-ESE SUPPLY FITTING
EXPOSED FLUSH VALVE
BEDPAN CLEANSER WITH PEDAL MIXING VALVE

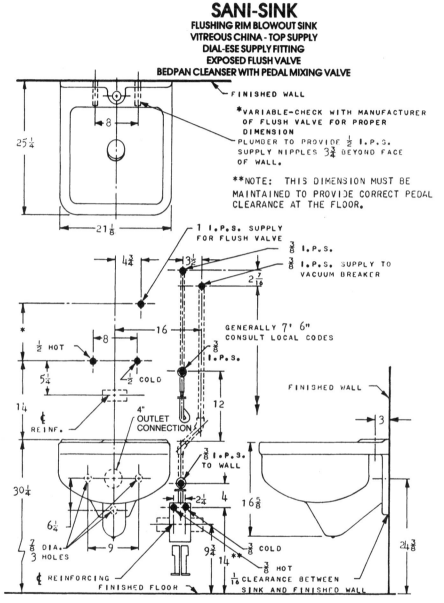

Figure 27.29 Detail of flushing rim service sink in Fig. 27.28. (*Courtesy of Crane Plumbing.*)

Laboratory fittings

Laboratory fittings are very similar to other types of plumbing fittings. There is, however, one very noticeable difference. Laboratory fittings typically have serrated nozzles. This allows the connection of hoses for water, gas, air, and vacuum use. The basic working parts of these fittings are comparable to other types of plumbing fittings.

Other Special Fixtures and Fittings

Other special fixtures and fittings exist in the plumbing world (Figs. 27.30 through 27.35). If you work in commercial kitchens, you are likely to work with prerinse sprayers. While these devices are not common in residential plumbing, they are a way of life in commercial kitchens. However, the troubleshooting skills needed to work with them are no more complex than those used on other types of plumbing.

Whether you are called upon to troubleshoot a garbage can washer, a clinical sink, or a commercial garbage disposer, your basic troubleshooting skills will get you through the job. There will be times when you may need detailed specifications for a given piece of equipment, but the principles and procedures for troubleshooting specialty fixtures are no different from those described throughout this book. Once you have your troubleshooting abilities refined, you can find any type of problem, under any conditions, almost all of the time.

NORWICH
Vitreous China Shampoo Lavatory

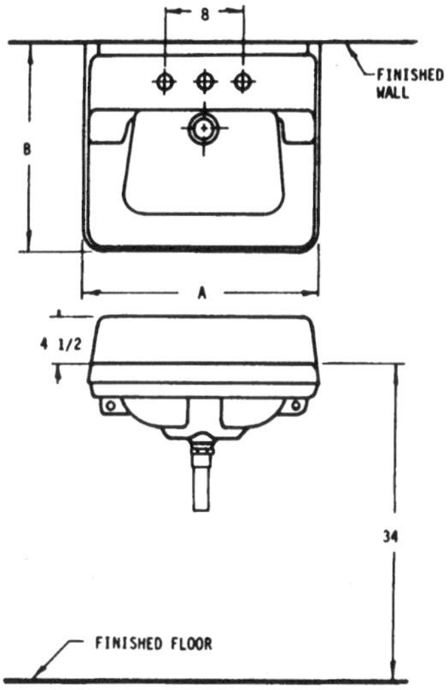

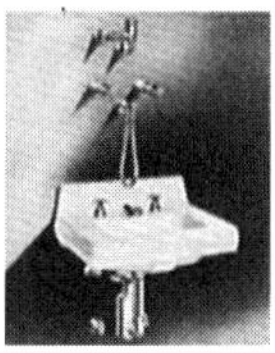

Norwich 1H-209 On Chair Carrier or Wall Hung

$\underline{A}$
18
20
24

$\underline{B}$
15
18
21

Lavatory:	1H-209 **Norwich** vitreous china lavatory with back, retangular basin, splash lip, front overflow and two soap depressions.
Support:	Concealed hanger supplied. Drilled for concealed arm carrier.
Trim:	
Trap:	
Supplies:	
Color:	
Size:	18" x 15", 20" x 18", 24" x 21"

Exposed metal trim is polished chromium plated, unless otherwise described.

Figure 27.30   Shampoo sink. (*Courtesy of Crane Plumbing.*)

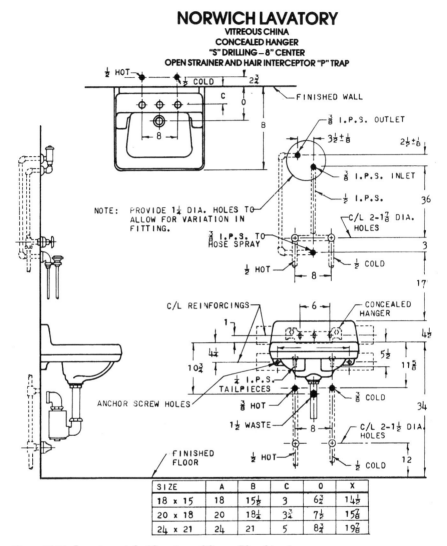

Figure 27.31 Lavatory sink. (*Courtesy of Crane Plumbing.*)

VITREOUS CHINA
Service Sink

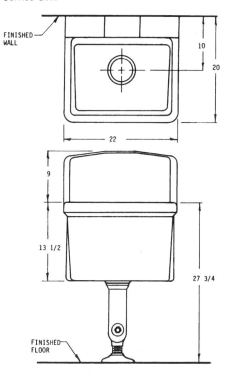

Drilled Back 7-503
Without Rim Guard

Sink: 7-503 **vitreous china** , service sink with 8" high back and hole for supporting screw.

Supports:

Trim:

Trap:

Rim Guard:

Exposed metal trim is polished chromium plated, unless otherwise described.

Figure 27.32   Detail of service sink in Fig. 27.31. (*Courtesy of Crane Plumbing.*)

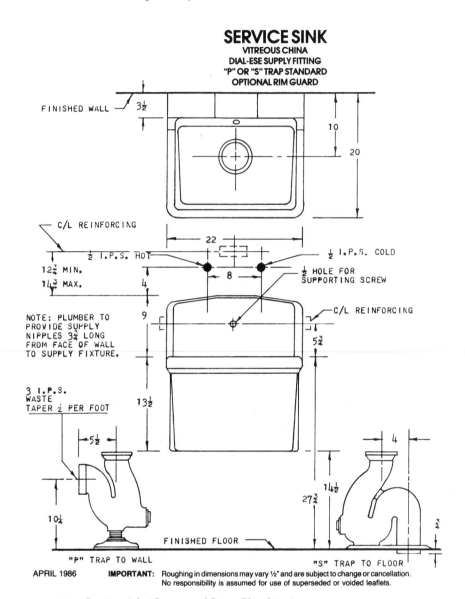

Figure 27.33 Service sink. (*Courtesy of Crane Plumbing.*)

NEU-MEDIC™
Vitreous China Plaster Sink

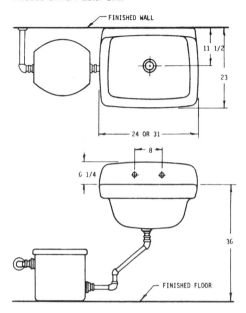

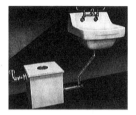

Neu-Medic 5H-258 Plaster Sink Vitreous China

Sink: 5H-258 **Neu-Medic** vitreous china plaster sink with 6¼" high back and return ends.

Supports:

Trim:

Supplies:

Waste:

Trap:

Exposed metal trim is polished chromium plated, unless otherwise described.

Figure 27.34   Detail of plaster sink in Fig. 27.33. (*Courtesy of Crane Plumbing.*)

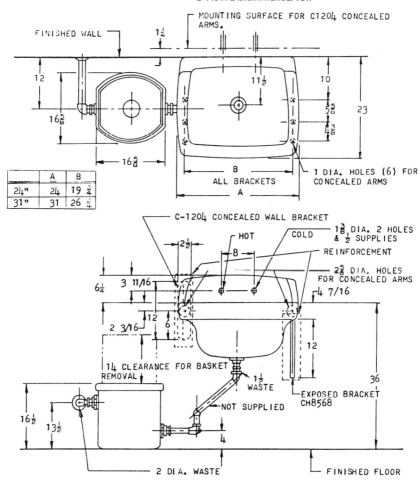

Figure 27.35   Plaster sink. (*Courtesy of Crane Plumbing.*)

28

Troubleshooting
Well Systems

Troubleshooting well systems is very difficult for some plumbers. Rural locations depend on wells and water pumps for their potable water, and plumbers moving from a city to a country location can be stymied by the many facets of well systems.

It is not only plumbers from cities who have trouble working with well systems. Those who work with new construction projects have little opportunity to get accustomed to servicing and repairing well systems. Plumbers who are forced into service work by a bad economy and a slowdown in building need a crash course in well systems.

There are three basic types of wells: drilled, driven, and dug wells. Drilled and dug wells are the most common.

As for water pumps, there are two types that are most common: submersible and jet pumps. Jet pumps can work with a one-pipe system or a two-pipe system, depending upon the depth of the well it is serving.

In addition to the well and the water pump, there are many other components involved in a well system. Control boxes, pressure gauges, tank tees, pressure tanks, and air-volume controls are just some of what a plumber has to know about in order to troubleshoot well systems.

A good electrical meter is a necessity for troubleshooting pumps, and a plumber must be able to evaluate meter readings to determine the causes of various types of problems.

Wells

We will begin our education on well systems with the wells themselves. Let's look at each type of well and see how the different types may affect the troubleshooting procedures used.

Driven wells

Driven wells typically are shallow and have small diameters. These wells are used for livestock and occasionally for residential use, but they are not dependable water sources during dry spells. It is not unusual for driven wells to run out of water.

If you are called to a property with a pump that runs and doesn't produce water, the problem may be that the water level has dropped below the pick-up point of the well pipe.

Most driven wells are equipped with a one-pipe, shallow-well jet pump.

Dug wells

Dug wells are much more common for residential uses than driven wells are. In areas with a high water table, such as many areas in the South, dug wells are abundant. These wells have a large diameter and are rarely over 40 feet deep. Many dug wells are less than 30 feet deep.

Modern dug wells are lined with concrete cylinders and capped with heavy concrete tops. This type of well can use a submersible pump, a two-pipe jet pump, or a one-pipe jet pump, depending on the lift of the water. If the lift of water is 25 feet or less, a one-pipe jet pump is normally installed. For deeper wells, either a two-pipe jet pump or a submersible pump is used.

Dug wells are susceptible to pollution and unwanted fill-in conditions. It is not uncommon for mud, sand, pebbles, or other debris to clog the end of a well pipe in a dug well.

Depending upon the depth of the water reservoir, dug wells can run dry during hot, dry months. If you have a pump that is not producing water from a dug well as it should be, check for fill-in conditions and insufficient water depth.

Drilled wells

Drilled wells are the most dependable type of well. Since these wells are drilled deep into the earth, they are usually unaffected by common dry spells.

Drilled wells usually have a modest diameter, a steel casing, and a depth of 100 feet or more. Submersible pumps are the pump of choice for most drilled wells. A two-pipe jet pump can be used with drilled wells with depths of less than 100 feet, but submersible pumps are typically more efficient.

Well-System Components

The number of well-system components a plumber must be familiar with is substantial. Before we dig into the specifics of troubleshooting well systems, let's get acquainted with the many components you may have to deal with.

Shallow-well pumps

Shallow-well pumps use only one pipe to pull water from a well, but the depth of the well should not exceed 25 feet. These pumps work well for shallow wells, but they cannot function when the lift requirements are much more than 25 feet. This type of pump is often called a jet pump. These pumps are installed outside of the well.

Deep-well pumps

Deep-well pumps are very similar in appearance to shallow-well pumps. However, these pumps require a two-pipe system to work properly. One pipe pushes water down into the piping system, and the second pipe pulls the water up to the pressure tank. This type of jet pump is suitable for wells with depths ranging from 25 to 100 feet. These pumps are installed outside of the well.

Submersible pumps

Submersible pumps can be used with any well where the water will cover the pump. Submersible pumps are suspended under the water level in wells.

Rather than pulling water up from the well, submersible pumps push the water up a single pipe to a pressure tank. The design of submersible pumps makes them ideal for deep wells.

The type of pump chosen for a given system is determined largely by the depth of the well. In all cases the pump should not be rated at a higher delivery rate than the recovery rate of the well. If it is, the pump can deplete the water supply in the well. When a well's recovery rate is sufficient, a pump with a 5-gallon-a-minute rating is normally used.

Pressure tanks

Pressure tanks are where water pumped from a well is stored. The use of a pressure tank reduces the wear on a pump and extends its life. Without a pressure tank, a pump would have to cut on and off with each demand for water. The pressure tank allows water to be used to a certain point before the pump must cut on to replenish the water supply.

Most modern pressure tanks are built with an internal diaphragm (Fig. 28.1) to prevent waterlogging. Older tanks don't have this feature and are much more likely to loose their air cushion, resulting in a pump that must cycle more often than it should.

Air-volume controls

Air-volume controls work to maintain the proper volume of air in a pressure tank. Float-type air-volume controls regulate the proper air volume by opening a valve and providing air from the outside atmosphere when the water level rises too high.

Diaphragm-type air-volume controls are more common than float-type controls. These devices allow air to enter the pressure tank each time the pump stops if there is not enough air.

Pressure switches

Pressure switches open and close the electrical circuit for a pump based on preset points. When the pressure in the pressure tank drops to a predetermined point, the pressure switch cuts on and demands water from the pump. As the water pressure reaches the preset limit, the pressure switch cuts the pump off.

Typical settings for pressure switches have them cut on when the pressure drops to 20 pounds per square inch and off when the pressure reaches 40 pounds per square inch. The switch can be set with a cuton of 30 pounds per square inch and a cutoff of 50 pounds per square inch or at any other reasonable settings. There is typically a 20-pound difference between the cutin and the cutout setting.

Pressure gauge

A pressure gauge is used to determine the pressure being produced by the well system (Fig. 28.2).

Relief valve

Relief valves are needed with well systems to prevent excessive pressure buildups. These safety devices should be installed on the discharge side of a pump. As a rule, the relief valve blowoff setting is usually about 20 pounds per square inch higher than the cutout setting on the pressure switch.

1. WELL-X-TROL has a sealed-in air chamber that is pre-pressurized before it leaves our factory. Air and water do not mix eliminating any chance of "waterlogging" through loss of air to system water.

2. When the pump starts, water enters the WELL-X-TROL as system pressure passes the minimum pressure precharge. Only usable water is stored.

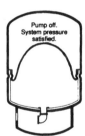

3. When the pressure in the chamber reaches maximum system pressure, the pump stops. The WELL-X-TROL is filled to maximum capacity.

4. When water is demanded, pressure in the air chamber forces water into the system. Since WELL-X-TROL does not waterlog and consistently delivers the maximum usable water, minimum pump starts are assured.

Figure 28.1 Diaphragm-type well pressure tank. (*Courtesy of Amtrol, Inc.*)

Figure 28.2 Pressure gauge and pressure switch.

Foot valves

Foot valves act as both a strainer and a check valve. They are installed on the submerged end of suction pipes in wells. The strainer prevents sand and pebbles from entering the pipe, and the check-valve action maintains the prime in the suction line when the pump is not running.

Control boxes

Control boxes (Figs. 28.3 and 28.4) are used with submersible pumps and are the heart of the electrical system. The wiring coming from the pump enters the control box and is routed out to the pressure switch.

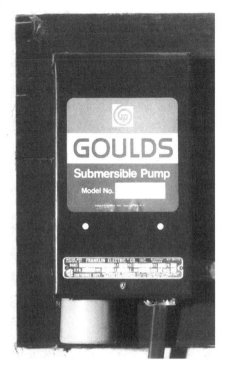

Figure 28.3   Control box.

Figure 28.4   Typical deep-well setup of controls.

Other elements of a well system

Other elements of a well system are numerous. There is, of course, the well piping and various fittings, drains, and a multitude of other gadgets and gizmos. We will discuss them as we come to them in our troubleshooting tour.

Troubleshooting Jet Pumps

Troubleshooting jet pumps (Figs. 28.5 through 28.7 and Tables 28.1 and 28.2) is the first hard look we are going to take at pump problems. We will progress in a logical, step-by-step manner that will be easy to understand. Before we dive right into pump problems, let's examine how to use an amprobe and an ohmmeter. These meter readings are essential to the effective troubleshooting of pumps.

Amprobe

An amprobe is a multirange meter that combines the use of an ammeter and a voltmeter. The scales of the voltmeter range from 150 to 600 volts. On the ammeter you will find scales from 5 to 40 amps and from 15 to 100 amps.

The ammeter should be held with its tongs around the wire being tested, with the rotary scale set at the 100-amp range. With this done, rotate the scale back to the smaller numbers until you arrive at an exact reading of the amperage.

When an amprobe is used as a voltmeter, the two leads are connected to the bottom of the meter, with the rotary scale set at 600 volts. When the voltage reading is less than 150 volts, the rotary scale should be set down to 150 volts to obtain a more accurate reading.

Ohmmeter

An ohmmeter measures electrical resistance, and each unit of measure is 1 ohm. If you look at the settings on an ohmmeter, you will notice that there are six of them. The readings are as follows:

- RX1
- RX100
- RX10K
- RX10
- RX1000
- RX100K

The round knob found in the center of an ohmmeter is used to

1. What well conditions might possibly limit the capacity of the pump?	Rate of flow from the source of supply, the diameter of a cased deep well and the pumping level of the water in a cased deep well.
2. How does the diameter of a cased deep well and pumping level of the water affect the capacity?	Limits the size pumping equipment which can be used.
3. If there are no limiting factors, how is capacity determined?	Maximum number of outlets or faucets likely to be in use at the same time.
4. What is suction?	A partial vacuum created in suction chamber of pump obtained by removing pressure due to atmosphere, thereby allowing greater pressure outside to force something (air, gas, water) into the container.
5. What is atmospheric pressure?	The atmosphere surrounding the earth presses against the earth and all objects on it, producing what we call atmospheric pressure.
6. How much is the pressure due to atmosphere?	This pressure varies with elevation or altitude. It is greatest at sea level (14.7 pounds per square inch) and gradually decreases as elevation above sea level is increased. At the rate of approximately 1 foot per 100 feet of elevation.
7. What is maximum theoretical suction lift?	Since suction lift is actually that height to which atmospheric pressure will force water into a vacuum, theoretically we can use the maximum amount of this pressure 14.7 pounds per square inch at sea level which will raise water 33.9 feet. From this, we obtain the conversion factor of 1 pound per square inch of pressure equals 2.31-feet head.
8. How does friction loss affect suction conditions?	The resistance of the suction pipe walls to the flow of water uses up part of the work which can be done by atmospheric pressure. Therefore, the amount of loss due to friction in the suction pipe must be added to the vertical elevation which must be overcome, and the total of the two must not exceed 25 feet at sea level. This 25 feet must be reduced 1 foot for every 1000-feet elevation above sea level, which corrects for a lessened atmospheric pressure with increased elevation.
9. When and why do we use a deep-well jet pump?	The resistance of the suction pipe walls to below the pump because this is the maximum practical suction lift which can be obtained with a shallow-well pump at sea level.

Figure 28.5 Questions and answers about pumps. (*Courtesy of A. Y. McDonald Mfg. Co.*)

10. What do we mean by water system?	A pump with all necessary accessories, fittings, etc., necessary for its completely automatic operation.
11. What is the purpose of a foot valve?	It is used on the end of a suction pipe to prevent the water in the system from running back into the source of supply when the pump isn't operating.
12. Name the two basic parts of a jet assembly.	Nozzle and diffuser.
13. What is the function of the nozzle?	The nozzle converts the pressure of the driving water into velocity. The velocity thus created causes a vacuum in the jet assembly or suction chamber.
14. What is the purpose of the diffuser?	The diffuser converts the velocity from the nozzle back into pressure.
15. What do we mean by "driving water"?	That water which is supplied under pressure to drive the jet.
16. What is the source of the driving water?	The driving water is continuously recirculated in a closed system.
17. What is the purpose of the centrifugal pump?	The centrifugal pump provides the energy to circulate the driving water. It also boosts the pressure of the discharged capacity.
18. Where is the jet assembly usually located in a shallow-well jet system?	Bolted to the casing of the centrifugal pump.
19. What is the principal factor which determines if a shallow-well jet system can be used?	Total suction lift.
20. When is a deep-well jet system used?	When the total suction lift exceeds that which can be overcome by atmospheric pressure.
21. Can a foot valve be omitted from a deep-well jet system? Why or why not?	No, because there are no valves in the jet assembly, and the foot valve is necessary to hold water in the system when it is primed. Also, when the centrifugal pump isn't running, the foot valve prevents the water from running back into the well.

Figure 28.6 Questions and answers about pumps. (*Courtesy of A. Y. McDonald Mfg. Co.*)

adjust the meter to zero when the two leads are clipped together. This must be done whenever the range selection for resistance is altered.

It is important to note that an ohmmeter should only be used after the electrical power to the wiring being tested has been turned off.

22. What is the function of a check valve in the top of a submersible pump?	To hold the pressure in the line when the pump isn't running.
23. A submersible pump is made up of two basic parts. What are they?	Pump end and motor.
24. Why did the name submersible pump come into being?	Because the whole unit, pump and motor, is designed to be operated under water.
25. Can a submersible pump be installed in a 2-inch well?	No, they require a 4-inch well or larger for most domestic use. Larger pumps with larger capacities require 6-inch wells or larger.
26. A stage in a submersible pump is made up of three parts. What are they?	Impeller, diffuser, and bowl.
27. Does a submersible pump have only one pipe connection?	Yes, the discharge pipe.
28. What are two reasons we should always consider using a submersible first?	It will pump more water at higher pressure with less horsepower. Easier installation.
29. The amount of pressure a pump is capable of making is controlled by what?	The diameter of the impeller.
30. What do the width of an impeller and guide vane control?	The amount of water or capacity the pump is capable of pumping.

Figure 28.7 Questions and answers about pumps. (*Courtesy of A. Y. McDonald Mfg. Co.*)

A pump that will not run

A pump that will not run can be suffering from one of many failures. The first thing to check is the fuse or circuit breaker to the circuit. If the fuse is blown, replace it. When the circuit breaker has tripped, reset it.

When the fuse or circuit breaker is not at fault, check for broken or loose wiring connections. Bad connections account for many pump failures.

It is possible the pump won't run because of a motor overload protection device. If the protection contacts are open, the pump will not function. This is usually a temporary condition that corrects itself.

If the pump is attempting to operate at the wrong voltage, it may not run. To test for this condition the power must be on, so be careful. Test the voltage with a volt-ammeter. With the leads attached to the meter and the meter set in the proper voltage range, touch the black lead to the white wire and the red lead to the black wire in the disconnect box near the pump. Test both the incoming and outgoing wiring.

TABLE 28.1   Troubleshooting Suggestions

Cause of trouble	Motor does not start Checking procedure	Corrective action
No power or incorrect voltage.	Using voltmeter, check the line terminals. Voltage must be ± 10% of rated voltage.	Contact power company if voltage is incorrect.
Fuses blown or circuit fuse breakers tripped.	Check fuses for recommended size and check for loose, dirty, or corroded connections in fuse receptacle. Check for tripped circuit breaker.	Replace with proper or reset circuit breaker.
Defective pressure switch.	Check voltage at contact points. Improper contact of switch points can cause voltage less than line voltage.	Replace pressure switch or clean points.
Control box malfunction.	For detailed procedure, see pages 28.11 and 28.14 to 28.15.	Repair or replace.
Defective wiring.	Check for loose or corroded connections. Check motor lead terminals with voltmeter for power.	Correct faulty wiring or connections.
Bound pump	Locked rotor conditions can result from misalignment between pump and motor or a sand bound pump. Amp readings 3 to 6 times higher than normal will be indicated.	If pump will not start with several trials, it must be pulled and the cause corrected. New installations should always be run without turning off until water clears.
Defective cable or motor.	For detailed procedure, see pages 28.3, 28.4, 28.6, and 28.7.	Repair or replace.
	Motor starts too often	
Pressure switch.	Check setting on pressure switch and examine for defects.	Reset limit or replace switch.
Check valve, stuck open.	Damaged or defective check valve will not hold pressure.	Replace if defective.
Waterlogged tank (air supply).	Check air-charging system for proper operation.	Clean or replace.
Leak in system.	Check system for leaks.	Replace damaged pipes or repair leaks.

source: A. Y. McDonald Manufacturing Co

Your next step in the testing process should be at the pressure switch. The black lead should be placed on the black wire, and the red lead should be put on the white wire for this test. There should be a plate on the pump that identifies the proper working voltage. Your

TABLE 28.2 Troubleshooting Suggestions

	Motor runs continuously	
Causes of trouble	Checking procedure	Corrective action
Pressure switch.	Switch contacts may be "welded" in closed position. Pressure switch may be set too high.	Clean contacts, replace switch, or readjust setting.
Low-level well.	Pump may exceed well capacity. Shut off pump, wait for well to recover. Check static and drawdown level from well head.	Throttle pump output or reset pump to lower level. Do not lower if sand may clog pump.
Leak in system.	Check system for leaks.	Replace damaged pipes or repair leaks.
Worn pump.	Symptoms of worn pump are similar to that of drop pipe leak or low water level in well. Reduce pressure switch setting. If pump shuts off, worn parts may be at fault. Sand is usually present in tank.	Pull pump and replace worn impellers, casing, or other close fitting parts.
Loose or broken motor shaft.	No or little water will be delivered if coupling between motor and pump shaft is loose or if a jammed pump has caused the motor shaft to shear off.	Check for damaged shafts if coupling is loose, and replace worn or defective units.
Pump screen blocked.	Restricted flow may indicate a clogged intake screen on pump. Pump may be installed in mud or sand.	Clean screen and reset at less depth. It may be necessary to clean well.
Check valve stuck closed.	No water will be delivered if check valve is in closed position.	Replace if defective.
Control box malfunction.	See pages 28.11, 28.14, and 28.15 for single phase.	Repair or replace.
	Motor runs but overload protector tips	
Incorrect voltage.	Using voltmeter, check the line terminals. Voltage must be within ± 10% of rated voltage.	Contact power company if voltage is incorrect.
Overheated protectors.	Direct sunlight or other heat source can make control box hot, causing protectors to trip. The box must not be hot to touch.	Shade box, provide ventilation, or move box away from heat source.
Defective control box.	For detailed procedures, see pages 28.11, 28.14, and 28.15.	Repair or replace.
Defective motor or cable.	For detailed procedures, see pages 28.3, 28.4, 28.6, and 28.7.	Repair or replace.
Worn pump or motor.	Check running current. See pages 28.15 to 28.17.	Replace pump and/or motor.

SOURCE: A. Y. McDonald Manufacturing Co.

test should reveal voltage that is within 10 percent of the recommended rating.

An additional problem that you may encounter is a pump that is mechanically bound. You can check this by removing the end cap and turning the motor shaft by hand. It should rotate freely.

A bad pressure switch can keep a pump from running. Remove the cover from the pressure switch, and you will see two springs, one tall and one short. These springs are depressed and held in place by individual nuts.

The short spring is preset at the factory and should not need adjustment. This adjustment controls the cutout sequence for the pump. If you turn the nut down, the cutout pressure will be increased. Loosening the nut will lower the cutout pressure.

The long spring can be adjusted to change the cutin and cutout pressure for the pump. If you want to set a higher cutin pressure, turn the nut tighter to depress the spring further. To reduce the cutin pressure you should loosen the nut to allow more height on the spring. If the pressure switch fails to respond to the adjustments, it should be replaced.

It is also possible that the tubing or fittings on the pressure switch are plugged. Take the tubing and fittings apart and inspect them. Remove any obstructions and reinstall them.

The last possibility for the pump failure is a bad motor. Use an ohmmeter to check the motor after the power to the pump has been turned off.

Start checking the motor by disconnecting the motor leads. We will call these leads L1 and L2. The instructions you are about to receive are for Goulds pumps with motors rated at 230 volts. When you are conducting the test on different types of pumps, you should refer to the manufacturer's recommendations.

Set the ohmmeter to RX100 and adjust the meter to zero. Put one of the meter's leads on a ground screw. The other lead should be systematically touched to all terminals on the terminal board, switch, capacitor, and protector. If the needle on your ohmmeter doesn't move as these tests are made, the ground check of the motor is okay.

The next check to be conducted is for winding continuity. Set the ohmmeter to RX1 and adjust it to zero. You will need a thick piece of paper for this test; it should be placed between the motor switch points and the discharge capacitor. You should read the resistance between L1 and A to see that it is the same as the resistance between A and yellow. The reading between yellow to red should be the same as L1 to the same red terminal.

The next test is for the contact points of the switch. Set the ohmmeter to RX1 and adjust it to zero. Remove the leads from the switch and attach the meter leads to each side of the switch; you should see a reading of zero. If you flip the governor weight to the run position, the reading on your meter should be infinity.

Now let's check the overload protector. Set your meter to RX1 and adjust it to zero. With the overload leads disconnected, check the resistance between terminals 1 and 2 and then between 2 and 3. If reading of more than 1 occurs, replace the overload protector.

The capacitor can also be tested with an ohmmeter. Set the meter to RX1000 and adjust it to zero. With the leads disconnected from the capacitor, attach the meter leads to each terminal. When you do this, you should see the meter's needle go to the right and drift slowly to the left. To confirm your reading, switch positions with the meter leads and see if you get the same results. A reading that moves toward zero or a needle that doesn't move at all indicates a bad capacitor.

Pump runs but gives no water

When a pump runs but gives no water, you have seven possible problems to check out. Let's take a look at each troubleshooting phase in its logical order.

The first consideration should be that of the pump's prime. If the pump or the pump's pipes are not completely primed, water will not be delivered.

For a shallow-well pump you should remove the priming plug and fill the pump completely with water. You may want to disconnect the well pipe at the pump and make sure it is holding water. You could spend considerable time pouring water into a priming hole only to find out the pipe was not holding the water.

For deep-well jet pumps, you must check the pressure-control valves. The setting must match the horsepower and jet assembly used, so refer to the manufacturer's recommendations.

Turning the adjustment screw to the left will reduce pressure, and turning it to the right will increase pressure. When the pressure-control valve is set too high, the air-volume control cannot work. If the pressure setting is too low, the pump may shut itself off.

If the foot valve or the end of the suction pipe has become obstructed or is suspended above the water level, the pump cannot produce water. Sometimes shaking the suction pipe will clear the foot valve and get the pump back into normal operation. If you are working with a two-pipe system, you will have to pull the pipes and do a visual inspection. However, if the pump you are working on is a one-pipe pump, you can use a vacuum gauge to determine if the suction pipe is blocked.

If you install a vacuum gauge in the shallow-well adapter on the pump, you can take a suction reading. When the pump is running, the gauge will not register any vacuum if the end of the pipe is not below the water level or if there is a leak in the suction pipe.

An extremely high vacuum reading, such as 22 inches or more, indicates that the end of the pipe or the foot valve is blocked or buried in mud. It can also indicate that the suction lift exceeds the capabilities of the pump.

A common problem when the pump runs without delivering water is a leak on the suction side of the pump. You can pressurize the system and inspect it for these leaks.

The air-volume control can be at fault for a pump that runs dry. If you disconnect the tubing and plug the hole in the pump, you can tell if the air-volume control has a punctured diaphragm. If plugging the pump corrects the problem, you must replace the air-volume control.

Sometimes the jet assembly will become plugged up. When this happens with a shallow-well pump, you can insert a wire through the $\frac{1}{2}$-inch plug in the shallow-well adapter to clear the obstruction. With a deep-well jet pump, you must pull the piping out of the well and clean the jet assembly.

An incorrect nozzle or diffuser combination can result in a pump that runs but that produces no water. Check the ratings in the manufacturer's literature to be sure the existing equipment matches them.

The foot valve or an in-line check valve could be stuck in the closed position. This type of situation requires a physical inspection and the probable replacement of the faulty part.

Pump cycles too often

When a pump cycles on and off too often, it can wear itself out prematurely. This type of problem can be caused by several things. For example, leaks in the piping or pressure tank would cause frequent cycling of the pump.

The pressure switch may be responsible for a pump that cuts on and off to often. If the cutin setting on the pressure gauge is set too high, the pump will work harder than it should.

If the pressure tank becomes waterlogged (filled with too much water and not enough air), the pump will cycle frequently. If the tank is waterlogged, it will have to be recharged with air. This would also lead you to suspect that the air-volume control is defective.

An insufficient vacuum could cause the pump to run too often. If the vacuum does not hold at 3 inches for 15 seconds, it might be the problem.

The last thing to consider is the suction lift. It's possible that the pump is getting too much water and creating a flooded suction. This can be remedied by installing and partially closing a valve in the suction pipe.

Won't develop pressure

Sometimes a pump will produce water but will not build the desired pressure in the holding tank. Leaks in the piping or pressure tank can cause this condition.

If the jet or the screen on the foot valve is partially obstructed, the same problem may result.

A defective air-volume control may prevent the pump from building suitable pressure. You can test for this by removing the air-volume control and plugging the hole where it was removed. If this solves the problem, you know the air-volume control is bad.

A worn impeller hub or guide vane bore could result in a pump that will not build enough pressure. The proper clearance should be 0.012 on a side, or 0.025 diametrically.

With a shallow-well system, the problem could be being caused by the suction lift being too high. You can test for this with a vacuum gauge. The vacuum should not exceed 22 inches at sea level. For deep-well jet pumps you must check the rating tables to establish their maximum jet depth. Also with deep-well jet pumps, you should check the pressure-control valve to see that it is set properly.

Switch fails to cut out

If the pressure switch fails to cut out when the pump has developed sufficient pressure, you should check the settings on the pressure switch. Adjust the nut on the short spring and see if the switch responds; if it doesn't, replace the switch.

Another cause for this type of problem could be debris in the tubing or fittings between the switch and pump. Disconnect the tubing and fittings and inspect them for obstructions.

We have now covered the troubleshooting steps for jet pumps, but before we move on to submersible pumps, look over the illustrations I've given you that show engineering data, multistage jet pumps, and typical installation procedures (Figs. 28.8 through Fig. 28.12).

Troubleshooting Submersible Pumps

There are some major differences between troubleshooting submersible pumps and jet pumps. One of the most obvious differences is that jet pumps are installed outside of wells, and submersible pumps are installed below the water level of wells (Fig. 28.13).

WX-100 Series WELL-X-TROL with tank mounted at jet pump.

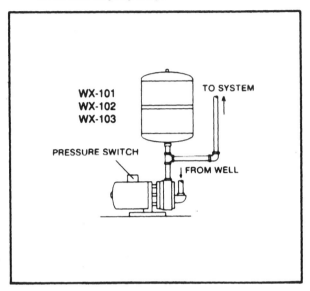

Figure 28.8 Pressure tank mounted at jet pump. (*Courtesy of Amtrol, Inc.*)

WX-100 Series WELL-X-TROL installed on-line with jet pump.

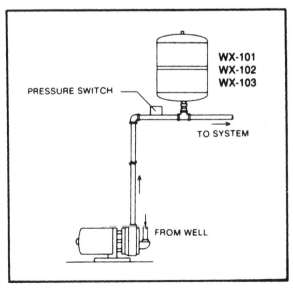

Figure 28.9 In-line pressure tank. (*Courtesy of Amtrol, Inc.*)

WX-103 or WX-200 with #162 pump stand.
WX-200 with #163 pump stand mounting jet
pump.

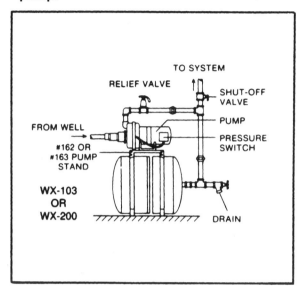

Figure 28.10 Jet pump mounted on pressure tank. (*Courtesy of Amtrol, Inc.*)

There are times when a submersible pump must be pulled out of a well, and this can be quite a chore. Even with today's lightweight well pipe, the strength and endurance needed to pull a submersible pump up from a deep well is considerable. Plumbers who work with submersible pumps regularly often have a pump puller to make removing the pumps easier.

When a submersible pump is pulled, you must allow for the length of the well pipe when planning the direction to pull from and where the pipe and pump will lay once removed from the well. It is not unusual to have between 100 and 200 feet of well pipe to deal with, and some wells are even deeper.

It is important when pulling a pump, or lowering one back into a well, that the electrical wiring does not rub against the well casing. If the insulation on the wiring is cut, the pump will not work properly. Let's look now at some specific troubleshooting situations.

Pumps that won't start

Pumps that won't start may be victims of a blown fuse or tripped circuit breaker. If these conditions are okay, turn your attention to the voltage.

WX-201 through WX-350 WELL-X-TROL installed on-line with jet pump.

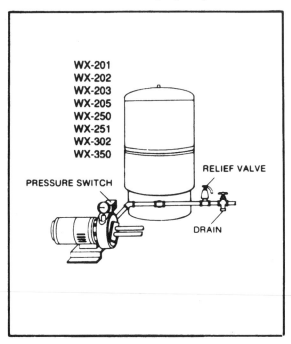

Figure 28.11 Jet pump mounted on floor, next to pressure tank. (*Courtesy of Amtrol, Inc.*)

In the following scenarios we will be dealing with Goulds pumps and Q-D-type control boxes.

To check the voltage, remove the cover of the control box to break all motor connections. Be advised: Wires L1 and L2 are still connected to electrical power. These are the wires running to the control box from the power source.

Press the red lead from your voltmeter to the white wire and the black lead to the black wire. Keep in mind that any major electrical appliance that might be running at the same time the pump would be, like a clothes dryer, should be turned on while you are conducting your voltage test.

Once you have a voltage reading, compare it to the manufacturer's recommended ratings. For example, with a Goulds pump that is rated for 115 volts, the measured volts should range from 105 to 125. A pump with a rating of 208 volts should measure a range from 188 to 228 volts. A pump rated at 230 volts should measure between 210 and 250 volts.

WX-201 through WX-203 using #161 pump stand. WX-205 through WX-350 using #165 pump stand. Shallow well jet pump mounted on tank.

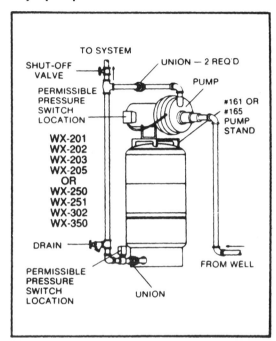

Figure 28.12   Jet pump mounted on pressure tank. (*Courtesy of Amtrol, Inc.*)

If the voltage is okay, check the points on the pressure switch. If the switch is defective, replace it.

The third likely cause of this condition is a loose electrical connection in the control box, the cable, or the motor. Troubleshooting for this condition requires extensive work with your meters.

To begin the electrical troubleshooting, we will look for electrical shorts by measuring the insulation resistance. Use an ohmmeter for this test; the power to the wires you are testing should be turned off.

Set the ohmmeter scale to RX100K and adjust it to zero. You will be testing the wires coming out of the well, from the pump, at the well head. Put one of the ohmmeter's lead to any one of the pump wires and place the other ohmmeter lead on the well casing or a metal pipe. As you test the wires for resistance, you will need to know what the various readings mean, so let's examine this issue.

You will be dealing with normal ohm and megohm values.

WX-201 through WX-350 WELL-X-TROL installed on-line using submersible pump.

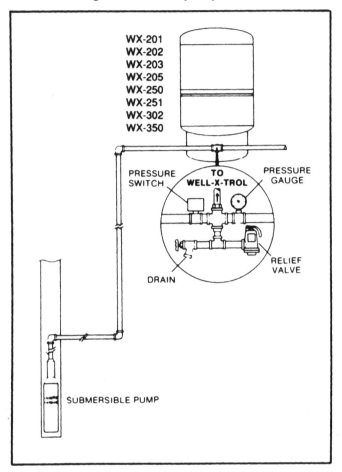

Figure 28.13   In-line pressure tank with submersible pump. (*Courtesy of Amtrol, Inc.*)

Insulation resistance does not vary with ratings. Regardless of the motor, horsepower, voltage, or phase rating, the insulation resistance remains the same.

A new motor that has not been installed should have an ohm value of 20,000,000 or more and a megohm value of 20. A motor that has been used but is capable of being reinstalled should produce an ohm reading of 10,000,000 or more and a megohm reading of 10.

For a motor that is installed in the well, which will be the case in most troubleshooting, the readings will be different. A new motor installed with its drop cable should give an ohm reading of 2,000,000 or more and a megohm value of 2. An installed motor in a well that is not new but is in good condition will present an ohm reading of between 500,000 and 2,000,000. Its megohm value will be between 0.5 and 2.

A motor that gives a reading in ohms of between 20,000 and 500,000 and a megohm reading of between 0.02 and 0.5 may have damaged leads or may have been hit by lightning; however, don't pull the pump yet.

You should pull the pump when the ohm reading ranges from 10,000 to 20,000 and the megohm value drops to between 0.01 and 0.02. These readings indicate a motor that is damaged or cables that are damaged. While a motor in this condition may run, it probably won't run for long.

When a motor has failed completely or the insulation on the cables has been destroyed, the ohm reading will be less than 10,000 and the megohm value will be between 0 and 0.01.

With this phase of the electrical troubleshooting done, we are ready to check the winding resistance. You will have to refer to charts as reference for correct resistance values, and you will have to make adjustments if you are reading the resistance through the drop cables. I'll explain more about this in a moment.

If the ohm value is normal during your test, the motor windings are not grounded and the cable insulation is intact. When the ohm readings are below normal, you will have discovered that either the insulation on the cables is damaged or the motor windings are grounded.

To measure winding resistance with the pump still installed in the well, you will have to allow for the size and length of the drop cable. Assuming you are working with copper wire, you can use the following figures to obtain the resistance of cable for each 100 feet in length and ohms per pair of leads:

Cable size	Resistance
14	0.5150
12	0.3238
10	0.2036
8	0.1281
6	0.08056
4	0.0506
2	0.0318

If aluminum wire is being tested, the readings will be higher. Divide the ohm readings above by 0.61 to determine the actual resistance of aluminum wiring.

If you pull the pump and check the resistance for the motor only (not with the drop cables being tested), you will use different ratings. You should refer to a chart supplied by the manufacturer of the motor for the proper ratings.

When all the ohm readings are normal, the motor windings are fine. If any of the ohm values are below normal, the motor is shorted. An ohm value that is higher than normal indicates that the winding or cable is open or that there is a poor cable joint or connection. Should you encounter some ohm values that are higher than normal while others are lower than normal, you have found a situation where the motor leads are mixed up and need to be attached in their proper order.

If you want to check an electrical cable or a cable splice, you will need to disconnect the cable and have a container of water to submerge it in; a bathtub will work. Start by submerging the entire cable, except for the two ends, in water. Set your ohmmeter to RX100K and adjust it to zero. Put one of the meter leads on a cable wire and the other to a ground. Test each wire in the cable with this same procedure.

If at any time the meter's needle goes to zero, remove the splice connection from the water and watch the needle. If the needle falls back to give no reading, the leak is in the splice.

Once the splice is ruled out, you have to test sections of the cable in a similar manner. In other words, once you have activity on the meter, you should slowly remove sections of the cable until the meter settles back into a no-reading position. When this happens, you have found the section that is defective. At this point, the leak can be covered with waterproof electrical tape and reinstalled, or you can replace the cable.

A pump that will not run

A pump that will not run can require extensive troubleshooting. Start with the obvious and make sure the fuse is not blown and the circuit breaker is not tripped. Also check to see that the fuse is of the proper size.

Incorrect voltage can cause a pump to fail. You can check the voltage as described in the electrical troubleshooting section above.

Loose connections, damaged cable insulation, and bad splices, as discussed above, can prevent a pump from running.

The control box can have a lot of influence on whether or not a pump will run. If the wrong control box has been installed or if the box is located in an area where temperatures rise to over 122 degrees F, the pump may not run.

When a pump will not run, you should check the control box out carefully. We will be working with a quick-disconnect-type box. Start by checking the capacitor with an ohmmeter. First, discharge the capacitor before testing. You can do this by putting the metal end of a screwdriver between the capacitor's clips. Set the meter to RX1000 and connect the leads to the black and orange wires out of the capacitor case. You should see the needle start toward zero and then swing back to infinity. Should you have to recheck the capacitor, reverse the ohmmeter leads.

The next check involves the relay coil. If the box has a potential relay (three terminals), set your meter on RX1000 and connect the leads to red and yellow wires. The reading should be between 700 and 1800 ohms for 115-volt boxes. A 230-volt box should read between 4500 and 7000 ohms.

If the box has a current relay coil (four terminals), set the meter on RX1 and connect the leads to black wires at terminals 1 and 3. The reading should be less than 1 ohm.

In order to check the contact points, set your meter on RX1 and connect to the orange and red wires in a three-terminal box. The reading should be zero. For a four-terminal box, set the meter at RX1000 and connect to the orange and red wires. The reading should be near infinity.

Now you are ready to check the overload protector with your ohmmeter. Set the meter at RX1, and connect the leads to the black wire and the blue wire. The reading should be at a maximum of 0.5.

If you are checking the overload protector for a control box designed for $1\frac{1}{2}$ horsepower or more, set your meter at RX1 and connect the leads to terminals 1 and 3 on each overload protector. The maximum reading should not exceed 0.5 ohms.

A defective pressure switch or an obstruction in the tubing and fittings for the pressure switch could cause the pump not to run.

As a final option, the pump may have to be pulled and checked to see if it is bound. There should be a high amperage reading if this is the case.

Pump runs but doesn't produce water

When a submersible pump runs but doesn't produce water, there are several things that could be wrong. The first thing to determine is if the pump is submerged in water. If you find that the pump is submerged, you must begin your regular troubleshooting.

Loose connections or wires connected incorrectly in the control box could be at fault. The problem could be related to the voltage. A leak in the piping system could easily cause the pump to run without producing adequate water. A check valve could be stuck in the closed position. If the pump was just installed, the check valve may be installed backward.

Other options include a worn pump or motor, a clogged suction screen or impeller, and a broken pump shaft or coupling. You will have to pull the pump if any of these options are suspected.

Enough tank pressure

If you don't have enough tank pressure, check the setting on the pressure switch. If that's okay, check the voltage. Next, check for leaks in the piping system, and as a last resort, check the pump for excessive wear.

Frequent cycling

Frequent cycling is often caused by a waterlogged tank, as was described in the section on jet pumps. Of course, an improper setting on the pressure switch can cause a pump to cut on too often, and leaks in the piping can be responsible for the trouble.

You may find the problem is being caused by a check valve that has stuck in an open position. Occasionally the pressure tank will be sized improperly and cause problems. The tank should allow a minimum of 1 minute of running time for each cycle.

Get Familiar with Your Meters

Before you attempt to troubleshoot pumps, get familiar with your meters. Meter readings play a vital role in the successful troubleshooting of pumps, so you must know how to take, read, and interpret your meter.

Once you master your meter, you are well on your way to being a good troubleshooter for pump problems. Ideally, you should equip yourself with charts and tables from various pump manufacturers for reference in the field.

29

Troubleshooting Water Distribution Systems

Troubleshooting water distribution systems involves many problem possibilities. These systems can be simple residential installations or complex commercial jobs. Since a water distribution system consists primarily of pipe and fittings, many plumbers take the troubleshooting process for these piping layouts for granted.

It would seem logical that finding trouble with a water distribution system wouldn't be difficult, but it can be. If you've been in the trade for a number of years, you know I'm speaking the truth. Plumbers new to the trade may doubt the level of difficulty in troubleshooting water distribution systems, but they shouldn't.

I've been in the trade about 20 years, and I've spent many hours looking for problems in water distribution systems. Sometimes I've found them quickly, and sometimes I've been totally perplexed. Over all of these years I've wasted a lot of time because I didn't know where to look for the problems. With the experience I've gained, the hard way, I've learned where to look for plumbing problems and what possibilities to rule out quickly. This saves me time and my customers money.

With the help of this chapter, you will learn what to look for, where to look for it, and when to look for it when troubleshooting water distribution systems. As we move through this chapter, we will have to cover a lot of ground. There are a myriad of possibilities for problems with water distribution systems, and we will cover most, if not all, of them in the following pages.

Are you wondering what types of problems might crop up with a water distribution system? Well, I don't want to give away all of the secrets so soon, but I will give you a few hints of what we will be looking for.

Have you ever had a problem with copper tubing that developed pinhole leaks for no apparent reason? When was the last time a customer asked you to solve the problem of banging water pipes? Has water pressure ever been a complaint you have had to deal with? Do you know what to do when a house has frequent problems with faucets that leak? Well, we are going to answer all of these questions and a lot more as we travel through this chapter.

Noisy Pipes

Noisy pipes can make living around a plumbing system very annoying. Sometimes the pipes bang, sometimes they squeak, and sometimes they chatter. This condition can be so severe that living with the noise is almost unbearable. What causes noisy pipes? Water hammer is one reason pipes play their unfavorable tunes, but it is not the only reason.

Not all pipe noises are the same. The type of noise you hear gives a strong indication to the type of action that will be required to solve the problem. You must listen closely to the sounds being made in order to diagnose the problems properly.

What will you be listening for? The tone and type of noise being made will be all that you may have to go on in solving the problem of a water hammer. Let's look at the various noises individually and see what they mean and how you can correct the problem.

Water hammer

Water hammer is the most common cause of noisy pipes in a water distribution system. Banging pipes are a sure sign of water hammer. Can water hammer be stopped? Yes; there are ways to eliminate the actions and effects of water hammer, but implementing the procedure is not always easy.

What causes water hammer? Water hammer occurs most often with quick-closing valves, like ballcocks and washing-machine valves, but it can be a problem with other fixtures. The condition can be worsened when the water distribution pipes are installed with long, straight runs. When the water is shut off quickly, it bangs into the fittings at the end of the pipe run or at the fixture and produce the hammering or banging noise. The shock wave can produce some loud noises.

If a plumbing system is under higher than average pressure, it can be more likely to suffer from water hammer. There are several ways to approach the problem to eliminate the banging. To illustrate these options, I would like to put the problems into the form of real-world situations. I plan to share my past experiences with you in an entertaining and informative way.

A troublesome toilet

This first example of a water hammer problem involve a troublesome toilet. The residents of the home where the toilet was located hated the banging noise their water pipes made on most occasions when the toilet was flushed.

The toilet was located in the second-floor bathroom, and when it was flushed, it would rattle the pipes all through the home. There came a day when the homeowners decided they could not live with the problem any longer. They were fed up with the annoying banging of the pipes every time the toilet was flushed.

The couple called in a plumber to troubleshoot their problem. The plumber looked over the situation and went about his work in trying to locate the cause of the problem. He knew they were suffering from a water hammer situation, but he wasn't sure how to handle the call. In fact, the plumber gave up, washed his hands of the job, and left.

Distraught, the homeowners called my company. One of my plumbers responded to the call and quickly assessed that the upstairs toilet was the culprit. He recommended that the couple allow him to install an air chamber in the wall, near the toilet.

At first, the couple was reluctant to give their permission to my plumber to cut into their bathroom wall. Under the circumstances, however, there were not many viable options, so the homeowners gave their consent.

My plumber made a modest cut in the bathroom wall, around and above the closet supply, and installed an air chamber. This particular homeowner was one who wanted to know every move that was being made, and he was not too sure the plumber was doing anything that would really improve the situation.

After the air chamber was installed, my plumber activated the system and began to flush the toilet. After several sequences of flushing the pipes remained quiet; there was no banging. The homeowner had been doubtful of my plumber's decision, but he did admit the work solved the problem, and he was very happy.

Air chambers are frequently all that are needed to reduce or eliminate banging pipes (Fig. 29.1). My plumber was correct in his diagnosis of the problem and in effecting an efficient cure for the problem. It is possible, however, that the problem could have been solved by installing a master unit for a water hammer arrester in the basement of the home, eliminating the need to cut into the bathroom wall. There is, however, no guarantee that a master unit would have controlled the situation on the second floor.

The plumber made a wise and prudent decision. By installing the air chamber at the fixture, satisfaction was practically guaranteed. While the master unit may have worked, there was some risk that it

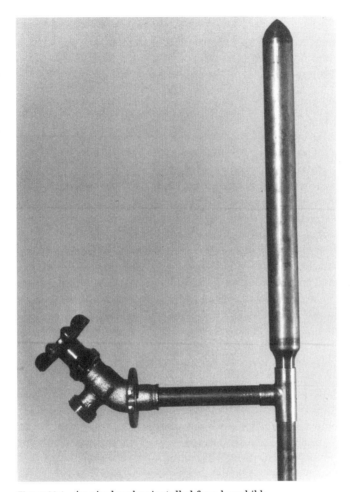

Figure 29.1   An air chamber installed for a hose bibb.

wouldn't. If the homeowner had objected to opening the bathroom wall, I'm sure the plumber would have installed a master unit in an attempt to control the problem.

Existing air chambers

Sometimes existing air chambers become waterlogged and fail to function properly. This problem is not uncommon, but it is a bother. If you have a plumbing system equipped with air chambers that is being affected by water hammer, you must recharge the exiting air chambers with a fresh supply of air. This is a simple process.

To recharge air chambers, you must first drain the water pipes of most of their reserve water. This can be done by opening a faucet or valve that is at the low end of the system and one that is at the upper end.

Once the water has drained out of the water distribution pipes you can close the faucets or valves and turn the water supply back on. As the new water fills the system, air will be replenished in the air chambers. As easy as this procedure is, it is all that is required to recharge waterlogged air chambers.

Unsecured pipes

Unsecured pipes are prime targets when you are having a noise problem. If the pipes that make up the water distribution system are not secured properly, they are likely to vibrate and make all sorts of noise. Not only is this annoying to the ears, it can be damaging to the pipes. Pipes that are not secure in their hangers can vibrate to the point that they wear holes in themselves. If enough stress is present, the connection joints may even be broken.

Local plumbing codes dictate how far apart the hangers for pipe may be. If the hangers fall within the guidelines of the local code and secure the pipes tightly, no problems should exist. However, when the hangers are farther apart than they should be, are the wrong size, or are not attached firmly, trouble can develop.

Pipes that are not secured in the manner described by the plumbing code may create loud banging sounds. This noise imitates the noise made by a system that is experiencing a water hammer. Squeaking and chattering noises may also be present when pipes are not secured properly.

The problem of poorly secured pipes is easy to fix, if you have access to the pipes. Unfortunately, the pipes and their hangers are often concealed by finished walls. Sometimes the problem pipes can be accessed in basements or crawlspaces, but as often as not, they are inaccessible, unless walls and ceilings are destroyed.

Once you have access to the pipes that are not secured in the proper fashion, all you have to do is add hangers or tighten the existing hangers. This is fine if you can get to the pipes easily, but it is a hard-sell to a homeowner who doesn't want the walls and ceilings of a home cut open. There is, however, no other way to eradicate the problem.

That squeaking noise

That squeaking noise you often hear from water pipes is almost always a hot-water pipe. Whether the water distribution pipe is copper or plastic, it tends to expand when it gets hot. As hot water moves through the pipe, the pipe expands and creates friction against the pipe hanger.

The expansion aspect of hot-water pipes is not practical to eliminate, but you can do something about the squeaking noise. The simple solution, when you have access to the pipe and hanger, is to install an insulator between the pipe and the support. This can be a piece of foam, a piece of rubber, or any other suitable insulator. Your only goal is to prevent the pipe from rubbing against the hanger when expansion in the pipe occurs.

Chattering

The chattering heard in plumbing pipes is not caused by cold temperatures; it is caused by problems in the faucets of fixtures. While this is more of a faucet and valve problem than a water distribution problem, we will cover it here since the noise is often associated with water pipes.

When you hear a chattering sound in a plumbing system, you should inspect the faucet stems and washers at nearby faucets. You will probably find that the faucet washer has become loose and is being vibrated, or fluttered as some plumbers say, by the water pressure between it and the faucet seat.

To solve the problem of chattering pipes, all you should have to do is tighten the washers in the faucets. Once the washers are tight, the noise should disappear.

Muffling the system at the water service

When you have a plumbing system that is banging because of water hammer, you can try muffling the system at the water service. This procedure doesn't always work, but sometimes it does.

Access to piping in many buildings is not readily available. Under these conditions the only easy way to attack a water hammer is at the water service or main water distribution pipe. By installing air chambers or water hammer arresters on sections of the available pipe, you may be able to eliminate the symptoms of water hammer throughout the building.

To install an air chamber or water arrester on the water service or main water distribution pipe is not a big job. Once the water is cut off, you simply have to cut in some tee fittings and install the air chambers or arresters. The risers that accept these devices should be at least 2 feet high, when possible.

Depending upon the severity of the water hammer problem, several devices may need to be installed at different locations along the water piping. Typically, one at the beginning of the piping, one at the end of the piping, and devices installed at the ends of branches will greatly reduce, if not eliminate, the problems associated with water hammers.

Adding offsets

Adding offsets to long runs of straight piping is another way to reduce the effects of water hammer. Long runs of piping invite the slamming and banging noise of a water hammer. If you cut out the straight sections and rework them with some offsets, you increase your odds of beating the banging noises.

Pinhole Leaks in Copper Tubing

Pinhole leaks in copper tubing are not uncommon in rural areas. These leaks can occur anywhere within the water distribution system, and they can cause a lot of damage to building materials, floors, and other items located in the buildings. These small leaks may drip a little water, or they may be large enough to produce a steady spray.

What causes pinhole leaks in copper tubing and pipe? Acidic water is the major cause of such problems. Many plumbing systems that derive their water supply from wells suffer from some degree of acidic water. When the pH rating of the water is low enough, the water can eat holes in the copper water pipes. It can also work to destroy faucet stems, pump parts, water tanks, fittings, and other elements of the plumbing system.

When a thin-wall copper tubing, like type M copper, is used to carry acidic water, it can become very tender. Squeezing the pipe with a pair of pliers can result in a crushed pipe and quite a leak. There are usually signs of acidic water that show up before the copper tubing is damaged. The evidence of acid being in the water is usually in the form of a blue-green stain in the plumbing fixtures.

Acid in the water distribution system will eat away at the bibb screws in faucets and other metal in the plumbing system. The deteriorating action of the acid can, in time, literally eat a plumbing system up from the inside out.

Low Water Pressure

Low water pressure often becomes a problem in some plumbing systems. Trying to take a shower in a home with low water pressure is very frustrating, and waiting several minutes for a water closet to refill its tank can be nerve-racking. There are many reasons, to be sure, why low water pressure will induce customers to call in professional plumbers.

What causes low water pressure? Ah, that's a loaded question; there are many possible causes for low water pressure. For buildings using pump systems, it could be a problem with the pressure tank or the pressure switch. Buildings on city water supplies may have pressure-reducing valves that need to be adjusted. Properties that are plumbed with old galvanized steel piping could be suffering from rust obstructions in the pipes. There are many reasons why a plumbing system may not have the desired water pressure. To expand on this, let's look at most of the reasons under a microscope.

Pressure tanks

Properties that obtain their water from private water sources depend on pressure tanks to give them the working pressure they want from their plumbing systems. If the pressure tanks become waterlogged, they cannot produce the type of water pressure they are designed to provide.

A waterlogged pressure tank gives a symptom that is hard for a serious troubleshooter to miss. When a pressure tank is waterlogged, the well pump cuts on very frequently, often every time a faucet is opened.

Sometimes waterlogged pressure tanks provide adequate pressure, even though they are forcing the pump to work much harder than it is intended to. At other times, there is a noticeable loss in pressure.

If you are faced with a building that has unsatisfactory water pressure and a well pump that cuts on frequently, you should take a close look at the pressure tank.

When a pressure tank is suspected of being waterlogged, it should be drained, recharged with air pressure, and then refilled with water. This is a simple process, and it can solve your pressure problems.

Pressure switches

Properties that are served by private wells depend not only on pressure tanks for their water pressure, but on pressure switches as well. If the cutin pressure on a well system is not set properly, it is possible for pressure to drop to unacceptable levels before the pump will produce more water.

A fast visual inspection of the pressure gauge on the well system will tell you if the system is maintaining a satisfactory working pressure. It is important, however, to make sure there is demand from a plumbing fixture while you are observing the pressure gauge.

If no plumbing fixtures are being called upon for water, the pressure gauge will remain static at its highest level. To be sure the system is maintaining an acceptable working pressure, you must put a demand on the system.

A pressure gauge that shows a sharp fall in pressure before the pump cuts on indicates that the cutin pressure is set too low. Going into the box of the pressure switch and adjusting the spring-loaded nut will correct this problem.

Pressure-reducing valves

Pressure-reducing valves that are not set properly can cause trouble in the form of low water pressure. If these valves are not adjusted to the proper settings, they can reduce water to little more than a trickle.

Sometimes the adjustment screws on pressure-reducing valves are turned down too tightly. If you are dealing with a low-pressure situation where a pressure-reducing valve is involved, check the level of the screw setting. Turning the adjustment screw counterclockwise will increase the water pressure on the system.

Pressure-reducing valves are not often the cause of a pressure problem, but they can be. It may be that the adjustment screw is set improperly, or it could be that the entire valve is bad. If you cut the water off at the street connection, you can check the building pressure by removing the pressure-reducing valve and installing a pressure gauge on the piping. This will eliminate, or confirm, any doubts you may be having about the pressure-reducing valve.

Galvanized pipe

If you have been in the plumbing trade long enough to have worked with much galvanized pipe, you know how badly it can rust, corrode, and build up obstructions. Any time you are working with a water distribution system that is made up of galvanized pipe, you shouldn't be surprised to find low water pressure.

I have cut out sections of galvanized pipe where the open pipe diameter wasn't large enough to allow the insertion of a common drinking straw. When the open diameter of the pipe is constricted, water pressure must drop. Any long-time plumber will tell you that galvanized water pipe is a nightmare waiting to happen.

Most plumbing systems plumbed with galvanized pipe are equipped with unions on various sections of the piping. If you loosen these unions, you are likely to find the cause of your low pressure. A quick look at the interior of the pipe will probably be all it will take to convince you.

The only real solution to low pressure from blocked galvanized pipe is the replacement of the water piping. This can be a big and expensive job, but it is the only way to solve the problem positively.

Undersized piping

Undersized piping can create problems with water pressure. If the pipes delivering water to fixtures are too small, the fixtures will not receive an adequate volume of water at a desired pressure. As a matter of fact, I was on just such a job earlier today.

I was called to a meeting hall to determine why their water pressure was not as good as the pressure in neighboring buildings. The building is served by a municipal water supply, so that narrowed the list of possibilities.

When I got to the job, I went into the kitchen to test the pressure. Before I ever made it to the kitchen sink I saw the problem.

The water heater for the building is located in the kitchen, in plain view. When I looked over at the heater, I almost couldn't believe what I was seeing. The piping at the heater from both the inlet and outlet connections was up to code, but once the pipes rose a foot or so above the water tank, all bets were off. The $\frac{3}{4}$-inch copper tubing was connected, with flare fittings, to $\frac{1}{4}$-inch copper tubing.

The $\frac{1}{4}$-inch tubing provided the incoming water to the water heater and carried the hot water out to the building's plumbing fixtures. I removed a few ceiling tiles and traced the path of the tubing.

This building is equipped with two bathrooms and a kitchen, and every fixture is supplied with water from the $\frac{1}{4}$-inch tubing. Now I'm not talking about $\frac{1}{4}$-inch branch feeds; I'm saying that all of the water distribution pipe is run with $\frac{1}{4}$-inch tubing.

It is obvious that fixtures designed to be fed by $\frac{3}{4}$- and $\frac{1}{2}$-inch tubing will not perform as well when their supply is reduced to a quarter of an inch.

In a case like this, the only option is to replace the illegal water piping with tubing that will meet code requirements.

Many times the downsizing of piping will not be as drastic as the job I've just described, but inferior piping is common in many rural areas. If you are responding to a complaint of low water pressure, check the pipe sizing.

Hard water

Hard water can be a cause of reduced water pressure. The scale that hard water allows to build up on the inside of water pipes, fixtures, and tanks can reduce water pressure by a noticeable amount. When you have a job that is giving you trouble with low pressure and no sign of a cause, inspect the interior of some piping to see if it is being blocked by a scale buildup.

Clogged filters

Clogged in-line filters are notorious for their ability to restrict water flow. People have these little filters put in to trap sediment, and they do. They trap the sediment so well that they eventually clog up and block the normal flow of water, thereby reducing water pressure.

When you respond to a low-pressure call, ask the property owner if any filters are installed on the water distribution system. If one is, check the filter to see if it needs to be replaced.

Stopped-up aerators

One of the most common, and simplest to fix, causes of low water pressure is stopped-up aerators. The little screens in the aerators stop up frequently when a house is served by a private water supply. Iron particles and mineral deposits are usually the cause for stoppages.

Aerators can be removed and will sometimes clean up, but many times they must be replaced. As long as you have an assortment of aerators on your service truck, this is one problem that is easy to find and to fix.

Too Much Water Pressure

Too much water pressure in a water distribution system can be dangerous and destructive. When fixtures are under too much pressure, they do not perform well over extended periods of time. The O-rings, stems, and other components of the plumbing system are not usually meant to work with pressures exceeding 80 pounds per square inch.

Extreme water pressure can be dangerous for the users of the plumbing system. For example, I once worked for a homeowner who had cut herself badly at the kitchen sink as a result of high water pressure.

The woman was holding a drinking glass under the kitchen faucet when she turned the water on to fill the glass. The water rushed out with such force that it knocked the glass out of her hand. The glass shattered on impact with the sink and pieces of the broken glass sliced into the woman's hand and arm.

High water pressure is easy to control. Water pressure can be reduced with a pressure-reducing valve and, in the case of well systems, with adjustments to the pressure switch.

Troubleshooting high water pressure is simple; all you have to do is look at the pressure gauge on the well system or install a pressure gauge on a faucet, usually a hose bibb. If the pressure is higher than 60 pounds per square inch, it should normally be lowered. Most residential properties operate well with a water pressure of between 40 and 50 pounds per square inch.

If you have a well system, all you have to do is to go into the pressure switch and adjust the cutout setting. In the case of city water service, you will have to install a pressure-reducing valve. If a pressure-reducing valve is already in place, you can lower the system pressure by turning the adjustment screw clockwise.

Water Leaks

Water leaks are probably the most common type of complaint plumbers receive about water distribution systems. Most of the time leaks are easy to find, and they are not usually too difficult for experience plumbers to repair. There are times, however, when the leaks are not so easy to find or to fix.

As the list of acceptable plumbing materials for water distribution systems has grown, so has the diversity of the types of piping and tubing used in water systems. You never know when you will be dealing with copper, polybutylene (PB), polyethylene (PE), chlorinated polyvinyl chloride (CPVC), galvanized steel, or maybe even brass. Just having enough variety of fittings and supplies on your service truck to deal with all of these types of materials can be a job.

Since there are so many possibilities for the makeup of a water distribution system, let's take the time to look at each type of material on its own. This will allow us to see exactly what problems are unique to the different types of pipes.

Copper leaks

Copper leaks are the most common type of leak found in the piping of water distribution systems. Copper tubing and pipe has been used for many years, and it is still quite popular for new installations.

While any experienced plumber knows how to solder joints on copper pipe, there are times when the job doesn't go by the book. A little water trapped in a copper pipe can make soldering a joint a hair-pulling experience if you don't know how to handle the situation. There are also those times when the fittings that must be soldered are located in places where the flame from a torch poses some potentially serious problems.

This book is not designed to tell you exactly how to fix problems on a step-by-step basis. Since the expected readers are professionals, it is assumed that they will know the basics of making standard repairs. However, since some soldering situations are problems in themselves, we will cover some tricks of the trade to help you out in difficult times. After all, this is a problem-solving book.

Finding leaks in copper

Finding leaks in copper pipe and tubing is not usually difficult. By the time plumbers get called in to fix a leak, someone normally knows where the leak is. There are times, though, when the location of the leak is not known.

In other chapters throughout the book we have talked about ways to find hidden leaks. You've seen how to find leaks in the risers to shower heads and how to track down water that is running across a ceiling. Those tactics can be applied to leaks in the water distribution system.

Big leaks are easy to find. You can either see or hear them. It is the tiny leak that does its damage over an extended period of time that is difficult to put your finger on. With these leaks it is sometimes necessary to start at the evidence of the leak and work your way back to the origin of the water. This can mean cutting out walls and ceilings, but there are times when there just isn't any other way to pinpoint the problem.

Copper leaks often spray water in many directions. This can also make finding the leak difficult. If a solder joint weakens to the point that water can spray out of it, the water may travel quite a distance before it splashes down. These leaks are not hard to find when they are exposed, but if they are concealed, you can find yourself several feet away from the leak when you cut into a wall or ceiling that is showing water damage.

The good thing about spraying leaks is that you can usually hear them once you have made an access hole. This is a big advantage over trying to find a mysterious drainage leak. The spraying water will lead you to the leak in a hurry.

Copper pipe and the joints made on it can leak for a number of reasons. Pinholes in the pipe can occur from acidic water. The pipe can swell and split or blow out of a fitting if it has frozen. Stress can break joints loose, and sometimes joints that were never soldered properly will blow completely out of a fitting. Bad solder joints can also begin to drip slowly.

Fixing copper leaks

Fixing copper water lines is usually not a complex procedure, but what will you do if there is water in the lines? Water in the piping can be of two types; it can be standing water that is trapped, and it can be moving water that is leaking past a closed valve. In either case, the water makes it hard to get a good solder joint.

Inexperienced plumbers often don't know how to overcome the problem of water in the piping that they need to solder. Some inexperienced plumbers will try to make a solder joint and be fooled by the actions of the solder. This type of situation will result in a new leak, but the leak may not show up immediately; let me explain.

When water is present in the area being soldered, several things can happen. If there is enough water in the pipe, the joint will not get hot enough to melt solder. This is frustrating, but at least the plumber is aware that the solder joint can't be made without some type of action being taken against the standing water.

In some cases the pipe and fitting will get hot enough to allow solder to melt, but the fitting will not obtain a temperature suitable for a solid joint. When this happens, solder will melt and roll out around the fitting, but it is not being sucked into the fitting as it should be. To inexperienced eyes, this type of joint can look okay, but it's not.

When the water is turned back on, the defective joint may leak immediately, if you're lucky. If the joint leaks right away, the plumber knows the job is not done. However, sometimes these fouled joints will not leak immediately, and this means trouble.

When the weak joint doesn't leak during the initial inspection, it may be left and possibly concealed by a wall or ceiling repair. In time, and it usually won't be long, the bad joint will begin to leak. It may drip, or it may blow out. Either way, the plumber's insurance company won't be happy.

When the bad joint fails and is inspected by the next plumber, it will be obvious the joint was not made properly. There will not be evidence of solder deep in the fitting because it was never there. This will usually be considered neglectful on the professional plumber's part.

A third way that water in the piping will drive an inexperienced plumber crazy is with the results of steam building up in the pipe. As the area around the joint is heated, the water will turn to steam. The steam will vent itself, usually through some portion of the fitting being soldered. This steam may escape without being seen. When this happens, solder runs around the fitting as it should, except that a void is created where the steam is blowing out. If the plumber can't see, hear, or sense the steam, the joint will look good. Once the water is turned on, the joint will no longer look so good, it will leak and the process will have to be repeated.

Inexperienced plumbers will think they just had a leak because of poor soldering skills. They will go back through the same process and wind up with the same results. Until they figure out what is happening, and what to do about it, they will just spin their wheels trying to solder around the steam. There are ways to beat all of these problems, and I'm about to show them to you.

Trapped water.   Trapped water is easier to overcome than moving water. Trapped water can sometimes be removed by bending the cut ends of horizontal pipes down. Opening fixtures above and below the work area will often remove trapped water from pipes. When the problem pipe is installed vertically, a regular drinking straw can be inserted into the pipe and the water blown out. If you have the equipment and time, you can use an air compressor to blow trapped water out of pipes. When none of these options work, you have to get a bit more creative.

The easiest way to beat trapped water in pipes that just won't drain is the use of bleed fittings. These are cast fittings that are made with removable drain caps on them. Heating mechanics call them vent fittings, I call them bleed fittings, and many people just call them drain fittings.

Drain fittings are available as couplings and ells (Figs. 29.2 and 29.3). The vent on a coupling is on the side of the fitting, at about the center point. Drain ells have their weep holes on the back of the ell, where the ell makes its turn.

If you install a drain fitting in close proximity to the trapped water, you can steam the water out of the pipe as you solder the fitting. Remove the cover cap and the little black seal that covers the drain opening. Make sure the drain opening is not pointing toward you.

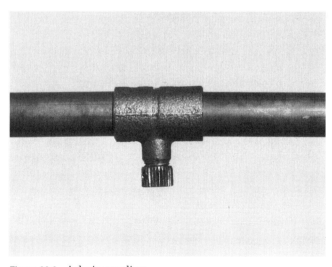

Figure 29.2   A drain coupling.

Figure 29.3 A drain ell.

As you heat the pipe and fitting, the trapped water will turn to steam and vent through the weep hole in the fitting. The hot water can spit out of the hole and the steam can come out fast and hot. This is why you don't want the hole pointed in your direction.

With the water and steam coming out of the vent hole, you can solder the joint with minimal problems. Unless you have a huge amount of water in the pipe, the soldering process shouldn't take long, and it should go about the same as it would if the water wasn't in the pipe.

Once the joints have cooled, you can replace the black seal and the cap to finish your watertight joint. If you get caught without bleed fittings on your truck, a stop-and-waste valve can be substituted for the drain fitting. The little weep drain on the side of the valve will work in the same way described for the drain fittings. Make sure, however, that the valve is in its open position before you begin to solder.

Bread. Bread is an old standby for seasoned plumbers who are troubled by water in the pipes they are trying to solder. Inserting bread into the pipe will block the water long enough for a good joint to be soldered. When the water pressure is returned to the pipe, the bread will break down and come out of a faucet. This procedure works well, but there are a few traps involved with it.

I've used bread countless times to control water that was inhibiting my soldering. On occasion, I've regretted it. Sometimes I've created more problems for myself by using bread. I would recommend that you use drain fittings rather than bread, but I'll tell you what to look out for when bread is used, just in case you don't have a choice.

First, always remove the crust from loaf bread before stuffing the pipe. The crust is much more dense than the heart of the bread, and it doesn't dissolve as well.

When bread has been used in a pipe, try to remove it through the spout of a bathtub. If you must get it out through a faucet, remove the aerator before you cut the water on. The broken-down bread will clog the screen of the aerator in the blink of an eye.

Don't flush a toilet to remove bread from the line. The bread particles may become lodged in the fill valve and cause more trouble.

Avoid packing the bread in the pipes too tightly. I once used a pencil to push and pack bread into a pipe that had moving water in it. The bread blocked the water and allowed me to make my solder joint, but it didn't come out once the water was turned on. Oh, it came through after awhile, and blocked up the supply to a toilet.

Moving water.   Moving water makes the job of soldering joints an adventure. Old valves don't always hold water back completely. There will be many times when a trickle, or more, of water will creep past the valve and make soldering a bad experience.

When you have moving water in a pipe, you will have to use some special way to get the soldering job done. Bleed fittings will let you do your work if the water is not too abundant.

There is a special tool that I've seen advertised, but I can't remember its name, that is designed to make soldering wet pipes possible. As I recall, the tool has special fittings that are inserted into the pipe and expanded. The plug holds back the water while a valve is installed on the pipe. When the valve is soldered onto the end of the pipe, the tool is removed and the valve is closed, allowing you to work from the valve without the threat of water.

The tool is nifty but not needed. You can do the same thing with the old standby, bread. Stuff the pipe with bread, and pack it if you have to. Once the water is stopped, solder a gate valve onto the end of the pipe. Don't use a stop-and-waste valve because you may need access to the bread in order to get it out.

Once the valve is soldered properly and cooled, close it and cut the water on. Open the valve to let the bread out of the system, and remember that water will be coming out right behind the bread. If the bread is packed too tightly for the water to come through, poke holes in it with a piece of wire, like a coat hanger. The bread will break up and come out of the pipe. Once the pipe is clear, close the valve and go on about your business.

Use a heat shield. When you are soldering in close quarters, use a heat shield and have a fire extinguisher close by. If your work is dangerously close to combustible materials, wet them down with water before putting your torch into the area.

Heat shields that attach to the tip of a torch are available, but these units are small and are not always enough to consider the conditions safe. I carry a piece of duct work on my truck that can be used to block flames from combustibles, and I've used aluminum foil in years past. Both work well, but you should never depend too heavily on any heat shield to provide positive protection; keep a fire extinguisher at hand.

I knew a plumber once (he didn't work for me) who almost set an entire apartment building on fire. He was repairing a copper pipe that had split due to freezing. During his work, the insulation in the wall caught on fire. The plumber had limited experience, and he tried to put the fire out at the point where it had started. The wall, however, was acting as a chimney, and the fire was spreading upward, quickly.

The plumber's boss was in the next room making some repairs and came running in. As soon as he saw what was happening, he used his hammer to open the wall where it met the ceiling, several feet above the work area. I understand it was touch and go, but the supervisor was able to contain the fire in the one apartment and get it put out.

The apartment building was old and not built to present-day fire codes. If the experienced supervisor had not been on the job, the whole building probably would have gone up in flames. Keep this story in mind, and keep that fire extinguisher close by. And remember to head fire off if it is spreading, not to fight it just where you can see it.

PB leaks

PB leaks are not common, except at connections that were not made properly. Because of its flexibility and durability, PB pipe rarely gives plumbers problems with leaks. It doesn't even normally burst under freezing conditions. There are, however, several times when the pipe is not put together properly, and this can result in leaks.

Bad crimps. Bad crimps account for some leaks with PB pipe. If a crimping tool gets out of adjustment, and they do after extended use, the crimp ring may not be seated properly. If this is the case, the bad connection must be removed and a new one made.

Fittings not inserted. It is not often that an insert fitting is not inserted far enough, but I've seen it happen. This type of problem is obvious and normally not difficult to repair. The replacement of the bad connection is all that is required.

Stainless steel clamps.   Some plumbers are not familiar with PB pipe, and they sometimes use stainless steel clamps to hold their connections together. This rarely works for long, if at all. If you have a leak at a PB connection where a stainless steel clamp has been used, remove the clamp and crimp in a proper connection.

Compression ferrules.   Compression ferrules may account for most of the after-installation leaks in PB piping systems. Compression fittings can be used on PB pipe, but brass ferrules should be avoided (Fig. 29.4). Nylon ferrules are the proper type to use with compression fittings and PB pipe.

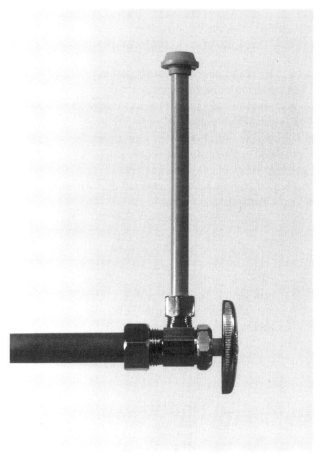

Figure 29.4   A polybutylene closet supply.

If you have a PB pipe or supply tube leaking around the point of connection where a compression fitting has been used, inspect the ferrules. If brass ferrules have been used, replace them with nylon ferrules. The brass ferrules can cut into the PB pipe if the compression nut is tightened too much.

I'm not saying that brass ferrules can't be used on PB pipe. In fact, I've used them on PB supply tubes many times without any problems. But, if the plumber puts too many turns on the compression nut, the PB will be cut. Whenever possible, use nylon ferrules.

PE leaks

PE leaks are not much of a consideration when talking about water distribution systems. PE pipe is used frequently for water services, but its use is very limited in water distribution systems.

Cracked fittings and loose clamps are the most frequent causes of leaks in PE piping. The fittings can be replaced, and the clamps can be tightened or replaced.

On the occasions when PE pipe develops pinhole leaks, there are a few easy ways to fix the problem. The best way to deal with the situation is to cut out the bad section of the piping and replace it. When this isn't convenient or possible, you can use repair clamps.

Repair clamps can be used to patch small holes in all types of piping. You can buy the clamps, or in some cases you make your own. In the case of PE pipe, rubber tape and a stainless steel pipe clamp will make a good repair on small holes. All you have to do is wrap rubber tape around the hole and clamp it into place.

CPVC leaks

CPVC leaks are common, and in my opinion, are a real pain to deal with. I've worked with CPVC off and on for going on 20 years, and I've never liked it. The one house that I plumbed with CPVC was enough to turn me off from using it in future jobs. Since that first house, early in my career, my only association with CPVC has been in repairing it.

If you have worked with CPVC much, you know it is brittle, takes a long time for its joints to set up, and is famous for its after-the-fact leaks.

When you are troubleshooting a water distribution system made from CPVC piping, there is a lot to look for. You have to keep your eyes open for little cracks in the pipe. Because of its makeup, CPVC will crack without a lot of provocation. These hairline cracks can be hard to find.

In addition to cracks in the piping, you have to look for cracks in the threaded fittings. Fittings that have been cross-threaded are also common in CPVC systems. And to top it all off, CPVC pipe that is not supported properly can vibrate itself to the point of weakening joints that will leak.

When you find a leak in CPVC pipe or fittings, don't attempt to reglue the fitting. Cut it out and make a good connection with new materials. Attempting to patch old CPVC will more often than not result in frustration and continued problems.

You don't have to solder CPVC, but water in the line will still mess up your new joints. The water will seep into the cement and create a void that will leak.

I know some plumbers just cut the pipe, slap a little glue on it, and stick it together with its fitting, but I don't think you should operate this way. CPVC is so finicky that I believe you should make your connections by the book.

Cut the ends of the pipe squarely and rough them up with some sandpaper. Using a cleaner and a primer prior to applying the cement is also a good idea. When you glue and connect the pipe with the fitting, turn it if you can to make sure the cement gets good coverage.

Don't turn the pipe loose right away. If you release your grip too soon, the pipe is likely to push out of the fitting to some extent, weakening the joint. Hold the connection in place as long as your patience will allow. CPVC takes a long time to set up, so don't move it or cut the water on too soon. New joints should set for at least an hour before being subject to water pressure.

Galvanized steel and brass

Galvanized steel and brass pipes aren't used much for water distribution these days, but they do still exist in some buildings. These types of pipe present some special problems.

Leaking threads. Leaking threads are a common problem with old piping. The threads are the weakest point in the piping, and they tend to be the first part of the system to go bad. Acidic and corrosive water can work on these threads to make them leak prematurely.

When you have a leak at the threads going into a fitting, a repair clamp is not going to help you. Under these conditions, the leaking section of pipe must be removed and replaced. This can become quite a job.

What starts out as a single leak at one fitting can quickly become a plumber's nightmare. As you cut and turn on the bad section of pipe, you are likely to weaken or break the threads at some other connection. This chain reaction can go on and on until you practically have to replace an entire section of the water distribution system.

Many young plumbers see leaks at threads and believe they can correct the problem by tightening the pipe. In old piping, this is only a pipe dream, no pun intended. Twisting the old pipe tighter is not likely to solve problem, but it may worsen it.

If you have leaks at the threads, you might as well come to terms with the fact that the section is going to have to be replaced. Unlike drain pipe, you can't use rubber couplings to put water piping back together. You have to remove sections of the pipe until you can get to good threads. Then you can convert to some type of modern piping to replace the bad section. Be prepared to have to replace more than you plan to.

Pinholes.   Pinholes in these hard pipes can be repaired with repair clamps. Rubber tape and a pipe clamp will work, but a real repair clamp is best.

Brass imitates copper.   Brass pipe imitates copper to the point that some plumbers will think they are working with copper. In fact, I hate to admit it, but I've even mistaken brass for copper. It is easy to do in poor lighting.

Brass water pipe can be cut with copper tubing cutters. The cut takes a little longer, but unless you suspect the pipe is brass, it can fool you. However, once you try to get a copper fitting on the end of the pipe, you'll know something is wrong; the fitting won't fit.

If you think you're working with copper and can't figure out why your fittings won't work, look for an existing joint and see if it is a threaded connection. If it is, you're probably dealing with brass pipe.

If the fittings are soldered and your standard fittings won't fit the pipe, somebody plumbed the job with refrigeration tubing. When this is the case, you're going to need refrigeration fittings to get the job done. This one tip can save you a lot of frustration.

Compression leaks

Compression leaks are frequent problems with water distribution systems. The compression fittings are usually easy to fix. All that is normally required is the tightening of the compression nut. These leaks happen when the nuts were not tight to begin with or when the pipe has vibrated enough to loosen the fitting. It is also possible that the connection was hit with something and loosened.

I responded to a call last week that would fool a lot of plumbers, so I think it is worth telling you about. The call came in from the owner of a motel. His maintenance mechanic had investigated a leak under a lavatory and determined that a plumber should be called in. The owner explained to me that the leak was at the threads of a straight stop under the lavatory. He said the leak was below the cutoff and the water supply to the whole motel would have to be cut off.

I went to the job personally and inspected the leak. The piping came up through the floor. A long, chrome nipple ran from the floor to the straight stop. A steady stream of water was running down the nipple, and at first glance, it did appear to be coming from the threaded connection.

Both the owner of the motel, who is handy, and the maintenance person had inspected the leak. They were convinced the problem was with the threaded connection. If I had wanted to take advantage of the customer, I could have easily played along with their opinions and made a little job into a much larger one.

I could have inconvenienced everyone in the motel by cutting off the main water supply and worked for some time in replacing the nipple and the cutoff valve. Instead, I did the right thing.

What did I do? I didn't rely on the information I had been given. I looked at the problem closely, as any good troubleshooter would, and I saw what I believed to be the real problem. Do you know what it was?

When I looked at the stop valve, I saw a bubble of water sitting on top of the compression nut that held the supply tube in place. I cut the supply valve off and wiped all the water off the nipple. The water didn't continue to run down the nipple. Obviously, if the leak had been at the threads on the bottom of the valve, the leak would have continued.

I had been reasonably sure that the leak was at the supply-tube connection, but my little test proved me to be right. I tightened the compression nut on the supply tube and cut the valve back on. The leak was gone.

Instead of spending an hour or so replacing the nipple, only to find that it wasn't the problem to begin with, I completed the job in less than 10 minutes.

The maintenance person was a bit embarrassed by my ease in fixing the problem, and the owner was delighted. They had both been nearby during the repair and knew that I had not cut the main water supply off or spent much time on the job.

They asked how I had fixed the problem so quickly, and I told them. While I was on the job, the owner had me rebuild a tub valve and replace a relief valve in a water heater. My honesty, professionalism, and troubleshooting skills were appreciated so much that I now have the account not only for that motel but for another one as well. The owner has even asked me to plumb 10 new units he will be building in a month or so. It is surprising what a little job can lead to when you do it right.

I can easily understand how the two people thought the leak was at the threads. Even a plumber who was in a hurry might have assumed the same thing. By troubleshooting the job properly, the root cause was found and the job was simple to do. This is well worth remembering in your work.

Frozen Pipes

Frozen pipes can be a plumber's bread-and-butter money in the winter. They can also be troublesome to work with, hard to find, and difficult to fix and are potentially dangerous if they are not worked with in the proper manner.

Steel pipe, and sometimes copper pipe, can be thawed with the use of a welding machine. If the leads are attached at opposite ends of the pipe, with the frozen section in the middle, the welding rig can produce enough juice to thaw them. I've seen plumbers do this many times over the years, but the process can be dangerous; fires can be started.

Special thawing machines are also available for dealing with frozen pipes. They work on pretty much the same principle as the welding machine. I prefer to avoid using electricity to thaw frozen pipes.

When I was a plumber in Virginia, I used to get numerous calls for thawing and repairing frozen pipes. Normally, the job entailed only a single pipe, often that of an outside hose bibb. A heat gun, a hair dryer, or a torch made quick work of thawing these individual sections of pipe.

In Maine, the freeze-ups are often considerably larger than the ones I dealt with in Virginia. Here it is not uncommon for whole systems to freeze. To thaw these pipes, I usually use a portable heater. A large space heater, like those used on construction sites, will bring a building up to a thawing temperature quickly and safely. Once the building is warm and the pipes are thawed, necessary repairs can be made.

If you haven't worked with many frozen pipes, you may not be aware of the way the freezing action can swell copper pipe. Even though the split resulting from the freeze-up is in one place, the pipe might be swollen for several feet on either side of the split. This makes it impossible to get a fitting on the end of the swollen pipe. To handle this dilemma, you will have to keep moving back on the pipe until you find a piece that has not swollen from the cold.

Knowing What to Look For

Knowing what to look for is a key aspect of any type of troubleshooting. If you know what to look for, you have a much better chance of finding it. Problems with water distribution systems can involve many possibilities, but they are not hard to work through when you use good troubleshooting skills. This chapter has shown you the ropes on water pipes, so let's move onto the next chapter and see what's in store for us there.

Troubleshooting Drainage and Vent Systems

Troubleshooting drainage and vent systems can involve a lot more than just looking for leaks and stoppages in the drainage system. While leaks and stoppages are the two most common complaints with drainage, waste, and vent (DWV) systems, they are far from being the only problems people experience with their DWV systems.

Let me give you a quick quiz to see how much you already know about troubleshooting DWV systems. I'll pose the questions to you now and answer them as we move through the chapter. Get a note pad and jot down your answers to the questions. We'll see how you did you did on the quiz a little later.

1. This question deals with odors in a home. This house has three bathrooms, two with tub/shower combinations and one with a stall shower. The house has two stories with two baths upstairs and one, the one with the shower, downstairs. The homeowner has noticed that when she is working in her kitchen there is a bad odor in the area. There are several possible causes for this problem, but what would you look for first?

2. This question has to do with a noisy drain pipe in the wall between the kitchen and the living room of a home. The homeowner despises having guests listen to water rush down this drain every time the upstairs toilet is flushed. He is willing to go to any reasonable expense to correct the problem. What would you suggest?

3. The third question in our quiz has to do with the drainage from a commercial dishwasher. The business owner calls you and explains that the plastic pipe that carries the drainage from his dishwasher is leaking again. He goes on to explain that this is the third time he has had the same problem, and he wants you to fix the piping so that it won't happen again. What will you do?

4. This one has you looking for the reason why the fixtures in the bathroom of a new home are not draining properly. The bathtub and the lavatory in this new house are both draining slowly. You have snaked the drains, but they still won't create the drainage whirlpool that they should. What might the problem be in this case?

5. This question finds you wondering why a kitchen sink that had been working satisfactorily is suddenly backing up. All the homeowner can tell you is that a garbage disposer was recently installed on the sink, and that seems to be when the problem started. You should be able to make a good guess about the cause of the problem without even looking under the sink. What do you think the problem is?

6. A homeowner has called you and asked if it is possible for their vent pipes to freeze in the winter time. How will you answer the customer?

7. This question has to do with a foul odor in the basement of a home. The basement contains most of the home's DWV system. There is also a floor drain and a sump for a sump pump in the basement. What is the most likely cause of the foul odor?

8. This question has to do with a kitchen drain that stops up frequently. The homeowner has explained to you that she has to have the drain snaked out about every month. It seems the plumbers always hit the clog within 15 feet of the sink. The fixture drains okay for a while, but then it stops up again. What do you suspect the answer to this problem is?

9. This question has to do with a homeowner who went into his attic to get down some items he had in storage. While the man was in the attic, he noticed a strange smell. At first, he thought maybe a small animal had died in the attic, but he couldn't find any evidence of it, so he called you. What are you going to look for first?

10. The final question in our quiz revolves around a house with old plumbing in it. The house has a half-basement under it. The rest of the foundation, where the building drain passes into it, is a tight crawl space. The building drain of the house is stopped up. The old cleanout in the basement doesn't want to come out, so you elect to snake the drain from the first-floor toilet. You send your snake down the drain, and it hits something solid. The snake cable kinks and won't go any further. The cable is acting like it is hitting a ball of tree roots or a broken pipe.

When you can't get the snake to go in any further, you retrieve it and note how many feet of cable you had in the drain. After going into the basement and guesstimating the distances of the piping, you believe the snake was hung up in the pipe section that is in the crawl space. Since the problem is in the crawlspace, you rule out the possibility of tree roots. A quick inspection with a flashlight seems to rule out a collapsed pipe. What else might the problem be?

I'm about to give you the answers to the above questions. Check your notes to see how you did. If you did well, congratulations. Should your score be lower than you would like, at least you learned your lessons here, instead of in the field.

1. The most appropriate answer for question 1 is a dry trap in the downstairs shower. This question was based on a real-life problem that my mother-in-law was experiencing. She asked me one day if I had any idea where the odor in her kitchen was coming from. I knew she had a shower just down the hall from the kitchen, and I didn't think she used it often. After asking if the shower got much use, she confirmed my suspicion. Acting on my instructions, she ran the shower for a while to replenish the seal in the trap. The problem has never come back.

2. Many people don't like having noisy drains in the vicinity of their living rooms. Unfortunately, there is nothing simple that can be done about this problem once the pipes are installed and the walls are finished off.

Since this homeowner is willing to go to some expense, there are two viable options. The plastic drain could be replaced with cast-iron pipe. This would help to deaden the sound. Insulation could then be wrapped around the cast-iron pipe to muffle it even more. It would also be conceivable to just wrap insulation around the existing drain. Of course, the wall with the pipe in it would have to be opened and repaired to allow for this type of work.

3. The best way to solve the problem in question 3 is to replace the plastic pipe with either DWV copper or cast-iron pipe. The intense heat from the commercial dishwasher is more than the plastic pipe and associated joints can handle.

4. The answer to question 4 is one that I doubt many got right. If you missed this one, don't feel bad. I have seen the problem occur, but the reason for it is one that should not exist. What was the cause of the problem? The plumber who installed and tested the new piping forgot to remove the test caps from the vents on the roof. This may sound weird, but I've run into the problem twice.

5. I hope you got question 5 right. Why would the drain suddenly be backing up? My guess would always be that the kitchen drain was plumbed with old galvanized pipe. The pipe was probably closing up

with rust, grease, and assorted gunk. I expect there was a small hole in the center of the closing pipe that would allow water from the sink to drain, but with the disposer on the pipe, the small opening left in the drain pipe couldn't handle it. This type of problem happens so often that I won't install a garbage disposer on a sink that drains into galvanized pipe unless the property owner releases me from liability for stoppages.

6. Is it possible for vents to freeze as was asked in question 6? Yes, it is possible. In extremely cold conditions, vents can condensate and the moisture can freeze on the inside of the pipe. If the temperatures stay low for a long time, the ice can eventually reach all the way across the pipe and seal the vent.

7. How did you answer question 7? If you guessed that the trap for the floor drain had dried up, you're right. Water in traps will evaporate over time. If the drains are not used periodically, the water seal will dissipate and odors can escape. In some cases, trap primers must be installed to make sure the condition does not occur.

8. After reading question 8, did you guess the problem to be galvanized piping that is closing up? If you did, you're right on the money. As galvanized drains close up and are snaked out, generally only a small hole is punched through the obstruction. Since few plumbers run cutting heads down sink drains, their spring heads do nothing more than open small passages in the drain. This works for a while, but then new grease and gunk joins the existing obstruction to block the pipe again.

9. What do you think the problem is in question 9? More than likely the problem is a vent pipe that was never extended through the roof. It could be a pipe that was broken after installation, but it's probably just one that didn't get hooked up. If you are faced with this type of situation and don't see any pipes that are not installed properly, look under the insulation. Many times the insulation will be covering the pipe that is responsible for the problem.

10. If you're a young plumber, you may have had a problem with question 10. Older plumbers probably knew right off what to expect. The problem would probably be an old house trap.

In the old days, it was common to install a house trap in the building drain or sewer in close proximity to where the drain left the foundation of the property. Sometimes the traps are installed inside the foundation, and sometimes they are installed underground, outside of the foundation. When they were used, house traps were usually the only traps installed on the drainage system. Large snakes frequently can't get through these traps. If you have a mystifying problem with your snake in an unusual section of drainage pipe, you can make a pretty safe bet that you will find a house trap installed.

Now that you've had a chance to test your knowledge of DWV systems, let's move on into the chapter and see what else you can learn.

Clogged Vents

Clogged vents can cause fixtures to drain slowly. You've seen one example where the test caps were never removed from a home's vents, but that is not the only way that vents get plugged up.

Squirrels and birds can disable a plumbing vent very quickly, especially vents with small diameters. Birds sometimes start building nests on top of vent pipes, and squirrels have been known to use the pipes as a cache for nuts.

Aside from wildlife, Mother Nature can put a strain on plumbing vents. If a vent is positioned under trees, all sorts of things can get into the pipe. Leaves can fall into the vent, pine cones could plug the pipe, and a combination of nuts, leaves, and twigs can restrict the opening of the vent pipe.

If you have a drain that is draining slowly after a thorough snaking, you should check to see first if the fixture is vented. If it is, check the vent for stoppages. If necessary, you can snake the vent from the roof to break up stoppages.

When a fixture is not vented, you might consider installing a mechanical vent on the fixture to allow it to drain better. Some code officers frown on this practice, so make sure you are within the limits of code compliance before you put a mechanical vent on the system.

Vents Too Close to Windows

Vents that are too close to windows that open, doors, or roof soffits can be responsible for odors in a building. The vents should be at least 10 feet away from such openings. If the vents are closer than that, they should extend at least 2 feet above the opening. Otherwise, the fumes from the vent pipe can be sucked into the building through the opening.

Frozen Vents

You learned in the quiz questions that vents can freeze. Many plumbers don't realize this. When a plumbing system suddenly fails to drain as well as it normally does in cold weather, look to the vents to see if they are frozen. If they are, an acid-based drain cleaner will normally eat through the ice and clear up the problem, at least temporarily.

Vents in General

Vents in general don't present many problems for plumbers. Unless they are nonexistent or stopped up, vents don't normally require any attention.

Copper Drainage Systems

Copper drainage systems are usually not much trouble. The copper provides many years of good service, and it is not normally a contributor to stoppages. In fact, in 20 years, I can't recall a time when a copper DWV system caused any problems. Oh sure, they get stopped up, but all drains can. In general, copper is a good above-ground drainage material that is very dependable.

Galvanized Pipe

Galvanized pipe has to be one of the worst materials ever used in plumbing. This pipe is famous for its ability to rust and to catch every imaginable thing that goes down it. Grease and hair are especially common stoppages found in galvanized drains.

As galvanized pipes age, they begin to develop buildups that slowly restrict their openings. Eventually, the buildup blocks the pipe completely. Snaking the drain will punch holes in the obstructions, but the stoppage will recur in a matter of weeks or months.

Old galvanized drains are also known to rust out at their threads, causing leaks at their joints. The best solution to galvanized drain problems is the replacement of the old piping with more modern plumbing materials. Rubber couplings make it easy to adapt new drainage materials to the old piping.

Cast-Iron Pipe

Cast-iron pipe has been used for DWV systems for longer than I've been a plumber and then some. Typically, cast iron gives good, dependable, long-term service. Whether the system is plumbed with service weight pipe or no-hub pipe, the cast iron lasts a long time.

There are some drawbacks to cast-iron drains, however. The interior of cast-iron pipe can be rough, especially as it starts to rust. These rough surfaces catch a lot of debris as it is being drained down the pipes. This leads to stoppages, but they are not as bad as the ones found in galvanized pipes, and they don't recur as quickly.

One of the biggest problem with cast-iron drainage systems is the removal of cleanout plugs. The old brass plugs that have been in the cleanouts for years can be next to impossible to remove.

I've had occasions when a 24-inch pipe wrench with a 2-foot cheater bar wouldn't turn the cleanout plug. If you've done much work with pipe wrenches and cheater bars, you know the kind of leverage that is being applied under these conditions. Even so, some cleanout plugs just won't budge, and some of them are in locations where you just can't get enough leverage on them to make the turn.

When this happens, your options are limited. Some plumbers drill the brass plugs out. Others use cold chisels and hammers to knock them out. And others just cut the pipe and put it back together with a rubber coupling to make cleaning the drain the next time easier. I usually opt for the latter.

I had a service call a few weeks ago that I would like to tell you about. The call came in from a customer who owns a few rental houses. One of his houses was suffering from a stopped-up drain. He told me that nothing in the house would drain and that the condition had existed for a few days.

I went to the house and talked with the tenant. Sure enough, all of the fixtures on the main floor of the one-level house were out of service. When I went into the basement, water was leaking through the floor at the base of the toilet. The building drain was made up of cast-iron pipe. Some of the fixture branches were piped with PVC pipe.

As I looked around, trying to decide which drain cleaner to use, I noticed a washing-machine hookup. It was located near where the building drain went through the foundation. I thought it strange that the fixtures above me were all backed up, but no water was coming out of the washing-machine receptor.

The sewer had a 4-inch diameter. The combination wye-and-eighth-bend fitting at the foundation wall reduced the building drain to a 3-inch diameter. The horizontal extension of the wye had been reduced to a $1\frac{1}{2}$-inch diameter and served only the washing-machine receptor. It had a cleanout in the end of it and was plumbed with PVC pipe. Tentatively, I removed the cleanout plug. Nothing came out. I didn't really think anything would, but you can never be sure with plumbing.

The fact that nothing flooded out of the cleanout or the laundry receptor told be that the blockage was not in the sewer, but in the 3-inch building drain.

The building drain had a developed length of about 24 feet and very few offsets. In fact, the only offset was one long-sweep ell. This looked as though it would be a very simple job, but boy was I wrong.

A section of the horizontal drain near the bulkhead door was cracked and leaking. I suspect the crack came from a previous stoppage that froze before it was cleared up. Wind probably whipped through the bulkhead door and froze the contents of the pipe, cracking the cast iron. At any rate, the dripping crack told me that the stoppage was downstream of the broken pipe.

Unfortunately, there was no cleanout at the change in direction of the pipe, only a long-sweep ell. The only cleanout upstream of the clog was a small one, an 1½-inch one, where a branch took a turn for the kitchen sink. This cleanout was plumbed in with PVC, so I decided to try to clear the stoppage through it.

I removed the cleanout plug and caught the backed-up liquid in a bucket. Since the pipe was hanging tight to the floor joists, I decided to use a hand-held snake for my first attempt. I put 25 feet of cable down the drain and the spring head brought back a significant amount of hair and toilet tissue. Hoping the drain was cleared, I had the homeowner run the sink upstairs. It didn't take long for the water to back up and come out the cleanout. Another tactic would be needed.

I don't like to run a big sewer machine in a drain that is over my head, and all of these pipes were high. There was a set of stairs going up near the drain for the lavatory and toilet. This gave me an idea.

There was a short section of PVC that connected the toilet and the lavatory to the cast-iron drain. I could have gone upstairs, pulled the toilet, and snaked down from above, but I wanted to avoid pulling the toilet and making a mess in the bathroom.

I cut the 1½-inch lavatory drain and set my big drain cleaner up on the steps. A paint can was placed on the lower step to provide a solid base for the machine. The snake with a spring head went down the drain easily. It also brought back hair and toilet tissue when it was retrieved.

The tenant ran the upstairs sink again. The drain was still plugged up. I tried various heads on the snake cable, but none of them cleared the stoppage. Confused, I tried a sewer bag with water pressure. Still nothing happened. Next, I put a flat-tape snake down the drain. I hit resistance and broke through it, but the upstairs sink still backed up.

It had been nearly 2 hours since I started working with this simple drain stoppage, and it was still plugged after I had taken my best shots at it. I knew the problem had to be in the horizontal section of the building drain, and I couldn't figure out why it wouldn't clear.

Assuming the landlord would want the cracked piece of drainage pipe replaced, I decided to break it out and snake from that point. I broke out the side of the pipe and was promptly met with all sorts of nasty contents. The stoppage was definitely downstream, but now there was less than 15 feet of pipe for it to be hiding in.

Since the pipe was over my head, I used the flat-tape snake on it. I met with strong resistance and couldn't get the snake through the stoppage. By this time, I'm thinking some child's rubber ball or duck got flushed down the drain.

I was hot, sweaty, and beginning to lose my patience. Never in 20 years had I run across such a stubborn stoppage that was this unexplainable.

I went to where the building drain offset downward to the sewer. There was a short section of pipe between the combination-wye-and-eighth-bend and the eighth-bend that brought the pipe into a horizontal position. Not having a set of cast-iron cutters with me, I broke out this section of the piping with a framing hammer. Nothing came out of the pipe. Aha, the stoppage was between the two open sections of piping. There was no way it could escape me now.

I put a sewer bag in the pipe and turned on the water pressure. Nothing happened. I couldn't believe it. Next, I went back after the clog with the tape snake. Knowing that I was hitting the stoppage and not a fitting, I put all of my considerable weight behind the snake. When the clog broke free, it splattered all over the basement floor, near the sump pump.

If the tape snake had not gotten the clog, I was going to remove the entire section of piping and dissect it to see what was going on, even if it were at my own expense. Though much of the clog blew out into the basement, a lot of it was still in the pipe. The snake pushed a huge amount of unidentified crud out of the pipe. It was clearly the worst residential clog I've ever seen. I don't understand how so much stuff could develop inside a pipe before someone called a plumber. It was as if the whole 3-inch diameter was filled with debris.

Anyway, I replaced the cracked pipe and put the rest of the pipe back together. The problem was solved, but it took nearly 3 hours. The electric drain cleaners were apparently just punching holes through the stoppage. I wasn't able to use a large cutting head because I was accessing the pipe through an $1\frac{1}{2}$-inch drain. The end result was a problem that you won't learn about in apprenticeship classes or by reading most books. It was one of those rare occurrences that in my case has only happened once in 20 years, but it happened. Typically, cast-iron drainage piping doesn't produce the kinds of problems you have just read about. It is normally a good, dependable DWV pipe.

Plastic Pipe

Plastic pipe, as you probably know, is the most common pipe used for DWV systems today. It may be ABS or PVC, but both are good and dependable. PVC is more brittle than ABS and is more likely to be cracked or broken, but under normal conditions, neither pipe gives much cause for trouble.

Drum Traps

Drum traps are illegal for most uses under most plumbing codes. With the exception of combination-waste-and-vent systems, drum traps are hardly ever used. They are, however, prevalent in Maine, where few fixtures are individually vented.

If you are trying to snake a tub or shower drain and your snake begins to kink up quickly, you may be wrapped up in a drum trap. Because of their design, drums traps will not allow a snake to pass through them. You probably won't run into many drum traps, unless you work in Maine, but it does help to know that they may exist and that they impede the progress of a snake.

Lead Traps and Closet Bends

Lead traps and closet bends are not found too often these days, but there are some still in use. If you are called in to a building where a ceiling is showing water stains that are probably drainage related, the problem may be with a lead trap or closet bend.

The lead used in these fittings gives out after awhile, and leaks develop. There is no need to attempt to repair old lead traps and closet bends; it is best to replace them.

Mystery Leaks on the Floor

Mystery leaks on the floor in a commercial building can be caused by indirect wastes that are plugged up. Sometimes indirect waste pipes are blocked almost completely. It would be easy to diagnose the problem if the pipe were blocked completely, but when there is enough of an opening for water to slowly slip by, pinpointing the problem can be more difficult. Let me give you a case-history example about an ice machine in a motel.

I was a field supervisor for a plumbing company when this story took place. A motel manager called the company I worked for and requested service for a leak at her ice machine. A plumber was dispatched, and he found a puddle of water under the drainage piping of the ice machine.

The drain was made of $\frac{3}{4}$-inch copper tubing with sweat joints. The drain terminated over an indirect waste with a $\frac{3}{4}$-inch copper ell dumping into an $1\frac{1}{2}$-inch open drain.

The plumber searched for the leak and couldn't find it. He told the manager to mop up the floor and keep an eye on it. If water reappeared, he would come back. The manager had the floor mopped, and the next day the puddle was back.

The manager called the company and requested a different plumber. Since I was the field supervisor and the public relations person for the company, I was dispatched to the call.

When I arrived there was a shallow, but wide puddle of water on the tiled floor. I inspected the copper drainage tubing and could find no evidence of a leak. The pipe was dry and the fittings all looked to be in good shape and well soldered.

There was no way to tell that the indirect waste was stopped up by looking at it, since the trap holds water at all times. I had never experienced anything like this, but it made sense to me that the only reason for the puddle had to have something to do with the indirect drain.

I wasn't sure if some other fixture was causing the drain to back up and spill its contents or if the puddle was being made by the discharge of the ice machine. There was, however, an easy way to find out.

The motel manager gave me a pitcher full of water that I poured into the indirect waste, and guess what? Yep, the pipe overflowed, creating a puddle. I snaked the drain and the problem never recurred, to the best of my knowledge.

In this case, the occasional dripping of the ice machine was not enough to make the pipe overflow, but when the machine went into its heavy discharge cycle, the drainage was more than the blocked pipe could handle, which created the puddle.

This is a very good example of how a simple troubleshooting technique solved a problem and made a customer who had doubts about the credibility of the plumbing company happy.

Material Mistakes

Material mistakes account for some strange problems in the DWV systems of buildings. Good plumbers tend to think along the lines of the plumbing code when they are creating a mental picture of a plumbing problem. They basically assume the job was done to code, but this is sometimes far from the truth.

Whether you are troubleshooting a DWV system or any other plumbing problem, don't assume anything. I know it can be hard to think outside of the code, but there are times when you must, as police officers have to think like criminals to catch the bad guys. You have to think like an irresponsible plumber or a rank amateur sometimes to figure out plumbing problems. Let me give you a few examples of what I'm talking about.

The sand trap

This first example could aptly be dubbed the sand trap. The story has to do with a shower drain. A friend of mine who owns his own plumbing business was in my office recently. We were discussing a project that we would be working on together, and he told me this story.

The plumber was called in to work on a shower in the basement bathroom. During the course of his work, he noticed that something was not quite right with the drain in the shower. He removed the strainer plate for a closer look and was amazed at what he saw. Looking through the drain of the shower, he saw sand. His first thought was that the trap had somehow become filled with sand. During his work with the drain, he discovered that it turned freely in the shower base. With a little more investigation, the plumber found that the trap had not been filled with sand. It turns out that the sand was the trap.

Whoever installed the shower never bothered to connect it to a trap or a drain. The fixture simply emptied its waste water into the sand beneath the concrete slab. This is clearly a situation few plumbers would ever imagine.

The building drain that wouldn't hold water

This next case history is about the building drain that wouldn't hold water. The job began when the customer called my office and complained of strong odors in the corner bedroom of the home. It was summer, and I suspected that the customer's problem was an overworked septic field, but I couldn't be sure without an on-site inspection.

I went to the house and met with the homeowner. She started to explain her problem to me, but I could already smell the odor, and I hadn't even gone in the house.

The house was on a pier foundation. I walked around the left side of the home and saw two young children playing in the yard. They wore shorts and no shoes, and were splashing around in a puddle. This seemed odd since it hadn't rained in days.

The closer I got to the back corner of the house, and the children, the worse the smell became. I was starting to get a feeling that made me uncomfortable.

A bathroom had recently been added to the house, near the corner bedroom. Whoever installed the plumbing did so in a way that I'd never seen before. Looking under the house, I could see the problem, but I could hardly believe what I was seeing.

The drainage hanging from the floor joists was piped with schedule 40 PVC. The PVC dropped straight down to the sewer that ran from under the house to the septic tank.

It was the sewer pipe that was creating the problem. You see, someone install slotted drain pipe for the sewer between the house and the septic tank. Much of the sewer had never been buried in the ground. The holes were on top, but the septic tank was full, and raw sewage was seeping out of the slotted sewer pipe. Sewage was puddled under the house, under the bedroom window, and, you guessed it, where the children were playing.

I asked the homeowner who had installed the plumbing, and she claimed that the builder who had built the home some years back had done the work. Can you envision anyone using slotted pipe for a sewer? I couldn't either.

Duct tape

Duct tape is a plumber's friend, and it is capable of many good uses, but it is not meant to be used as a repair clamp on a drainage line.

Many years ago I was called to a house by a homeowner who said he could hear water dripping under his home after flushing the toilet. He thought perhaps the wax seal under his toilet had gone bad. The house was built on a crawlspace foundation. My first step in troubleshooting the situation was to flush the toilet. No water seeped out around the base of the toilet, but I could hear water dripping under the home, just as the homeowner had described.

I took my flashlight and went under the home to inspect the problem. As I worked my way back to the bathroom area, I got a strong whiff of sewage. Once I was close enough to the problem to see the piping and the ground, I noticed a rather large puddle on the ground. Shining my light on the pipe, I could see the remains of duct tape wrapped around the cast-iron drain.

After a closer inspection, I found that the cast-iron pipe had a large hole in the side of it. The pipe appeared to have been hit with a hammer at sometime in the past. Apparently, whoever attempted to repair the hole used duct tape to seal the pipe. Maybe they had done this as a temporary measure until they could get materials to do the job right, I don't know. In any event, the duct tape had long since lost its ability to retain the contents of the pipe, and raw sewage was blowing out the side of the pipe every time the toilet was flushed.

I cut out the damaged pipe and replaced it with a section of ABS and some rubber repair couplings. The repair wasn't a big job, but it was a messy one.

When jobs are not installed with the proper materials, your troubleshooting skills are put to the ultimate test. As plumbers, we are trained to look for normal problems, not problems created by someone's use of illegal materials or installation methods.

If a drain from a lavatory is piped with 1-inch PE pipe, as some I've found have been, it is difficult to understand why the pipe is stopped up or why a snake is so difficult to feed into the drain. Until you see the pipe, it is natural to assume it is of a legal size and material. You cannot, however, assume anything when you're troubleshooting plumbing.

I hope this chapter has opened your eyes to some of the strange calls you may get regarding DWV systems. Furthermore, I hope that by reading this book you have gained new insight to the value of learning effective troubleshooting skills. If you learn and practice the techniques we have talked about, I'm sure your plumbing career will be more successful and satisfying.

This brings us to the end of the book. I wish you good luck in all your endeavors.

Glossary

Accessible   Refers, as it relates to plumbing, to a means of access. For example, a tub waste is considered accessible when there is an access panel that can be opened or removed to gain access to the tub waste.

Air break   Refers, in the drainage system, to an indirect-waste procedure. The indirect waste enters a receptor through open air.

Air gap (drainage)   The vertical distance that waste travels through open air, between the waste pipe and the indirect-waste receptor.

Air gap (potable water)   The vertical distance water travels through open air, between the water source and the flood-level rim of its receptor or fixture.

Air gap (the device)   A device used to connect the drainage of a dishwasher with the sanitary drainage system.

Antisiphon   When a device cannot be made to form a siphonic action.

Area drain   A drain used to receive surface water from grounds or parking areas, for example.

Aspirator   A device used to create a vacuum, like in a suction system for medical offices.

Back-flow   The backward flow of water or other liquids in the drainage or water system.

Back-flow preventer   A device used to prevent back-flow.

Back-siphonage   Essentially the same as back-flow; it is the reverse flow of water or liquids in a pipe, caused by siphonic action.

Back-water valve   A device installed on drainage systems to prevent a back-flow from the main sewer into the building where the back-water valve is installed.

Ballcock   An automatic fill device. Ballcocks are most commonly found in toilet tanks. They supply a regulated amount of water, on demand, and then cut off when the water level reaches a desired height.

Branch   A part of the plumbing system that is not a riser, main, or stack.

Branch interval   A means of measurement for vertical waste or soil stacks. A branch interval is equal to each floor level or story in a building, but they are always at least 8 ft in height.

Branch vent   Vents that connect individual vents with a vent stack or stack vent. They can be used to connect multiple vents to a vent stack or stack vent.

Building drain   The primary drainage pipe inside a building.

Building sewer   The pipe extending from the building drain to the main sewer. Building sewers usually begin between 2 and 5 ft outside of a building's foundation.

Building trap   A trap installed on the building drain to prevent air from circulating between the building drain and sewer.

Cistern   A covered container used to store water, normally nonpotable water.

Code   A set of regulations that govern the installation of plumbing.

Code officer   An individual responsible for enforcing the code.

Combination waste and vent system   A plumbing system in which few vertical vents are used. In these systems, the drainage pipes are often oversized to allow air circulation in the system.

Critical level   The point where a vacuum breaker may be submerged, before back-flow can occur.

Cross-connection   A connection or situation that may allow the contents of separate pipes to commingle.

Developed length   A method of measurement, based on the total liner footage of all pipes and fittings.

Drain   A pipe that conveys wastewater or water-borne wastes.

Drainage system   All plumbing that carries sewage, rainwater, and other liquid wastes to a disposal site. A drainage system does not include public sewers or sewage treatment and disposal sites.

Existing work   Work that was installed prior to the adoption of current code requirements.

Fixture supply   The water supply between a fixture and a water-distribution pipe.

Fixture unit   A unit of measure assigned to fixtures for both drainage and water, to be used in pipe sizing.

Flood-level rim   The point of a fixture where its contents will spill over the rim.

Grade   The downward fall of a pipe.

Groundworks   Plumbing installed below a finished grade or floor.

Hot water   Water with a temperature of 110°F or more.

House trap   Same as a building trap.

Interceptor   A device used to separate and retain substances not wanted in the sanitary drainage system.

Leader   An exterior pipe used to carry storm water from a roof or gutter.

Local vent   A vertical vent used with clinical sinks to transport vapors and odors to the outside air.

Main   Any primary pipe for water service, distribution, or drainage.

Main sewer   The public sewer.

Main vent   The primary vent for a plumbing system.

Nonpotable water   Water not safe for human consumption.

Offset   A change in direction.

Open air   The air outside of a building.

Pitch   See *Grade*.

Plumbing inspector   See *Code officer*.

Potable water   Water safe for human consumption in drinking, cooking, and domestic uses.

Private sewage disposal system   A sewage disposal system serving a private party, a septic system.

Private water supply   A water supply serving a private party, such as a well.

Readily accessible   Having direct and immediate access to an object. If an access panel must be removed before the object can be accessed, the object is not readily accessible.

Rim   The open edge of a fixture.

Riser   A vertical pipe in the water-distribution system that runs vertically for at least one story.

Rough-in   The installation of plumbing in areas that will be concealed once the building is completed.

Sewage   Liquid waste that contains animal or vegetable matter. The matter may be contained in solution or in suspension and may contain chemicals in solution.

Slope   See *Grade*.

Stack   A vertical pipe in the drainage system. A stack may be a vent, soil pipe, or waste pipe.

Stack vent   A vertical pipe that extends above the highest drainage point to vent the drainage system.

Storm water   Rainwater.

Sump vent   A vertical vent that rises to vent a sump, such as in a sewer ejector sump.

Tempered water   Water tempered to maintain a temperature between 85 and 110°F.

Underground plumbing   See *Groundworks*.

Vacuum   A pressure that is less than the pressure produced by the atmo-00sphere.0

Vacuum breaker   A device used to prevent a vacuum.

Vent stack   A vertical pipe in the drainage system that acts only as a vent and does not receive the discharge of plumbing fixtures.

Waste   A discharge from fixtures and equipment that does not contain fecal matter.

Water-distribution system   The collection of water pipes within a building that delivers water to fixtures and equipment.

Water-hammer arrestor   A device that defeats water hammer by absorbing pressure surges.

Water main   The public water service.

Water service   The pipe delivering water from a water source to the water-distribution system of a building.

Well   A water source from a hole in the ground.

Wet vent   A pipe that receives the drainage from plumbing and does double duty by venting part of the plumbing system.

Yoke vent   An upward connection from a soil or waste stack to a vent stack. Yoke vents are normally made with wye fittings to prevent pressure changes in the stack.

List of Forms on the Enclosed CD-ROM

Where Is Your Money Going?

Weekly Expense Report

Field Expenses

Office Expenses

Petty-Cash Records

Cost Projections

Job-Cost Log

Customer Relations

News Release

Work Estimate

Quote

Proposal

Remodeling Contract

Letter of Engagement

Receipt for Deposit

Addendum

Liability Waiver Form

Request for Substitutions

Change Order

Early Termination and Mutual Release of Contract

Customer Punch List

Contractor Punch List

Certificate of Completion and Acceptance

Permission to Quote

Job Performance Report

Employees

Employment Application

Employment Agreement

Employment Acknowledgment (Part-Time or Temporary)

Employee-File Checklist

Employee Expenses

Authorization to Release Employee Information

Employment Verification Letter

Request for Medical Records

Medical Examination Release

Weekly Work Schedule

Weekly Work History

Weekly Time Sheet

Time-Off Record

Employee Warning Notice

Employment Termination Notice

Subcontractors

Letter Soliciting Bids from Subcontractors

Bid Request

Phone Log

Subcontractor Questionnaire

Contractor Rating Sheet

Independent Contractor Agreement

Independent Contractor Acknowledgment

Subcontractor Agreement

Subcontractor Contract Addendum

Subcontractor Schedule

Subcontractor List

Inspection Log

Code Violation Notification

Notice of Breach of Contract

Long-Form Lien Waiver

Short-Form Lien Waiver

Certificate of Subcontractor Completion and Acceptance

Credit Concerns

Business Credit Application

Request for Credit History

Credit Application Rejection

Invoice

Statement

Accounts Receivable Log

Receipt for Payment

Request for Information

Dishonored Check Notice

Payment Inquiry

Second Notice

Final Collection Notice

Demand for Payment

Appointment of Collection Agent

Turnover for Collection

Settlement Offer

Debt Acknowledgment

Confirmation of Payment Plan

Promissory Note

Installment Payment Acknowledgment

Notice of Default

Notice of Dispute

Mistakes With Materials

Take-Off Form

Letter Soliciting Material Quotes

Material Order Log

Demand for Delivery

Rejection of Goods

Cancel Order

Cancellation of Backordered Goods

Payment on Specific Accounts

Inventory Control

Inventory Log

General Inventory Control for Trucks

Spot Check of Truck Inventory

Vehicles And Equipment

Vehicle Daily Checklist

Vehicle Fuel Records

Vehicle-Related Expenses

Vehicle Sign-Out Sheet

Vehicle Mileage Log

Vehicle Maintenance Log

Vehicle Reminder Log

Daily Equipment Checklist

Equipment Safety Inspection

Equipment Failure/Damage Report

Equipment Sign-Out Sheet

Sign-Out Sheet for Equipment Loaned for Personal Use

Equipment Leasing Records

Lost/Stolen Equipment Report

Index

About the Author

R. Dodge Woodson is a general contractor, plumbing contractor, and licensed master plumber with over twenty years of field and business experience. He runs his own plumbing and construction company in Maine, and teaches plumbing classes at Maine Technical College. Mr. Woodson is the author of numerous McGraw-Hill books, including the *National Plumbing Codes Handbook, Plumber's Troubleshooting Guide, Plumber's Quick-Reference Manual,* and *Builder's Guide to Residential Plumbing.*

SOFTWARE AND INFORMATION LICENSE

The software and information on this diskette (collectively referred to as the "Product") are the property of The McGraw-Hill Companies, Inc. ("McGraw-Hill") and are protected by both United States copyright law and international copyright treaty provision. You must treat this Product just like a book, except that you may copy it into a computer to be used and you may make archival copies of the Products for the sole purpose of backing up our software and protecting your investment from loss.

By saying "just like a book," McGraw-Hill means, for example, that the Product may be used by any number of people and may be freely moved from one computer location to another, so long as there is no possibility of the Product (or any part of the Product) being used at one location or on one computer while it is being used at another. Just as a book cannot be read by two different people in two different places at the same time, neither can the Product be used by two different people in two different places at the same time (unless, of course, McGraw-Hill's rights are being violated).

McGraw-Hill reserves the right to alter or modify the contents of the Product at any time.

This agreement is effective until terminated. The Agreement will terminate automatically without notice if you fail to comply with any provisions of this Agreement. In the event of termination by reason of your breach, you will destroy or erase all copies of the Product installed on any computer system or made for backup purposes and shall expunge the Product from your data storage facilities.

LIMITED WARRANTY

McGraw-Hill warrants the physical diskette(s) enclosed herein to be free of defects in materials and workmanship for a period of sixty days from the purchase date. If McGraw-Hill receives written notification within the warranty period of defects in materials or workmanship, and such notification is determined by McGraw-Hill to be correct, McGraw-Hill will replace the defective diskette(s). Send request to:

Customer Service
McGraw-Hill
Gahanna Industrial Park
860 Taylor Station Road
Blacklick, OH 43004-9615

The entire and exclusive liability and remedy for breach of this Limited Warranty shall be limited to replacement of defective diskette(s) and shall not include or extend to any claim for or right to cover any other damages, including but not limited to, loss of profit, data, or use of the software, or special, incidental, or consequential damages or other similar claims, even if McGraw-Hill has been specifically advised as to the possibility of such damages. In no event will McGraw-Hill's liability for any damages to you or any other person ever exceed the lower of suggested list price or actual price paid for the license to use the Product, regardless of any form of the claim.

THE MCGRAW-HILL COMPANIES, INC. SPECIFICALLY DISCLAIMS ALL OTHER WARRANTIES, EXPRESS OR IMPLIED, INCLUDING BUT NOT LIMITED TO, ANY IMPLIED WARRANTY OF MERCHANTABILITY OR FITNESS FOR A PARTICULAR PURPOSE. Specifically, McGraw-Hill makes no representation or warranty that the Product is fit for any particular purpose and any implied warranty of merchantability is limited to the sixty day duration of the Limited Warranty covering the physical diskette(s) only (and not the software or information) and is otherwise expressly and specifically disclaimed.

This Limited Warranty gives you specific legal rights; you may have others which may vary from state to state. Some states do not allow the exclusion of incidental or consequential damages, or the limitation on how long an implied warranty lasts, so some of the above may not apply to you.

This Agreement constitutes the entire agreement between the parties relating to use of the Product. The terms of any purchase order shall have no effect on the terms of this Agreement. Failure of McGraw-Hill to insist at any time on strict compliance with this Agreement shall not constitute a waiver of any rights under this Agreement. This Agreement shall be construed and governed in accordance with the laws of New York. If any provision of this Agreement is held to be contrary to law, that provision will be enforced to the maximum extent permissible and the remaining provisions will remain in force and effect.